MW00843272

Aerial Access Networks

This book presents state-of-the-art research on aerial communications coexisting with terrestrial networks from the physical, MAC, network, and application layer perspectives. It includes thorough discussion of control issues, access techniques, and resource sharing between cellular communication and aerial communications to accommodate larger volumes of traffic and to provide better service to users. Other challenges explored in this text are identification of services, radio resource allocation and resource management for aerial links, self-organizing aerial networks, aerial offloading, and performance evaluation of aerial communications. This volume will be a highly useful resource for students, researchers, and engineers interested in obtaining comprehensive information on the design, evaluation, and applications of aerial access networks and communications.

Lingyang Song is a Boya Distinguished professor at Peking University, China. His main research interests include wireless communication and networks, signal processing, and machine learning. He was a recipient of the IEEE Leonard G. Abraham Prize (2016) and the IEEE Asia Pacific Young Researcher Award (2012). He has been an IEEE Distinguished Lecturer since 2015.

Boya Di is an assistant professor at Peking University, China. Prior to that, she was a postdoctoral associate at Imperial College London. She is a recipient of the IEEE Communications Society Asia-Pacific Outstanding Paper Award (2021) and the IEEE Communications Society Asia-Pacific Outstanding Young Researcher Award (2022). She is currently an associate editor for *IEEE Transactions on Vehicular Technology.*

Hongliang Zhang is an assistant professor at Peking University, China. Prior to that, he was a postdoctoral associate at Princeton University, New Jersey. Zhang is a recipient of the IEEE Communications Society Heinrich Hertz Award for Best Communications Letter (2021) and IEEE Communications Society Asia-Pacific Outstanding Paper Award (2021).

Zhu Han is currently a professor at the University of Houston, Texas. Han has received a National Science Foundation (NSF) Career Award (2010), Fred W. Ellersick Prize (2011), EURASIP Best Paper Award (2015), and IEEE Leonard G. Abraham Prize (2016) and was a winner of the IEEE Kiyo Tomiyasu Award (2021) for outstanding early to mid-career contributions to technologies.

Aerial Access Networks

Integration of UAVs, HAPs, and Satellites

Edited by

LINGYANG SONG
Peking University

BOYA DI
Peking University

HONGLIANG ZHANG
Peking University

ZHU HAN
University of Houston

CAMBRIDGE
UNIVERSITY PRESS

Shaftesbury Road, Cambridge CB2 8EA, United Kingdom

One Liberty Plaza, 20th Floor, New York, NY 10006, USA

477 Williamstown Road, Port Melbourne, VIC 3207, Australia

314–321, 3rd Floor, Plot 3, Splendor Forum, Jasola District Centre, New Delhi – 110025, India

103 Penang Road, #05–06/07, Visioncrest Commercial, Singapore 238467

Cambridge University Press is part of Cambridge University Press & Assessment, a department of the University of Cambridge.

We share the University's mission to contribute to society through the pursuit of education, learning and research at the highest international levels of excellence.

www.cambridge.org
Information on this title: www.cambridge.org/9781108837934

DOI: 10.1017/9781108936538

First published 2023

A catalogue record for this publication is available from the British Library.

Library of Congress Cataloging-in-Publication Data
Names: Song, Lingyang, author. | Di, Boya, author. | Zhang, Hongliang, 1992- author. | Han, Zhu, 1974- author.
Title: Aerial access networks : integration of UAVs, HAPs, and satellites / Lingyang Song, Peking University, Beijing, Boya Di, Peking University, Beijing, Hongliang Zhang, Peking University, Beijing, Zhu Han, University of Houston.
Description: Cambridge, United Kingdom ; New York, NY, USA : Cambridge University Press, [2023] | Includes bibliographical references and index.
Identifiers: LCCN 2023008165 (print) | LCCN 2023008166 (ebook) | ISBN 9781108837934 (hardback) | ISBN 9781108936538 (ebook)
Subjects: LCSH: Aerial networks (Computer networks)
Classification: LCC TK5103.254 .A37 2023 (print) | LCC TK5103.254 (ebook) | DDC 004.6–dc23/eng/20230405
LC record available at https://lccn.loc.gov/2023008165
LC ebook record available at https://lccn.loc.gov/2023008166

ISBN 978-1-108-83793-4 Hardback

Contents

Part II UAV Communication Networks

Part V Integration of UAVs, HAPs, and Satellites

Preface

Since the 2010s, it has been widely acknowledged that traditional terrestrial wireless communications are experiencing an explosive growth in terms of both the number of users and the services to be supported. Future networks are expected to provide more resources to cope with the increasing traffic demands of various services. To support the emerging applications such as the Internet of Things (IoT), cloud computing, and big data, new standards and technologies are being proposed and implemented. However, limited by network capacity and coverage, the terrestrial communication systems alone cannot provide wireless access services with high data rates and reliability to any place on Earth, especially in environmentally harsh areas such as the ocean and mountains. It is thus imperative to exploit new network architectures to accommodate diverse services and applications with different quality of service (QoS) requirements in various scenarios.

Utilizing modern information network technologies and interconnecting space, air, and ground network segments, the aerial access network (AAN) has attracted much attention from academia and industry and has been recognized as a potential solution for future communication systems. Aerial access networks are heterogeneous networks that are engineered to utilize satellites, high-altitude platforms (HAPs), and unmanned aerial vehicles (UAVs) to build communication access platforms. Compared to terrestrial wireless networks, AANs are characterized by frequently changing network topologies and more vulnerable communication connections. Furthermore, AANs demand seamless integration of heterogeneous networks such that the network QoS can be improved. Thus, designing mechanisms and protocols for AANs poses many challenges. To solve these challenges, extensive research has been conducted.

Note that AANs are not intended to replace existing technologies but instead to work with them in a complementary and integrated fashion. However, design, analysis, and optimization of AANs require multidisciplinary knowledge, of wireless communications and networking, signal processing, artificial intelligence (e.g., for learning), decision theory, optimization, and economic theory. Therefore, a book containing the basic concepts and theories needed to address the research advances that enable aerial communications in cellular networks as well as state-of-the-art research and development will be useful for researchers and engineers, especially graduate and undergraduate students who are interested in obtaining comprehensive information on the design, evaluation, and applications of AANs. This is the primary motivation for this book.

This book has three main objectives:

1. Provide a general introduction to the physical, medium access control (MAC), and networking layer requirements of AAN integrated networks.
2. Introduce the key components and the corresponding techniques that enable AANs communications systems. Then, the related design, analysis, and optimization problems will be presented in a comprehensive way.
3. Present state-of-the-art AANs and possible applications. This will include classifications of the different schemes and the technical details in each scheme.

To achieve these objectives, we have divided the book into five parts. Part I introduces the basics of an AAN, which is a three-tier aerial access network containing UAVs, HAPs, and satellites. Parts II to IV elaborate on the key features and main techniques for the UAV, HAP, and satellite layers, respectively. Finally, we show how to integrate these three layers into a tightly coupled AAN and how the aerial network complements the terrestrial cellular communication systems in Part V. This is the first book on complicated aerial communications and networking that includes UAV, HAPs, and low-Earth-orbit (LEO) satellites. It presents state-of-the-art research on the physical, MAC, and networking layers for aerial communications coexisting with cellular networks.

1 Introduction

1.1 Overview

This book will provide state-of-the-art research on aerial communications coexisting with terrestrial networks from physical, MAC, network, and application layer perspectives. The book also includes the fundamental theories upon which aerial communication systems will be constructed. The book discusses in detail the issues of control, access techniques, and resource sharing (e.g., spectrum and energy) between cellular communication and aerial communications to accommodate larger volumes of traffic and to provide better service to users. Other challenges to be discussed in the book include identification of services for which aerial communications are useful, radio resource allocation and resource management for aerial links, self-organizing aerial networks, aerial offloading, and capacity and performance evaluation of aerial communications.

The key features of this book are

- a unified view of aerial access communications and networking;
- comprehensive review of the state-of-the-art research and key technologies for aerial communications networks;
- comprehensive of a wide range of techniques for the design, analysis, optimization, and applications of aerial communications networks;
- outlining the key research issues related to aerial communications and terrestrial networks; and
- standardization activities in the area of aerial communications.

The purpose of this book is to provide a systematical and comprehensive overview of the potentially promising models and methodologies in the era of aerial access network (AAN) empowering 6G from the perspective of the system architecture and networking design, enabling technologies, modeling and performance analysis, system performance optimization, and technical challenges and future directions. The book also presents techniques for analysis, design, optimization, and applications of aerial communications systems under given objectives and constraints.

1.1.1 Integrated System Architecture of AANs

With the development trend of AANs for the provision of wireless coverage extension and enhancement for 6G services, we first elaborate on the AAN architecture, which is designed to provide seamless and ubiquitous 6G services for more users and applications. The key functional components of the AAN system are presented to give flexibility and scalability to the management and control of the AANs. Notably, valuable suggestions and recommendations are proposed and analyzed in each part of the AAN architecture.

1.1.2 Modeling and Performance Analyses

We investigate the related modeling and performance analysis strategies of the combined multitiered systems in AANs to provide readers with a convenient reference. Specifically, we first analyze the complicated channel model in an aerial network of highly dynamic unmanned aerial vehicles (UAVs), high-altitude platforms (HAPs), and satellites the goal of which is to assist the terrestrial network to extend coverage. We then study modeling and performance analysis in UAV/HAP-terrestrial networks. We also explore the network models and performance analysis schemes in integrated satellite-terrestrial networks for service performance enhancement. Notably, we present some insight for the modeling and performance evaluation in future AANs with multinetwork tiers integration.

1.1.3 System Performance Optimization

To improve the system performance of AANs, we consider AANs with a single-tier network and multitiers heterogeneous network. Because the resources of the different components (i.e., HAP, UAV, and satellite tiers) are dynamic and have different characteristics, we study performance optimization strategies from the perspectives of single HAP, single UAV, single satellite, and integrated networks. Furthermore, regarding system performance improvement, we review technologies on four aspects: wireless communication services enhanced by applying UAV networks, user service performance optimization based on HAP networks, multiresource collaboration in satellite networks, and heterogeneous resource collaboration in AANs.

1.2 Motivation

The current development of 5G networks represents a breakthrough in the design of communication networks for their ability to provide a single platform enabling a variety of different services, such as enhanced mobile broadband communications, automated driving, and the huge number of connected Internet of Things (IoT) devices. Nevertheless, looking at the current development of technologies and new services, we can already envision the need to move beyond 5G (B5G) or sixth generation (6G) with a new architecture incorporating new services and technologies.

Providing "connectivity from the sky" is the new innovative trend in wireless communications for future communication systems. Benefitting from the inherent advantages in high throughput, large coverage, and resilience of satellite communications, more and more commercial companies and organizations have been working on satellite-related projects such as OneWeb [1] and SpaceX [2] in recent years. In addition, the third generation partnership project (3GPP) defines the deployment scenario of nonterrestrial network (NTN) and provides 6G commercial services for areas with an underdeveloped ground network infrastructure. The project is expected to achieve wide coverage by using aerospace access facilities such as UAVs and satellites. As a result, the AAN composed of low-altitude platforms (LAP), UAV, HAP, and satellites is an emerging architecture to support ubiquitous services. Thanks to the 3D wide service coverage capabilities and reduced vulnerability of space/airborne vehicles to physical attacks and natural disasters, NTNs have a number of benefits:

- Foster the roll out of wireless service in unserved areas where terrestrial networks are not available (such as in isolated/remote areas or on board aircrafts or vessels) or underserved (e.g., in suburban/rural areas) to upgrade the network performance in a cost-effective manner.
- Improve the wireless service reliability by providing service continuity for machine-to-machine (M2M) and IoT devices or for passengers on board moving platforms (e.g., aircraft, ships, high speed trains, and buses) or ensuring service availability anywhere especially for critical communications, future railway/maritime/aeronautical communications, and to.
- Enable cellular network scalability by providing efficient multicast/broadcast resources for data delivery at the network edges or user terminal.

These benefits relate to NTNs operating alone as well as integrated terrestrial and nonterrestrial networks. They will affect coverage, user bandwidth, system capacity, service reliability or availability, energy consumption, and connection density. For example, the incorporation of links with aerial access points placed on drones or very low-Earth-orbit (LEO) satellites is an effective way to provide coverage on demand and cope with the high variability of data rates as a function of space and time. These access points will most likely operate over lower frequency bands (up to tens of gigahertz). Their use, coupled with a dense deployment of terrestrial access points placed on locations such as lampposts and tall buildings, will allow future communication systems to enable true 3D connectivity.

Exploiting AANs to supplement terrestrial networks is a key approach to deal with massive data traffic demands and global ubiquitous communication in the future. There are about 30 satellite constellations around the world. The service capabilities of the system are gradually expanding from traditional mobile communication services to broadband interconnection services. Despite the fact that the number and size of satellite constellations are growing quickly, the link between constellation scale, structure, frequency usage guidelines, and constellation service in satellite constellation systems isn't fully understood. In addition, on-demand integration and movement of heterogeneous resources are critical technologies to enhancing network performance in a typical heterogeneous AAN upgraded for ground services. The limited and mobile network resources between different tiers, however, have a significant impact on the network performance because of the AAN's inherent self-organization, heterogeneity, and time-variability. For ensuring ubiquitous on-demand user services, it is crucial to consider the system architecture and design, resource management and frequency allocation, mobility management, networking protocol, and network performance analysis and optimization of heterogeneous AANs.

There have been state-of-the-art surveys on AANs. First, some surveys were conducted on single networks. Recent advances and future challenges of satellite and CubeSat communication were introduced in [3], respectively. In [4], the technology, standards, and open challenges of satellite IoT were introduced. The vision and framework of future HAP networks were provided and the unrealized potential of the HAP system was emphasized in [5]. The research status and future research directions of AANs including HAP and UAV cellular communications are detailed in [6]. In [7], the practical problems and security challenges in UAV cellular communications were discussed. Other works have focused on the integration of different kinds of networks. Satellite-ground communication networks were discussed from different aspects in [8] and [9]. In addition, some surveys of the AANs were carried out in [10] and [11], including recent research work, key technologies, application prospects, and requirements, architecture and challenges in 6G.

However, a full analysis of the aforementioned parts or all tiers of AANs for special applications is missing from the existing publications concerning AANs that are either constrained to or dedicated to particular topics. This indicates that the relevant studies in this subject are still dispersed and independent, and their connections are sporadic and asynchronous. As a result of AANs' inherent self-organization, heterogeneity, and time-variability, the objectives and methodology of the existing research has also varied substantially. In order to provide a concise guide, this book highlights the following crucial issues, including system architecture and networking design, network performance analysis, system optimization schemes, simulation evaluation, and system testing, in both single-tier and multitier AAN scenarios, which had not been well investigated. Conceptually, combining current systems to create a multitiered, hierarchical AAN is intended to give readers direction and a complete reference model for future topics.

1.3 Objectives and Organization

Utilizing modern information network technologies and interconnecting space, air, and ground network segments, the AANs have attracted much attention from academia and industry and have been recognized as a potential solution for the future communication systems. Such heterogeneous networks that are engineered to utilize satellites, HAPs, and UAVs to build communication access platforms. Compared to terrestrial wireless networks, AANs are characterized by frequently changing network topologies and vulnerable communication connections. Furthermore, AANs demand the seamless integration of heterogeneous networks such that the network quality of service (QoS) can be improved. Thus, designing mechanisms and protocols for AANs poses many challenges. To solve these challenges, extensive research has been conducted.

Note that AANs are not intended to replace existing technologies, but instead to work with them in a complementary and integrated fashion. Thus, AAN design, analysis, and optimization requires multidisciplinary knowledge – namely, knowledge of wireless communications and networking, signal processing, artificial intelligence (e.g., for learning), decision theory, optimization, and economic theory.

The three main objectives of this book are to (1) provide a general introduction to integrated AANs, (2) introduce key components and their corresponding techniques to enable AANs communications systems, and (3) present the state-of-the-art AANs and possible applications. To achieve these objectives (outlined in more detail in the Preface), this book is organized as follows:

1.3.1 Part I: Basics of Aerial Access Networks

We start with an introduction to wireless communications (including radio propagation and channel models) and by reviewing different wireless access technologies. To provide global connectivity for terrestrial users, AANs can work as a complementary component. The requirements of aerial services and the system architecture will also be discussed. A three-tier aerial access network consisting of UAVs, HAPs, and satellites will be introduced. In this chapter, we highlight the characteristics of each tier respectively and discuss the possible integrations and applications of these three tiers.

1.3.2 Part II: UAV Communication Networks

Emerging UAVs have been playing an increasing role in the military, public, and civil applications. Dedicated UAVs, also called drones, can be used as communication platforms (e.g., as wireless access points or relays nodes) to further assist the terrestrial communications. This type of application can be referred to as UAV-assisted cellular communications, which we describe in this part.

The key techniques to support the UAV communication networks will also be presented. For example, it is challenging to plan the time-variant placements/trajectories of the UAVs served as base station (BS)/relay due to complicated 3D propagation environments as well as many other practical constraints such as power and flying speed. In addition, spectrum sharing with existing cellular networks and user associations are other interesting topics to investigate.

1.3.3 Part III: HAP Communication Networks

High-altitude-platforms (HAPs) have the potential to deliver a range of communications services and other applications cost effectively due to their ability to carry a heavy payload. In addition, HAPs can provide larger coverage than UAVs and have a lower propagation delay with more controllable wireless links compared with satellite communications. This chapter explains a step change in performance and availability and the advantages of being able to deliver high capacity similar to that available from terrestrial systems and wide area type coverage similar to that available from satellites.

To integrate seamlessly with existing communication networks and achieve wide adoption among potential users, the HAP network has to be based on the most suitable existing or developing communication standards with necessary adaptations that take into consideration some specific requirements and the particular operating environment. Some applications such as data/computation offloading will be investigated. Specific emphasis is place on how this technology will integrate within a terrestrial/satellite infrastructure. Here we also describe the broadband services and the investigation into broadband architectures.

1.3.4 Part IV: Satellite Communication Networks

With recent significant advances in ultra-dense LEO satellite constellations, satellite access networks (SANs) have shown their significant potential to integrate with 5G and beyond to support ubiquitous global wireless access. This chapter proposes an enabling network architecture for a dense LEO-SAN in which the terrestrial and satellite communications are integrated to offer more reliable and flexible access.

Many key techniques such as effective interference management, diversity techniques, and cognitive radio schemes will be further elaborated on. An integrated architecture for satellite-terrestrial network will be proposed, where all types of satellites and aerial components play different roles given a software defined network (SDN) framework. Various applications of satellites for aerial-terrestrial integrated services are also studied.

1.3.5 Part V: Integration of UAVs, HAPs, and Satellites

The integration issue is another important aspect of providing seamless and high-rate wireless links for wireless devices with different QoS requirements. In this chapter, we further discuss the key problems in the integration of UAVs, HAPs, and satellites, including association, data/computation offloading, and routing, involved in providing seamless connectivity from the sky.

Part I

Basics of Aerial Access Networks

2 Aerial Integration

UAVs, HAPs, and Satellites

2.1 Overview of AANs

The rapid development of cellular communications has triggered an increasing demand for high-data-rate applications [12]. Future networks are expected to provide more resources than current ones to cope with the increasing data demands of various services and emerging applications, which in turn raise stringent requirements on achieving extremely broad coverage, ubiquitous connectivity, and high-capacity communications [13]. However, due to limited coverage, scarce spectrum resources and imbalanced infrastructure deployment, traditional terrestrial communication network capabilities are far from enough to satisfy these requirements. Hence, it is imperative to exploit new network architectures to accommodate diverse services and applications with different QoS requirements in various scenarios [14]. Utilizing modern information network technologies and interconnecting space and air networks, AANs are being developed with the promise of fulfilling the challenging visions for future networks owing to their inherent advantages in terms of large coverage, high throughput, and resilience [10]. Specifically, satellites in space segment networks can provide seamless global connectivity in rural, ocean, and mountain areas, whereas UAVs and HAPs in air segment networks can enhance the capacity for covered areas with high service demands. The space network segment and air network segment can work independently or interoperationally. By integrating heterogeneous networks within the two segments, it is easy to build a hierarchical broadband wireless network (Fig. 2.1).

2.1.1 Space Network

The space network is composed of satellites and constellations as well as their corresponding terrestrial infrastructures such as user terminals (UTs) and an Earth gateway. These satellites and constellations are in different orbits and have different characteristics. Based on their altitude, satellites can be classified into three categories: geostationary-Earth-orbit (GEO) satellites, medium-Earth-orbit (MEO) satellites, and low-Earth-orbit (LEO) satellites, providing different data services [15].

- GEO satellites operate in geostationary Earth orbit with an altitude of 35,800 km. The orbital period is 24 hours. These satellites cover a large geographical area, and

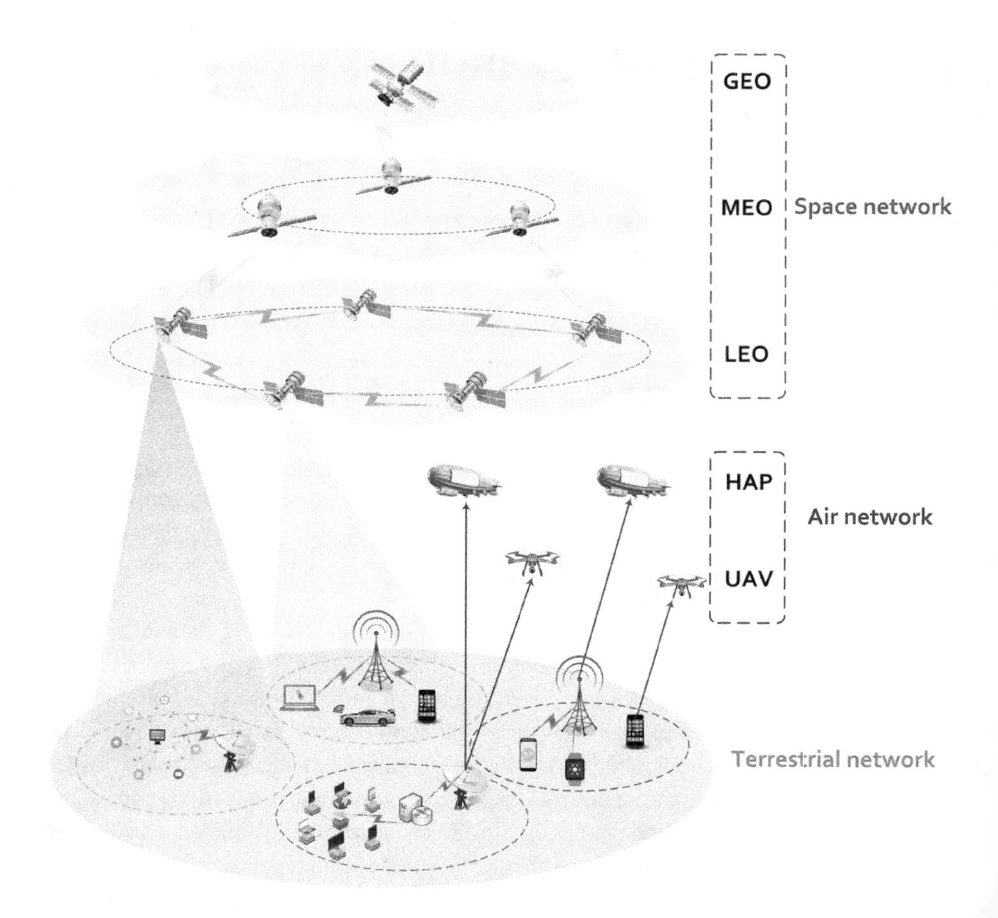

Figure 2.1 Architecture of aerial access networks

only three GEO satellites are needed to cover the Earth. GEO satellites mainly provide fixed satellite services and TV broadcast.

- MEO satellites operate in medium Earth orbit at an altitude of 5,000–12,000 km. The orbital period is 2–8 hours. The transmission delay of MEO satellites is less than that of GEO satellites, and their coverage area is larger than that of LEO satellites. MEO satellites mainly provide mobile data services.
- LEO satellites operate in low Earth orbit at an altitude of 500–1,500 km. The orbital period is 10–40 minutes. Transmission delay and path loss are the lowest for LEO satellite communications. LEO satellites mainly provide broadband multimedia and mobile data services.

2.1.2 Air Network

The air network is an aerial mobile system employing flying base stations (BSs) as carriers for information acquisition, transmission, and processing. Unmanned aerial vehicles and HAPs are the main infrastructures for providing broadband wireless communications to complement the terrestrial networks and space networks [16]. Specifically, compared to a terrestrial network, an air network is easily deployed and has large coverage so it can offer wireless access services on a regional basis. Compared to a space network, an air network has low latency and low cost, providing more flexible and reliable connectivity for urgent events. Next, we provide a brief introduction to UAVs and HAPs:

- UAVs are aircraft piloted by remote control or embedded programs without a human onboard. They operate at an altitude of 600 m to 18 km [17]. Compared to terrestrial communications or HAP communications, UAVs operating at low altitude are more cost effective and can be much more swiftly deployed. Typical applications of UAVs include sensing, communication, and surveillance.
- HAPs are quasi-stationary aerial platforms such as airships and balloons located at a height of 17–22 km above the Earth's surface in the stratospheric region of the atmosphere [6]. Compared to UAVs, HAPs have a larger coverage area and longer communication persistence. Typical applications of HAPs include real-time monitoring, communication relay, emergency recovery, and rocket launch platforms.

2.2 UAV Communication Networks

2.2.1 Basics of UAVs

During the 1930s, the US Navy began to experiment with radio-controlled UAVs. In the 1990s, micro UAVs started to be widely used in public and civilian applications. Recently, due to their ease of deployment, low acquisition and maintenance costs, high maneuverability, and ability to hover, UAVs have been widely used in civil and commercial applications [18]. However, UAV research has typically focused on navigation [19] and autonomy [20], which is required in military-oriented applications. In contrast, this book will focus on UAV communications over future cellular networks.

Unmanned aerial vehicle communication networks are one promising solution for data-rate enhancement. For one thing, UAVs can serve as flying access points (APs) to deliver reliable and cost-effective wireless communications, thereby increasing the number of APs. In addition, UAVs can serve as wireless relays to improve QoS for users, especially for cell-edge users. Moreover, another typical feature of future communication networks is the integration of massive numbers of connected Internet-of-Things (IoT) devices, which connect the physical world using sensors. The analyst firm Gartner estimated that over 2,020 billion connected things were expected to be in use by 2020 [21]. Benefiting from their ease of deployment, high autonomy, and ability to hover, UAVs equipped with sensors have come into our daily lives to execute a variety of sensing missions [22].

Table 2.1 Characteristics of UAVs

	Fixed-wing UAVs	Rotary-wing UAVs
Speed	Up to 500 km/h	Typically less than 60 km/h
Altitude	Up to 200 km	Typically less than 1 km
Flight time	Up to several hours	Typically less than 30 min
Applications	Carry cellular infrastructure	Sensing
	Airborne surveillance	Carry cellular BS

Generally, UAVs can be classified into two types: fixed-wings and rotary-wings.

- Rotary-wing UAVs allow vertical takeoff and landing and can hover over a fixed location. This high maneuverability makes them suitable to execute sensing tasks. In addition, rotary-wing UAVs can be used to deploy a BS because they can hover at desired locations with higher precision. However, rotary-wing UAVs have limited mobility and power.
- Fixed-wing UAVs can glide through the air, which makes them significantly more energy efficient and capable of carrying a heavy payload. Therefore, fixed-wing UAVs can carry cellular infrastructure to provide cellular coverage. Gliding also helps fixed-wing UAVs to fly at a faster speed, making them suitable for airborne surveillance. The disadvantage of fixed-wing UAVs is that they cannot hover over a fixed location. Fixed-wing UAVs are also more expensive than rotary-wing UAVs [23].

The characteristics of the UAVs are summarized in Table 2.1.

2.2.2 UAV Applications

Based on their characteristics, UAVs have two kinds of typical applications: flying infrastructure and flying users.

Flying Infrastructure

Due to their low operational cost and flexible deployment, UAVs have been adopted as a flying infrastructure to provide wireless communications. For example, Nokia's flying-cell (F-Cell) [24] and Facebook's Aquila [25] are two existing projects using UAVs to provide wireless coverage. In Facebook's Aquila project, UAVs with solar panels are sent 18–27 km into the stratosphere to provide widespread internet coverage, which can reach 40–80 km^2. In these applications, there exist two main scenarios:

- *UAV as BS:* UAVs can work as flying BSs in hotspots (areas where the data traffic load is heavy) for capacity enhancement or data offloading [26]. The UAV can also be used as a temporary BS for emergency communications due to its high mobility and ease of deployment.
- *UAV as relay:* UAVs can serve as relays for coverage extension or data aggregation [27]. The UAV can utilize an amplify-and-forward (AF) or decode-and-forward

(DF) protocol to relay the data. Unlike UAV BSs, which have a backhaul that is assumed to be perfect, the backhaul link for UAV relaying should also be considered.

Flying Users

In addition to flying infrastructures, UAVs can also be adopted as flying users for sensing purposes. Specifically, UAVs acting as flying users are equipped with a payload, such as a camera, thermometer, and air quality index (AQI) sensor, to execute real-time sensing tasks and transmit sensory data to the ground control station (GCS) or server. Considering the limited computation capability of UAVs, a terrestrial cellular network is a promising solution to support UAV sensing applications, which we refer to as the cellular Internet of UAVs [28], due to their larger coverage, higher reliability, and flexibility. In the cellular internet of UAVs, the UAVs transmit sensory data to the BS by two communication modes: UAV-to-network (U2N) communications, where UAVs transmit sensory data to the BS directly, and UAV-to-UAV (U2U) communications, where UAVs transmit to the BS through a UAV relay.

2.2.3 Challenges of UAV Communications

To guarantee the QoS of UAV communications, the following research challenges should be addressed:

- *Frequency allocation:* To avoid interference among different UAVs, proper frequency allocation and UAV transmission scheduling schemes should be well designed. Moreover, UAVs may move with varying speeds depending on the application, such that the topology of UAV networks is time-variant. The number of nodes, the relative positions of the nodes, and channel conditions also change, which complicates the frequency allocation scheme design.
- *Power control:* Energy is also a critical problem for UAVs because their onboard batteries are limited. Therefore, how best to balance the transmission energy and propulsion energy should be investigated. To increase the life of the network, it is also necessary to determine methods of conserving energy of power-starved UAVs.
- *Three-dimensional (3D) trajectory design or UAV placement:* As a UAV's altitude increases, the coverage area of the UAV BS becomes larger. However, the data rate reduces due to higher path loss. Therefore, the trajectory together with placement of the UAV should be well designed to achieve a trade-off between coverage and data rate.

2.3 High-Altitude Platform Communication Networks

2.3.1 Basics of HAPs

Since the 1990s, the investigations into AANs have not only been concerned with satellite communications but also increasingly with HAPs, which exploit the best features of both terrestrial and satellite communications [29]. Compared to satellite communications, HAP communications provide lower propagation delays and stronger signal strength. The upgrade and maintenance costs of HAPs are also lower than that of satellites [30]. In addition, HAPs can hover at high altitudes, which allows them to provide services for terrestrial users over an extensive area and makes them more favorable than terrestrial networks. Moreover, HAPs can fly on demand to serve different regions and their deployment is much more flexible than terrestrial BSs.

Generally, HAPs can be classified into two types:

- Aerostatic platforms float in the air due to their buoyancy. Balloons and airships are two typical aerostatic platforms. Specifically, balloons are unpowered platforms that use a gas such as diatomic hydrogen and helium to provide buoyancy. Airships are mainly powered by solar panels and have station-keeping capability provided by electric motors and propellers [31]. The volume of an aerostatic platform is generally large to compensate for the thin air in the lower stratosphere, which poses great challenges for takeoff and landing. In addition, the payload that an aerostatic platform can carry increases with its volume.
- Aerodynamic platforms exploit aerodynamic lift to fly and are powered by solar cells or fuel. They cannot stay in the air unless they move forward [29]. Hence, they have to fly on a circular path to maintain their quasi-stationary positions. Aerodynamic platforms are lightweight and typically have a wide wingspan. Their payload capacity is limited by their size. The operational cost is higher than aerostatic platforms. In addition, the circular flight of aerodynamic platforms places more stringent requirements on antenna technologies in terms of accurate beam steering.

2.3.2 HAP Applications

Based on their characteristics, HAPs can be applied to serve different scenarios such as telecommunications services, traffic and environmental monitoring, surveillance, intelligent transportation systems, and emergency services. In this subsection, we focus on typical HAP applications related to communications, such as broadband wireless access, multicast/broadcast services, and backhaul/fronthaul services.

Broadband Wireless Access

The International Telecommunications Union Radiocommunication Sector (ITU-R) has allocated several frequency bands with wide bandwidths for HAPs to provide different broadband multimedia services in a millimeter-wave band and International Mobile Telecommunications (IMT)-2000 services in third generation (3G) frequency bands [32]. The applicable services include audio/video streaming, high-definition video conference, high-definition television, distributed games, medical applications, distance education, and large file transfer [33]. In integrated terrestrial-HAP networks or terrestrial-HAP-satellite networks with large coverage areas, these services can be delivered efficiently through information relay without overloading the terrestrial networks, thereby reducing overall costs.

The advantages of HAPs in providing broadband wireless access have caught industry's attention. Specifically, the LOON project planned by Google adopts a network of interconnected solar-powered balloons equipped with lightweight redesigned LTE-based BSs. This project aims to provide internet coverage for terrestrial users in rural and remote regions [34]. The Communications from Aerial Platform Networks Delivering Broadband Communications for All (CAPANINA) project developed in Europe offers bidirectional high-capacity links between HAPs and ground terminals. The project aims to develop low-cost broadband technology for HAPs and provide fixed and mobile users with ubiquitous broadband wireless access in remote areas and inside high-speed vehicles [35].

Backhaul/Fronthaul Services

Nowadays, the shortage of available bandwidth and the limited backhaul/fronthaul capacity make it difficult for traditional terrestrial small cells to handle the explosive data transmission demands. High-altitude platforms are a promising solution to provide backhaul/fronthaul links for the ultra-dense deployed small cells. Specifically, HAPs support dual-band connectivity. First, they can communicate directly with terrestrial users over C-band. Second, they can provide ground UTs with data backhaul/fronthaul services over millimeter-wave bands [36]. The UT acts as an access point that can transmit users' data to the core network via HAP-based backhaul over a millimeter-wave band.

In addition to terrestrial networks, HAPs can also be utilized in the global wireless infrastructure to relay the huge amount of data transmitted from LEO satellites to Earth-based control centers. These platforms can break the weather-dependent satellite-terrestrial link to weather-independent, high-capacity optical links between satellites and HAPs [31]. Hence, it is possible to download data during a short time pass of a satellite. Although the channel condition between the HAP and Earth control center is subject to atmospheric weather conditions, the throughput requirements of the HAP-terrestrial link are lower than that of the satellite-terrestrial link.

2.3.3 Challenges of HAP Communications

To guarantee the QoS of HAP communications, the following research challenges should be addressed:

- *HAP deployment:* The deployment of a HAP will influence its coverage area and its users' QoS. Because the environment of the stratosphere layer is complicated, the HAP is inevitably affected by short-term airflow, leading to random perturbation with respect to location and angular movements. Therefore, the deployment of HAPs should be carefully designed while considering the probability distribution of the HAP perturbation.
- *Payload optimization:* The payload carried by a HAP depends on its application and size. The payload has a significant impact on the performance of the HAP. Specifically, although a heavy payload can improve its signal processing capability

a heavy payload results in higher power consumption and a shorter mission time [6]. Hence, the payload of HAPs must be optimized.
- *Physical layer function and protocol design:* Current physical layer functions and protocols are mainly designed for satellites and terrestrial applications. However, the characteristics of the HAP-terrestrial channel are completely different from a terrestrial channel or a terrestrial-satellite channel. Therefore, the physical layer functions and protocols should be optimized for use in HAPs.

2.4 Satellite Communication Networks

2.4.1 Basics of Satellites

Since the 1960s, because of their large coverage and bandwidth, several satellite systems have been proposed and implemented for various services such as communication, navigation, remote sensing, and scientific research for both civilian and military purposes. Due to their long communication distance, GEO and MEO satellite networks mainly provide low-date-rate services such as TV broadcasting services, which cannot satisfy users' increasing demand for high-data-rate applications. Recent advances in LEO satellite networks over the high-frequency band have provided an alternative solution for coverage extension and backhaul connectivity. Driven by SpaceX [37] and OneWeb [38], the ongoing LEO constellation projects plan to launch thousands of LEO satellites over the Earth, aiming to deploy an ultra-dense constellation and cooperate with the traditional operators to support seamless and high-capacity communication services. With promising developments in terms of feasibility, the projects use a pipeline production of small satellites to lower manufacturing costs [39]. Instead of the traditional small cell BS, a dedicated terrestrial-satellite UT equipped with steerable antennas acts as the AP in this case. The UT is easy to install on the roof or Evolved Node B (eNodeB) because of its miniaturized antennas. Each UT supports both the high-quality UT-satellite backhaul links over the Ka-band and the user-UT links over the C-band, enabling terrestrial small cell coverage for users. Compared to the traditional networks, the LEO network provides a great number of users with a high-capacity backhaul, vast coverage, and a more flexible access technique, which is less dependent on real environments.

2.4.2 Satellite Applications

Low-Earth-orbit satellite networks are being developed with the promise of supporting the explosive growth of wireless devices and various applications. In this subsection, we introduce two typical satellite applications related to communications: data offloading and data caching.

Data Offloading

Based on the superior communication characteristics of the LEO satellite network over the high-frequency band [40], the integrated ultra-dense LEO-based satellite-terrestrial network architecture has been proposed for data offloading. Specifically, the terrestrial operator considers offloading traditional small cell users to LEO-based small cells owned by the satellite operator for satellite-backhauled network access [41]. Data from the offloaded traditional small cell users is first transmitted to the LEO-based small cells over the dedicated C-band of the satellite operator. The LEO-based small cells then upload the received data to the core network via the satellite backhaul links.

Various aspects have been considered for data offloading in the satellite-terrestrial network, such as user fairness [42], energy efficiency [43], and sum rate maximization [44]. In [42], a two-step sequential carrier allocation strategy based on user fairness optimization has been presented to improve spectral efficiency in satellite-backhauled data offloading. In [43], a double-edge computation offloading algorithm has been proposed to optimize energy consumption in the satellite-terrestrial network. In [44], the maximization of the sum data rate in LEO-based satellite-terrestrial networks has been studied to achieve efficient satellite-backhauled data offloading while considering the intersatellite cochannel interference.

Data Caching

Traditionally, due to the limited backhaul capacity of traditional small cells, it is hard for users to download a large amount of real-time content from traditional small cells when they require files from the service provider through the Internet. The service provider needs to cache the files in the traditional small cells in advance, which is periodically updated based on user demand [45].

Benefiting from the high-capacity backhaul links, LEO satellites can be considered as an alternative to caching files. Specifically, for the files not cached in small cells, the service provider selects one or more satellites to multicast the required content to a number of UTs, and the UTs then forward the content to users via either a licensed or unlicensed spectrum. For services requiring a large amount of data such as multimedia streaming, a satellite-based multicast cache reduces the transmission delay compared to the terrestrial backhaul and covers a much larger area. The dense topology of satellites greatly improves the cache capacity and offers technical support to group-centric services. The intersatellite cochannel interference can also be reduced, and the occupied spectrum band is largely saved.

2.4.3 Challenges of Satellite Communications

To guarantee the QoS of LEO satellite communications, the following research challenges should be addressed:

- *Antenna technologies:* The high mobility of LEO satellites and the severe path loss compared to terrestrial networks put stringent requirements on antenna technologies in terms of accurate beam steering and high antenna gain. Traditional antennas integrated with UTs such as dish antennas and phased arrays either require heavy mechanics or costly phase shifters, making their implementation in practical systems prohibitive. Hence, LEO satellites call for novel antenna technologies with low power consumption and low hardware cost.

- *Interference management:* To support various reliable services for users, interference management is crucial to fully utilize the limited frequency and power resources in the integrated terrestrial network. Due to the entanglement of the terrestrial and satellite interfaces, the sources of interference, such as over cross-cell links and intersatellite links, are more complicated than in terrestrial networks. Moreover, the interference from the satellite segment influences the UT-satellite association and resource allocation and has an effect on the user association strategy, requiring a joint scheduling and resource-allocation scheme.
- *Data routing algorithm design:* Considering the propagation delay of LEO satellites and the limited frequency resources, it is not practical to construct communication links between each LEO satellite and the Earth gateway, especially in an ultra-dense network. Therefore, it is necessary to consider routing via inter-satellite communications to perform network control over satellites. The frequent relative motions among all LEO satellites make network control and routing algorithm design even more challenging.

Part II

UAV Communication Networks

3 UAVs Serving as Flying Infrastructures

Dedicated UAVs can be used as communication platforms, either as base stations or relays nodes, to further assist terrestrial communications. This type of application can be referred to as **UAV-assisted cellular communications**. Such communications have numerous use cases, including traffic offloading, wireless backhauling, swift service recovery after natural disasters, emergency response, rescue and search, information dissemination/broadcasting, and data collection from ground sensors for machine-type communications. However, unlike traditional cellular networks, it is challenging to determine the time-variant placements of the UAVs that serve as BSs or relays due to complicated 3D propagation environments as well as many other practical constraints such as power and flying speed. In addition, spectrum sharing with existing cellular networks is another interesting topic to investigate.

In this chapter, we first discuss the UAV offloading problem where the UAVs serve as BSs and then optimize the trajectory and transmit power in UAV relay networks.

3.1 UAVs Serving as Base Stations

From the aspect of wireless communications, one major advantage of utilizing UAVs is their high probability of maintaining line-of-sight (LoS) signals with other communication nodes, alleviating the problem brought on by severe shadowing in urban areas or mountainous terrain [46, 47]. In addition, properties such as low cost, high flexibility, and ease of scheduling make small-scale UAVs a favorable choice in civil applications, despite their disadvantages such as low battery capacity [48].

One of the major problems in the UAV-assisted wireless communications is optimally deploying UAVs in a way that best serves mobile users [48]. Many studies have dealt with this problem given different objectives and constraints. Among them, [49, 50] considered a scenario that uses only one UAV to provide coverage. The optimal height of a single UAV was deduced to maximize the coverage radius in [49]. In addition, [51–55] studied the coexistence of BSs and multi-UAVs, where data offloading becomes a major problem. Considering BSs in such scenarios, the benefit of deploying additional UAVs for offloading was discussed in [51–53]. The authors of [54] focused on the optimal cell partition strategy to minimize the users' average delay in a cellular network with multiple UAVs. In [55], the optimal resource allocation was presented, where one macro-cell base station (MBS), multiple small-cell base stations (SBSs), and multiple UAVs are involved.

Although UAV coverage and offloading problems have been discussed, few studies consider the situation where UAV operators could be selfish individuals with different objectives [17]. For instance, venue owners and scenic area managers may want to temporarily deploy their UAVs to better serve their visitors during periods with a temporary increase in the number of mobile users or due to the inconvenience of installing SBSs in remote areas [56]. In such cases, the deployment of multiple UAVs depends on each UAV operator, and a solution calculated by centralized algorithms is not likely to be optimal. In addition, the wireless channel allocation becomes a more critical problem because the bandwidth that the UAVs use to serve mobile users has to be explicitly authorized by the MBS manager. Therefore, further studies need to be done with respect to selfish UAV operators in UAV-assisted offloading cellular networks.

In this section, we focus on a scenario with one MBS that is operated by a MBS manager and multiple SBS-enabled UAVs owned by different UAV operators. To enable downlink transmissions of the UAVs, each UAV operator has to buy a certain amount of bandwidth that is authorized by the MBS manager. However, the total usable bandwidth of the MBS is limited, and selling part of the total bandwidth to the UAVs may harm the capacity of the MBS. Therefore, payments to the MBS manager should be made by UAV operators. Here, contract theory [57] can be applied as a tool to analyze the optimal contract that the MBS manager will design to maximize its revenue. Specifically, such a contract consists of a set of bandwidth options and corresponding prices. Because each UAV operator only chooses the most profitable option from the whole contract, the MBS manager has to guarantee that the contract is feasible (i.e., the option that a UAV operator chooses from the contract is exactly the one it was designed for.

The rest of this section is organized as follows. Section 3.1.1 presents our system model and formulates the optimal contract design problem. Section 3.1.2 theoretically deduces the optimal solution and provides our dynamic programming algorithm. Section 3.1.3 focuses on the height of the UAVs and discusses its impact on the revenue of the MBS manager. Section 3.1.4 shows the simulation results of the optimal contract. Finally, we conclude this section in Section 3.1.5.

3.1.1 System Model

We consider a scenario with one MBS and N UAVs, as shown in Fig. 3.1 [58]. The MBS is operated by a MBS manager, and the UAVs are run by different UAV operators. All the UAVs must stay at a legal height H, which is designated by the MBS manager, although the horizontal location of each UAV can be adjusted by its operator to cover as many local users as possible. Each UAV operator aims to provide better service for its local mobile users with its licensed spectrum, which is temporally bought from the MBS manager.

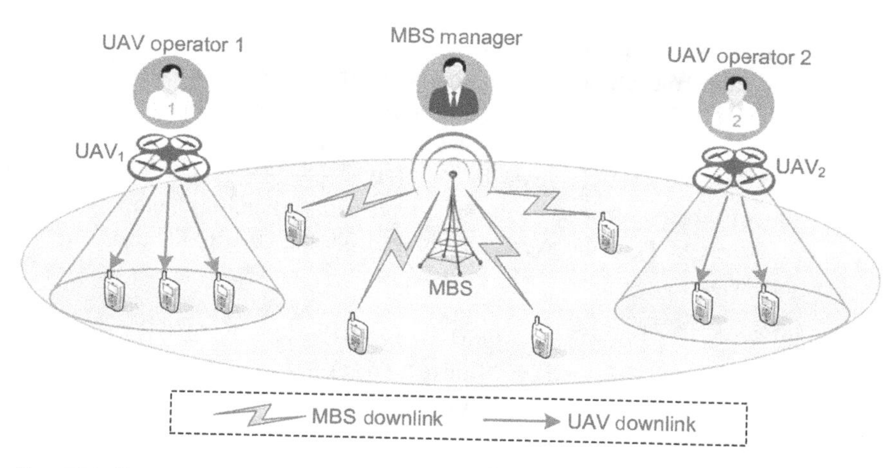

Figure 3.1 System model of UAV-assisted offloading in a cellular network with one MBS manager and multiple UAV operators

In the following, we first discuss the concerned mobility and energy consumption of the UAVs and then present the wireless downlink model of the MBS and the UAVs. After that, we introduce the utility of the UAV operators as well as the cost of the MBS manager. Finally, we formulate the contract design problem.

Mobility and Energy Consumption

Without a loss of generality, we consider the UAV offloading system in a series of short time slots.[1] In the sth time slot, the distribution of mobile users as well as the horizontal location of each UAV are assumed to be stable. In the $(s+1)$th time slot, the horizontal location of a UAV can be adjusted by its operator to cover as many mobile users as possible.

The total available energy for the nth UAV to stabilize or adjust its location is denoted by E_n. The energy consumption required by UAV_n stabilize itself for a whole time slot is given by e_n. The additional energy consumption of moving UAV_n for a distance of l between time slots is denoted by $q_n \cdot l$, where q_n is a constant for UAV_n. With E_n, e_n, q_n, and a specific movement behavior, we are able to obtain the number of time slots that UAV_n could sustain to provide wireless connections for its users.[2]

For the nth UAV operator, we also assume that there is a constant cost of deploying and retrieving UAV_n (unrelated to the number of time slots), denoted by C_n. To make it worth deploying its UAV, this UAV operator has to maximize its profit in each time slot during the deployment. In addition, the MBS manager also aims to maximize its own revenue in each time slot by properly designing the contract. Because the following parts of this subsection only correspond to the problem within one time slot, we omit the time slot number s for reading convenience.

Wireless Downlink Model

We adopt the elevation angle-based model in this section. The wireless channel between a UAV and a mobile user mainly consists of two parts: LoS and NLoS components. The probability of LoS for a user with elevation angle θ (in degrees) to a specific UAV is given by $P_{LoS}(\theta) = \frac{1}{1+a\exp(-b[\theta-a])}$, where a and b are the parameters that depend on the specific terrain (urban, rural, etc.).

Based on P_{LoS}, the average pathloss from the UAV to the user can be given by (in decibels)

$$\begin{cases} \bar{L}_{UAV}(\theta,d) = P_{LoS}(\theta)\cdot L_{LoS}(d) + \left[1-P_{LoS}(\theta)\right]\cdot L_{NLoS}(d), \\ L_{LoS}(d) = 20\log(4\pi fd/c) + \eta_{LoS}, \\ L_{NLoS}(d) = 20\log(4\pi fd/c) + \eta_{NLoS}, \end{cases} \tag{3.1}$$

where c is the speed of light, d is the distance between the UAV and the user, and f is the frequency of the channel. The pathloss of the LoS and NLoS components are $L_{LoS}(d)$ and $L_{NLoS}(d)$, respectively. The average additional loss that depends on the environment are η_{LoS}, η_{NLoS}. In contrast to the UAV-to-user wireless channel, the MBS-to-user channels are considered NLoS only, which gives us the average pathloss as[3]

$$\bar{L}_{MBS}(d) = 20\log(4\pi fd/c) + \eta_{NLoS}. \tag{3.2}$$

For simplicity, we assume that different channels have similar f and the difference can be ignored.

To see the signal quality that each user could experience, we use $\gamma_{MBS}(d)$ to denote the signal-to-noise ratio (SNR) for MSB users at the distance d from the MBS. And we have $\gamma_{MBS}(d) = \left[P_{MBS} - \bar{L}_{MBS}(d)\right]/N_0$, where P_{MBS} is the transmission power of the MBS and N_0 is the power of background noise. Similarly, we use $\gamma_{UAV}(d,\theta)$ to denote the SNR for the UAV users with elevation angle θ and distance d from a certain UAV, given as $\gamma_{UAV}(d,\theta) = \left[P_{UAV} - \bar{L}_{UAV}(d,\theta)\right]/N_0$, where P_{UAV} is the transmission power of the UAV.

It is also assumed that each user can automatically choose among the MBS and the UAVs to obtain the best SNR. Therefore, it is necessary to determine the region a certain UAV is able to provide better SNR than others (including the MBS and the other UAVs). We denote the region where UAV$_n$ provides better SNR as UAV$_n$'s effective offloading region, denoted by Ω_n.

Utility of the UAV Operators

Each mobile user in an effective offloading region is assumed to access the UAV randomly. We call the number of the users in Ω_n that want to connect to UAV$_n$ at any instant as the active user number of UAV$_n$, denoted by ε_n. We assume that ε_n obeys a poisson distribution[4] with a mean value of μ_n. Based on μ_n, we can classify the UAVs into multiple types. Specifically, we refer to UAV$_n$ as a λ-type UAV if $\mu_n = \lambda$, which means that there is an average λ users connecting to UAV$_n$ at any instant. The number of λ-type UAVs is denoted by N_λ, where $\sum_\lambda N_\lambda = N$. For simplicity, we use a random variable X_λ (instead of ε_n) to denote the active user number of a λ-type UAV. The probability of $X_\lambda = k$ is given by

$$P(X_\lambda = k) = \frac{(\lambda)^k}{k!} e^{-\lambda}, \qquad k = 0, 1, 2, \ldots \tag{3.3}$$

Without a loss of generality, we assume that each mobile user connecting to a UAV (or the MBS) is allocated one channel with a fixed bandwidth B in a frequency division pattern. Due to the variation of the active user number, there is always a probability that an UAV fails to serve the current active users. Therefore, the more channels are being obtained, the more utility the UAV can achieve. The utility function of obtaining w channels for a λ-type UAV is denoted by $U(\lambda, w)$. Because the utility of obtaining no channels is 0, we have

$$U(\lambda, w) = 0, \qquad w = 0. \tag{3.4}$$

Now assume that we have determined the value of $U(\lambda, w-1)$; the rest of the problem involves how to obtain $U(\lambda, w)$ by figuring out the marginal utility of obtaining the wth channel. Note that the newly added wth channel is only useful when there are more than $w-1$ active users at the given moment. Therefore, the marginal utility is $P(X_\lambda \geq w) \times 1$ (i.e., the probability of more than $w-1$ users are active at the moment). Thus, we have

$$U(\lambda, w) = U(\lambda, (w-1)) + P(X_\lambda \geq w), \qquad w \geq 1. \tag{3.5}$$

Based on (3.4) and (3.5), we can derive the general term of the λ-type UAV's utility as

$$U(\lambda, w) = \sum_{k=1}^{k=w} P(X_\lambda \geq k), \qquad w \geq 1. \tag{3.6}$$

Cost of the MBS Manager

It is assumed that the MBS will not reuse the spectrum that is already sold, which implies the MBS manager suffers a certain degree of loss as it sells the spectrum to UAV operators. The active user number of the MBS is also assumed to follow the Poisson distribution. We denote this random variable as X_{BS} and its mean value as λ_{BS}. Therefore, we have

$$P(X_{BS} = k) = \frac{(\lambda_{BS})^k}{k!} e^{-\lambda_{BS}}, \qquad k = 0, 1, 2, \ldots. \tag{3.7}$$

The total number of channels of the MBS is denoted by M, $M \in \mathbb{Z}^+$. Just like with UAVs, there is a utility of a certain number of channels for the MBS manager, $U_{BS}(m)$, representing the average number of users that m channels can serve, given as $U_{BS}(m) = 0$, for $m = 0$, and $U_{BS}(m) = U_{BS}(m-1) + P(X_{BS} \geq m)$, for $m \geq 1$. Based on the utility of the MBS manager, we define the cost function $C(m)$ as the utility loss of reducing the number of channels from M to $M-m$, given as

$$C(m) = U_{BS}(M) - U_{BS}(M-m) = \sum_{k=M-m+1}^{M} P(X_{BS} \geq k). \tag{3.8}$$

Contract Formulation

Because different types of UAVs have different demands, the MBS manager has to design a contract that contains a set of quality-price options for all the UAV operators, denoted by $\{(w(\lambda), p(\lambda)) \mid \forall \lambda \in \Lambda\}$, where Λ represents the set of all the UAV types. In this contract, the quality $w(\lambda)$ is the number of channels designed to sell to a λ-type UAV operator, and $p(\lambda)$ is the corresponding price designed to be charged. Each $(w(\lambda), p(\lambda))$ pair can be seen as a commodity with quality $w(\lambda)$ at price $p(\lambda)$.

However, each UAV operator is expected to choose the one that maximizes its own profit according to the whole contract. The contract is feasible if and only if any λ-type UAV operator considers the commodity $(w(\lambda), p(\lambda))$ its best choice. And to achieve this, the first requirement is the incentive compatible (IC) condition, implying that the commodity designed for a λ-type UAV operator in the contract is no worse than other commodities, given by

$$U(\lambda, w(\lambda)) - p(\lambda) \geq U(\lambda, w(\lambda')) - p(\lambda'), \qquad \forall \lambda' \neq \lambda. \tag{3.9}$$

If (3.9) is not satisfied, then a λ-type UAV operator may turn to another commodity, and the λ-type commodity is not properly designed. The second requirement is the individually rational (IR) condition, meaning that the λ-type UAV operator will not buy any of the commodities in the contract if all of the options lead to negative profits. In other words, the commodity designed for a λ-type UAV should lead to a nonnegative profit, even if this commodity is an empty commodity (with zero quality and zero price), given by

$$U(\lambda, w(\lambda)) - p(\lambda) \geq U(\lambda, 0) - 0 = 0, \tag{3.10}$$

where $U(\lambda, 0) - 0$ implies an empty commodity in the contract. This condition is added to avoid the case where the best commodity for a λ-type UAV is negative. In conclusion, a feasible contract has to satisfy the IC constraint and the IR constraint, and any contract that satisfies the IC and IR constraints is guaranteed to be feasible [60].

For the MBS manager, the overall revenue brought by the contract $\{w(\lambda), p(\lambda) \mid \forall \lambda \in \Lambda\}$ is

$$R = \sum_{\lambda \in \Lambda} \Big(N_\lambda \cdot p(\lambda) \Big) - C\left(\sum_{\lambda \in \Lambda} N_\lambda \cdot w(\lambda) \right), \tag{3.11}$$

where $N_\lambda \cdot p(\lambda)$ is the total payment obtained from λ-type UAV operators, and $\sum_{\lambda \in \Lambda} N_\lambda \cdot w(\lambda)$ is the total number of channels being sold. The objective of the MBS manager is to design proper $w(\lambda)$ and $p(\lambda)$ for any given $\lambda \in \Lambda$ in a way

Table 3.1 Notations in our model

Notation	Description
E_n	Total energy of UAV$_n$
e_n	Stabilization energy consumption of UAV$_n$ during a time slot
q_n	Mobility energy consumption of UAV$_n$ for a unit distance
C_n	Cost of deploying and retrieving UAV$_n$
a, b	Terrain parameters
η_{NOLS}, η_{LOS}	Additional pathloss parameters for non-LoS and LoS
P_{MBS}, P_{UAV}	Transmission power of the MBS and the UAVs
$\bar{L}_{UAV}(\theta, d), \bar{L}_{MBS}(d)$	Average pathloss to the user with elevation angle θ and distance d
Ω_n, S_n	Effective coverage region and effective coverage area of UAV$_n$
ε_n	Active user number of UAV$_n$ (random variable)
μ_n	Average active user number of UAV$_n$
λ	Type of a UAV (equal to average active users)
Λ	Set of the types of the UAVs
λ_{BS}	Average active user number of the MBS
N_λ	Number of λ-type UAVs
$U(\lambda, w)$	Utility of w channels for a λ-type UAV
M	Number of MBS's channels
$C(m)$	Cost of the MBS when selling m channels
$w(\lambda), p(\lambda)$	Number of channels and corresponding price designed for a λ-type UAV

that maximizes its own revenue with the preconsideration of each UAV operator's behavior, given as

$$
\begin{aligned}
\hat{R} = & \max_{\{w(\lambda)\},\{p(\lambda)\}} \sum_{\lambda\in\Lambda} \Big(N_\lambda \cdot p(\lambda)\Big) - C\Big(\sum_{\lambda\in\Lambda} N_\lambda \cdot w(\lambda)\Big), \\
s.t.\ & U(\lambda, w(\lambda)) - p(\lambda) \geq U(\lambda, w(\lambda')) - p(\lambda') \geq 0, && \forall \lambda, \lambda' \in \Lambda \text{ and } \lambda' \neq \lambda, \\
& U(\lambda, w(\lambda)) - p(\lambda) \geq 0, && \forall \lambda, \lambda' \in \Lambda \text{ and } \lambda' \neq \lambda, \\
& p(\lambda) \geq 0, \quad w(\lambda) = 0, 1, 2 \ldots && \forall \lambda \in \Lambda, \\
& \sum_{\lambda\in\Lambda} N_\lambda \cdot w(\lambda) \leq M,
\end{aligned}
\tag{3.12}
$$

where the first two constraints represent the IC and IR, and the last one indicates the limited number of channels possessed by the MBS. In the rest of this section, the quality assignment $w(\lambda)$ and the pricing strategy $p(\lambda)$ are the two most basic concerns. In addition, we call the contract that optimizes the problem in (3.12) as the MBS optimal contract. Before studying the contact design problem, we provide Table 3.1 to summarize the notations in our model.

3.1.2 Optimal Contract Design for UAV Offloading

For simplicity, we put all the types $\{\lambda\}$ in ascending order, given by $\{\lambda_1, \ldots \lambda_t, \ldots \lambda_T\}$, where T is the number of different types. We have $1 \leq t \leq T$ and $\lambda_{t_1} < \lambda_{t_2}$ if $t_1 < t_2$. Note that in this case we call λ_{t_1} a lower type and λ_{t_2} a higher type. In addition, we also simplify N_{λ_t} as N_t, $w(\lambda_t)$ as w_t, and $p(\lambda_t)$ as p_t in the following discussions.

Basic Properties

Before we analyze the property of the utility function $U(\lambda, w)$, we first provide a more basic conclusion with respect to a Poisson distribution property on which the utility function is defined.

LEMMA 3.1 *Given that X_λ and $X_{\lambda'}$ are two Poisson distribution random variables with mean values λ and λ', respectively, if $\lambda > \lambda' > 0$, then $P(X_\lambda \geq k) > P(X_{\lambda'} \geq k)$ for any $k \in \mathbb{Z}^+$.*

Proof Consider X_α a Poisson distribution random variable with mean value α. We have $P(X_\alpha \geq k) = 1 - P(X_\alpha < k) = 1 - e^{-\alpha} \sum_{i=0}^{k-1} \frac{\alpha^i}{i!}$. Because α can be a real number in its definition domain, we derive the derivative of $P(X_\alpha \geq k)$ with respect to α, given as

$$\frac{\partial P(X_\alpha \geq k)}{\partial \alpha} = e^{-\alpha} \sum_{i=0}^{k-1} \frac{\alpha^i}{i!} - e^{-\alpha} \frac{\partial}{\partial \alpha} \left(\sum_{i=0}^{k-1} \frac{\alpha^i}{i!} \right). \tag{3.13}$$

For $k = 1$, $\frac{\partial}{\partial \alpha} \left(\sum_{i=0}^{k-1} \frac{\alpha^i}{i!} \right) = 0$. And for $k > 1$, $\frac{\partial}{\partial \alpha} \left(\sum_{i=0}^{k-1} \frac{\alpha^i}{i!} \right) = \sum_{i=1}^{k-1} \frac{\alpha^{i-1}}{(i-1)!} = \sum_{i=0}^{k-2} \frac{\alpha^i}{i!}$. Therefore, we have $\frac{\partial P(X_\alpha \geq k)}{\partial \alpha} = e^{-\alpha} \frac{\alpha^{k-1}}{(k-1)!} > 0, \forall k \in \mathbb{Z}^+$, and $\alpha > 0$. For any given $\lambda > \lambda' > 0$, we can deduce that $P(X_\lambda \geq k) - P(X_{\lambda'} \geq k) = \int_{\lambda'}^{\lambda} \frac{\partial P(X_\alpha \geq k)}{\partial \alpha} d\alpha > 0$, $\forall k \in \mathbb{Z}^+$. □

This lemma is particularly singled out because it is used in many of the following propositions.

PROPOSITION 3.1 *The utility function $U(\lambda, w)$ monotonously increases with the type λ and the quality w, where $\lambda > 0$ and $w \in \mathbb{N}$. In addition, the marginal increase of $U(\lambda, w)$ with respect to w gets smaller as w increases, as shown in Fig. 3.2.*

Proof Consider a fixed value $w \in \mathbb{N}$ and $\lambda > \lambda' > 0$. If $w = 0$, we have $U(\lambda, w) = U(\lambda', w) = 0$ according to the definition. If $w > 0$, then $U(\lambda, w) - U(\lambda', w) = \sum_{k=1}^{k=w} \left[P(X_\lambda \geq w) - P(X_{\lambda'} \geq w) \right] > 0$ according to Lemma 3.1. Therefore, $U(\lambda, w)$ monotonously increases with λ.

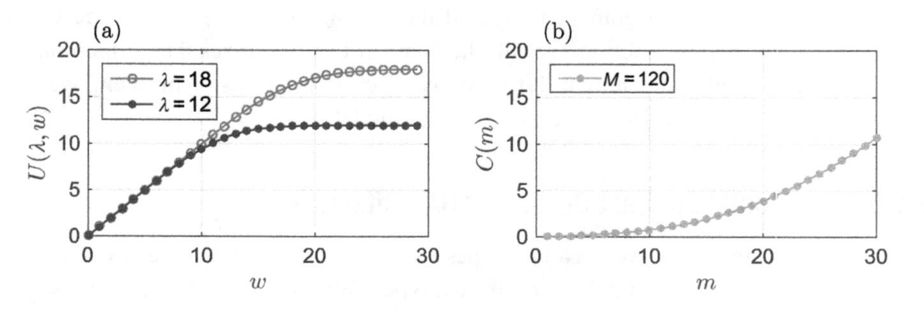

Figure 3.2 Simple illustration of the profiles of (a) UAV utility function and (b) MBS cost function

Now consider a fixed $\lambda > 0$ and $\forall w > w' \geq 0$, where $w, w' \in \mathbb{N}$. We have $U(\lambda, w) - U(\lambda, w') = P(X_\lambda \geq w) + \cdots + P(X_\lambda \geq w'+1) \geq P(X_\lambda \geq w) > 0$. Therefore, $U(\lambda, w)$ monotonously increases with w.

For a fixed $\lambda > 0$ and $\forall w \geq 1$, we have $U'(w) = U(\lambda, w) - U(\lambda, w-1) = P(X_\lambda \geq w)$. And for $w \geq 2$, we have $U''(w) = U'(w) - U'(w-1) = -P(X_\lambda = w-1) < 0$. Therefore, the marginal increase of $U(\lambda, w)$ with respect to w gets smaller as w increases. □

This proposition provides a basic property for us to design the optimal contract.

Based on Lemma 3.1 and Proposition 3.1, we exploit another important property of $U(\lambda, w)$, which says that a certain amount of quality improvement is more attractive to a higher type UAV than a lower type UAV. This property can be referred to as the increasing preference (IP) property, and we write it as the following proposition:

PROPOSITION 3.2 (IP property) *For any UAV types $\lambda > \lambda' > 0$ and channel qualities $w > w' \geq 0$, the following inequality holds: $U(\lambda, w) - U(\lambda, w') > U(\lambda', w) - U(\lambda', w')$.*

Proof According to the definitions of the utility function in (3.4) and (3.5), we have

$$U(\lambda, w) - U(\lambda, w') = P(X_\lambda \geq w) + \cdots + P(X_\lambda \geq w'+1), \tag{3.14}$$

$$U(\lambda', w) - U(\lambda', w') = P(X_{\lambda'} \geq w) + \cdots + P(X_{\lambda'} \geq w'+1). \tag{3.15}$$

Based on Lemma 3.1, each term in (3.14) is greater than each corresponding term in (3.15). Therefore, we can obtain $U(\lambda, w) - U(\lambda, w') > U(\lambda', w) - U(\lambda', w')$. □

With the help of this property, we are able to deduce the best pricing strategy in the following.

Optimal Pricing Strategy

In this part, we use a fixed quality assignment $\{w_t\}$ to analytically deduce the optimal pricing strategy $\{p_t\}$.

Based on the previous work on contract theory (such as in [60]), the IC and IR constraints and the IP property of the utility function in a contract design problem can directly lead to this conclusion:

PROPOSITION 3.3 *For the contract $\{(w_t, p_t)\}$ with the IC & IR constraints and the IP property, the following statements are simultaneously satisfied:*

- *The relation of types and qualities is $\lambda_i < \lambda_j \implies w_i \leq w_j$.*
- *The relation of qualities and prices is $w_i < w_j \iff p_i < p_j$.*

This conclusion contains the basic properties of a feasible contract. It indicates that a higher price has to be associated with a higher quality and that a higher quality means a higher price should be charged. Although different qualities are not allowed to be associated with the same price, it is possible that different types of UAVs are assigned the same quality and the same price.

LEMMA 3.2 *For the contract* $\{(w_t, p_t)\}$ *with the IC and IR constraints and the IP property, the following three conditions are the necessary and sufficient to determine a feasible pricing:*

- $0 \le w_1 \le w_2 \le \cdots \le w_T$,
- $0 \le p_1 \le U(\lambda_1, w_1)$, *and*
- $p_{k-1} + A \le p_k \le p_{k-1} + B$, *for* $k = 2, 3, \ldots, T$,
  where $A = [U(\lambda_{k-1}, w_k) - U(\lambda_{k-1}, w_{k-1})]$ *and* $B = [U(\lambda_k, w_k) - U(\lambda_k, w_{k-1})]$.

Proof **Necessity:** These three conditions can be deduced from the IC and IR constraints and the IP property as follows:

- Because $\{\lambda_1, \lambda_2, \ldots, \lambda_T\}$ is written in ascending order, we have $0 \le w_1 \le w_2 \le \cdots \le w_T$ and $0 \le p_1 \le p_2 \le \cdots \le p_T$ according to Proposition 3.3, where $w_i = w_{i+1}$ if and only if $p_i = i_{i+1}$.
- Considering the IR constraint of λ_1-type UAVs, we can directly obtain $0 \le p_1 \le U(\lambda_1, w_1)$. Here, if $w_x = 0$, then $U(\lambda_t, w_t) = 0$ and $p_t = 0$ for any $t \le x$.
- Considering the IC constraint for the k-type and the $(k-1)$-type where $k > 1$, the corresponding expressions are given by $U(\lambda_k, w_k) - p_k \ge U(\lambda_k, w_{k-1}) - p_{k-1}$, and $U(\lambda_{k-1}, w_{k-1}) - p_{k-1} \ge U(\lambda_{k-1}, w_k) - p_k$. As we focus on the possible scope of p_k, we can deduce that $p_{k-1} + [U(\lambda_{k-1}, w_k) - U(\lambda_{k-1}, w_{k-1})] \le p_k \le p_{k-1} + [U(\lambda_k, w_k) - U(\lambda_k, w_{k-1})]$.

Sufficiency: We have to prove that the prices $\{(p_t)\}$ determined by these conditions satisfy the IC and IR constraints. The basic idea is to use mathematical induction, from (w_1, p_1) to (w_T, p_T), by adding the quality-price terms one at a time into the whole contract. For simplicity, the contract that only contains the first k types of UAVs is denoted as $\Psi(k)$, where $\Psi(k) = \{(w_t, p_t)\}, 1 \le t \le k$. First, we can verify that $w_1 \ge 0$ and $0 < p_i < U(\lambda_1, w_1)$ provided by the conditions in Lemma 3.2 are feasible in $\Psi(1)$, because the IR constraint $U(\lambda_1, w_1) - p_i > 0$ is satisfied and the IC constraint is not useful in a single-type contract.

In the rest part of our proof, we show that if $\Psi(k)$ is feasible, then $\Psi(k+1)$ is also feasible, where $k + 1 \le T$. To this end, we first need to prove that the newly added λ_{k+1}-type complies with its IC and IR constraints, given by

$$\begin{cases} U(\lambda_{k+1}, w_{k+1}) - p_{k+1} \ge U(\lambda_{k+1}, w_i) - p_i, & \forall i = 1, 2, \ldots, k, \\ U(\lambda_{k+1}, w_{k+1}) - p_{k+1} \ge 0. \end{cases} \tag{3.16}$$

We next prove that the existing k types still comply with their IC constraints with the addition of λ_{k+1}-type, given by

$$U(\lambda_i, w_i) - p_i \ge U(\lambda_i, w_{k+1}) - p_{k+1}, \quad \forall i = 1, 2, \ldots, k. \tag{3.17}$$

First, we prove (3.16): Because $\Psi(k)$ is feasible, the IC constraint of λ_k-type should be satisfied, given by $U(\lambda_k, w_i) - p_i \leq U(\lambda_k, w_k) - p_k, \forall i = 1,2,\ldots,k$. Based on the right inequality in the third condition, we have $p_{k+1} \leq p_k + U(\lambda_{k+1}, w_{k+1}) - U(\lambda_{k+1}, w_k)$. By adding up these two inequalities, we have $U(\lambda_k, w_i) - p_i + p_{k+1} \leq U(\lambda_k, w_k) + U(\lambda_{k+1}, w_{k+1}) - U(\lambda_{k+1}, w_k), \forall i = 1,2,\ldots,k$. According to the IP property, we can obtain that $U(\lambda_k, w_k) - U(\lambda_k, w_i) \leq U(\lambda_{k+1}, w_k) - U(\lambda_{k+1}, w_i), \forall i = 1,2,\ldots,k$, because $\lambda_{k+1} > \lambda_k$ and $w_k \geq w_i$. Again, by combining these two inequalities together, we can prove the IC constraint of the λ_{k+1}-type, given by $U(\lambda_{k+1}, w_{k+1}) - p_{k+1} \geq U(\lambda_{k+1}, w_i) - p_i, \forall i = 1,2,\ldots,k$. The IR constraint of the λ_{k+1}-type can be easily deduced from the IC constraint because $U(\lambda_{k+1}, w_i) - p_i \geq U(\lambda_i, w_i) - p_i \geq 0, \forall i = 1,2,\ldots,k$. Therefore, we have $U(\lambda_{k+1}, w_{k+1}) - p_{k+1} \geq 0$.

Then, we prove (3.17): Because $\Psi(k)$ is feasible, the IC constraint of λ_i-type, $i = 1,2,\ldots,k$, should be satisfied, given by $U(\lambda_i, w_k) - p_k \leq U(\lambda_i, w_i) - p_i, \forall i = 1,2,\ldots,k$. Based on the left inequality in the third condition, we have $p_k + U(\lambda_k, w_{k+1}) - U(\lambda_k, w_k) \leq p_{k+1}$. By adding up the two previous inequalities, we have $U(\lambda_i, w_k) + U(\lambda_k, w_{k+1}) - U(\lambda_k, w_k) \leq U(\lambda_i, w_i) - p_i + p_{k+1}, \forall i = 1,2,\ldots,k$. According to the IP property, we can obtain that $U(\lambda_i, w_{k+1}) - U(\lambda_i, w_k) \leq U(\lambda_k, w_{k+1}) - U(\lambda_k, w_k), \forall i = 1,2,\ldots,k$, because $\lambda_k \geq \lambda_i$ and $w_{k+1} \geq w_k$. Again, by combining the two previous inequalities, we can prove the IC constraint of the existing types, λ_i, $\forall i = 1,2,\ldots,k$, given by $U(\lambda_i, w_i) - p_i \geq U(\lambda_i, w_{k+1}) - p_{k+1}$.

So far, we have proved that $\Psi(1)$ is feasible, and if $\Psi(k)$ is feasible, then $\Psi(k+1)$ is also feasible. We can conclude that the final contract $\Psi(T)$, which includes all the types is feasible. Therefore, these three necessary conditions are also sufficient conditions. □

It is an important guideline to design the prices for different types of UAVs. This implies that with a fixed quality assignment $\{w_t\}$ the proper scope of the price p_k depends on the value of p_{k-1}.

In the following, we provide the optimal pricing strategy of the MBS manager with a fixed quality assignment $\{w_t\}$. Here we call $\{w_t\}$ a feasible quality assignment if $w_1 \leq w_2 \leq \cdots \leq w_T$ and $\sum_{t=1}^{T} w_t \leq M$ (i.e., the first condition in Lemma 3.2 and the channel number constraint are satisfied). The maximum achievable revenue of the MBS manager with a fixed and feasible quality assignment $\{w_t\}$ is given by

$$R^*(\{w_t\}) = \max_{\{p_t\}} \left[\sum_{t=1}^{T} (N_t \cdot p_t) - C\left(\sum_{t=1}^{T} N_t \cdot w_t \right) \right]. \tag{3.18}$$

From this equation we can see that the key point is to maximize $\sum_{t=1}^{T}(N_t \cdot p_t)$ because the cost function is constant with a fixed quality assignment $\{w_t\}$. Accordingly, we provide the following proposition for the optimal pricing strategy:

PROPOSITION 3.4 (Optimal Pricing Strategy) *Given that $\{(w_t, p_t)\}$ is a feasible contract with a feasible quality assignment $\{w_t\}$, the unique optimal pricing strategy $\{\hat{p}_t\}$ is*

$$\begin{cases} \hat{p}_1 = U(\lambda_1, w_1), \\ \hat{p}_k = \hat{p}_{k-1} + U(\lambda_k, w_k) - U(\lambda_k, w_{k-1}), \quad \forall k = 2, 3, \ldots, T. \end{cases} \tag{3.19}$$

Proof By comparing (3.19) with Lemma 3.2, we can find that $\{\hat{p}_t\}$ is a feasible pricing strategy. In the following, we first prove that $\{\hat{p}_t\}$ is optimal and then prove that it is unique.

Optimality: Under the condition that quality assignment $\{w_t\}$ is fixed, $\{\hat{p}_t\}$ is optimal if and only if $\sum_{t=1}^{T}(N_t \cdot \hat{p}_t) \geq \sum_{t=1}^{T}(N_t \cdot p_t)$, where $\{p_t\}$ is any pricing strategy that satisfies the conditions in Lemma 3.2. Let's assume that there exists another better strategy $\{\tilde{p}_t\}$ for the MBS manager (i.e., $\sum_{t=1}^{T}(N_t \cdot \tilde{p}_t) \geq \sum_{t=1}^{T}(N_t \cdot \hat{p}_t)$). Because $N_t > 0$ for all $t = 1, 2, \ldots, T$, there is at least one $k \in \{1, 2, \ldots, T\}$ that satisfies $\tilde{p}_k > \hat{p}_k$. To guarantee that $\{\tilde{p}_t\}$ is still feasible, the following inequality must be complied with according to Lemma 3.2: $\tilde{p}_k \leq \tilde{p}_{k-1} + U(\lambda_k, w_k) - U(\lambda_k, w_{k-1})$, if $k > 1$. Because $\tilde{p}_k > \hat{p}_k$, we have $\hat{p}_k < \tilde{p}_{k-1} + U(\lambda_k, w_k) - U(\lambda_k, w_{k-1})$, if $k > 1$. By substituting (3.19) into this inequality, we have $\tilde{p}_{k-1} > \hat{p}_k + U(\lambda_k, w_k) - U(\lambda_k, w_{k-1}) = \hat{p}_{k-1}$, if $k > 1$. Repeating this process, we can finally obtain the result that $\tilde{p}_1 > \hat{p}_1 = U(\lambda_1, w_1)$, which contradicts Lemma 3.2, where p_1 should not exceed $U(\lambda_1, w_1)$. Due to this contradiction, the assumption that $\{\tilde{p}_t\}$ is better than $\{\hat{p}_t\}$ is impossible. Therefore, $\{\hat{p}_t\}$ is the optimal pricing strategy for the MBS manager.

Uniqueness: Assume that there exists another pricing strategy $\{\tilde{p}_t\} \neq \{\hat{p}_t\}$ such that $\sum_{t=1}^{T}(N_t \cdot \tilde{p}_t) = \sum_{t=1}^{T}(N_t \cdot \hat{p}_t)$. Because $N_t > 0$ for all $t = 1, 2, \ldots, T$, there is at least one $k \in \{1, 2, \ldots, T\}$ that satisfies $\tilde{p}_k \neq \hat{p}_k$. If $\tilde{p}_k > \hat{p}_k$, then the same contradiction occurs, as we discussed earlier. If $\tilde{p}_k < \hat{p}_k$, then there must exist another $\tilde{p}_l > \hat{p}_l$ to maintain $\sum_{t=1}^{T}(N_t \cdot \tilde{p}_t) = \sum_{t=1}^{T}(N_t \cdot \hat{p}_t)$. Either way, the contradiction is unavoidable, which implies that the optimal pricing strategy $\{\hat{p}_t\}$ is unique. □

We write the general formula of the optimal prices $\{\hat{p}_t\}$ as

$$\hat{p}_t = U(\lambda_1, w_1) + \sum_{i=1}^{t} \theta_i, \qquad \forall t = 2, \ldots, T, \tag{3.20}$$

where $\theta_1 = 1$ and $\theta_i = U(\lambda_i, w_i) - U(\lambda_i, w_{i-1})$ for $i = 2, \ldots, T$. The optimal pricing strategy is able to maximize R and achieve R^* with any given feasible quality assignment. However, it is still unsolved which $\{w_t\}$ is able to maximize R^* and achieve the overall maximum value $\hat{R}$.

Optimal Quality Assignment Problem

In this section, we analyze the optimal quality assignment problem and transform it into an easier form in preparation for the dynamic programming algorithm in the following.

The optimal quality assignment problem is given by

$$\begin{aligned} &\hat{R} = \max_{\{w_t\}} \left[R^*(\{w_t\}) \right], \\ &s.t. \sum_{t=1}^{T} N_t w_t \leq M, \ w_1 \leq w_2 \leq \cdots \leq w_T, \text{ and } w_t = 0, 1, 2, \ldots, \end{aligned} \tag{3.21}$$

where $R^*(\{w_t\})$ is the best revenue of a given quality assignment as given in (3.18). Based on the optimal pricing $\{\hat{p}_t\}$ in (3.20), we derive the expression of $R^*(\{w_t\})$ as

$$R^*(\{w_t\}) = \sum_{t=1}^{T} \left[C_t \cdot U(\lambda_t, w_t) - D_t \cdot U(\lambda_{t+1}, w_t) \right] - C\left(\sum_{t=1}^{T} N_t \cdot w_t \right), \tag{3.22}$$

where $C_t = \left(\sum_{i=t}^{T} N_i \right)$, $D_t = \left(\sum_{i=t+1}^{T} N_i \right)$ for $t < T$, and $D_T = 0$. Here, we are able to guarantee that $C_t > D_t \geq 0, \forall t = 1, 2, \ldots, T$, since $N_t > 0, \forall t = 1, 2, \ldots, T$. As we can observe from (3.22), w_i and w_j $(i \neq j)$ are separated from each other in the first term. This is a nonnegligible improvement to find the best $\{w_t\}$.

DEFINITION 3.3 *A set of functions* $\left\{ G_t(w_t) \middle| t = 1, 2, \ldots, T \right\}$, *with the quality* w_t *as the independent variable of* $G_t(\cdot)$, *with* C_t *and* D_t $(C_t > D_t \geq 0)$ *as the constants of* $G_t(\cdot)$, *is given by*

$$G_t(w_t) = C_t \cdot U(\lambda_t, w_t) - D_t \cdot U(\lambda_{t+1}, w_t), \quad w_t = 0, 1, 2, \ldots \quad \forall t = 1, 2, \ldots, T. \tag{3.23}$$

Based on (3.22) and Definition 3.3, we have $R^*(\{w_t\}) = \sum_{t=1}^{T} G_t(w_t) - C\left(\sum_{t=1}^{T} N_t \cdot w_t \right)$. The meaning of $G_t(w_t)$ is the independent gain of setting w_t for the λ_t-type UAVs regardless of the cost.

Given $\{G_t(w_t)\}$, we can rewrite the optimization problem in (3.21) as

$$\begin{aligned} \hat{R} = \max_{\{w_t\}} & \left[\sum_{t=1}^{T} G_t(w_t) - C\left(\sum_{t=1}^{T} N_t w_t \right) \right] \\ s.t. & \sum_{t=1}^{T} N_t w_t \leq M, \; w_1 \leq w_2 \leq \cdots \leq w_T, \text{ and } w_t = 0, 1, 2, \ldots. \end{aligned} \tag{3.24}$$

This problem can then be transformed into an equivalent one, given by

$$\begin{aligned} \hat{R} = & \max_{\{W = 0, 1, \ldots, M\}} \left\{ \max_{\{w_t\}} \left[\sum_{t=1}^{T} G_t(w_t) \right] - C(W) \right\}, \\ s.t. & \sum_{t=1}^{T} N_t w_t \leq W, \; w_1 \leq w_2 \leq \cdots \leq w_T, \text{ and } w_t = 0, 1, 2, \ldots, \end{aligned} \tag{3.25}$$

where the original problem is divided in to $M + 1$ subproblems (with different settings of W). Here we have $W \in \mathbb{Z}$ and $W \in [0, M]$, which can be comprehended as the possible value of $\sum_{t=1}^{T} N_t w_t$. From this formulation, we can see that the overall optimal revenue can be acquired by comparing the best revenue of $M + 1$ subproblems. Because $C(W)$ is fixed in each subproblem, in the following we only focus on how to maximize $\sum_{t=1}^{T} G_t(w_t)$, given as

$$\max_{\{w_t\}} \sum_{t=1}^{T} G_t(w_t),$$
$$s.t. \sum_{t=1}^{T} N_t w_t \leq W,\ w_1 \leq w_2 \leq \cdots \leq w_T,\ \text{and } w_t = 0,1,2,\ldots. \tag{3.26}$$

By calculating all the best results of (3.26) with different possible values of W, we are able to obtain the optimal solution of (3.25) by further taking into account $C(W)$.

Therefore, we regard (3.26) as the key problem to be solved. In the following, the proposed dynamic programming algorithm for this problem is presented.

Algorithm for the MBS Optimal Contract

In what follows, we first show how to consider (3.26) a distinctive form of the knapsack problem [61] and then provide our recurrence formula to calculate its maximum value $G_{\max}$. Next, we present the method to find the parameters $\{w_t\}$ that achieve $G_{\max}$ and finally provide an overview of whole solution including the optimal quality assignment $\{\hat{w}_t\}$ and the optimal pricing $\{\hat{p}_t\}$.

Special Knapsack Problem

First, we have to look at the constraints in the optimization parameters $\{w_i\}$. Because $w_i = 0,1,2,\ldots$ and $\sum_{t=1}^{T} N_t w_t \leq W$, we have $w_t \leq W$. To distinguish from the notation of *weight* in the following discussions, we use K instead of W as the common upper bound of $w_t, \forall t \in [1,T]$, where $K \leq W$. And we rewrite the constraint as $w_t \leq K$. Therefore for each t, there are in total $K+1$ optional values of w_t, given by $\{0,1,\ldots,K\}$. And the corresponding results of $G_t(w_t)$ are $\{G_t(0), G_t(1), G_t(2), \ldots, G_t(K)\}$, which represent the values of different object that we can choose. In addition, we interpret the constraint $\sum_{t=1}^{T} N_t w_t \leq W$ as the weight constraint in the knapsack problem, where W is the weight capacity of the bag and setting $w_t = k$ means taking up the weight of kN_t.

We list the values and the weight of different options in Table 3.2. Each row presents all the options of a type, and we should choose an option for each type. To be specific, the kth option in the tth row provides us with the value of $G_t(k)$ and the weight of kN_t. Due to the constraint of $w_1 \leq w_2 \leq \cdots \leq w_T$, we cannot choose the $(k+1)$th, $(k+2)$th, and so forth options in the tth row if we have already chosen the kth option in the $(t+1)$th row. Therefore, the algorithm introduced next basically starts from the last row and ends at the first row.

Table 3.2 All the optional objects to be selected

	Type	Optional values					Corresponding weights				
1	Type λ_1	$G_1(0)$	$G_1(1)$	$G_1(2)$	$\cdots$	$G_1(K)$	0	N_1	$2N_1$	$\cdots$	KN_1
2	Type λ_2	$G_1(0)$	$G_2(1)$	$G_2(2)$	$\cdots$	$G_2(K)$	0	N_2	$2N_2$	$\cdots$	KN_2
$\cdots$	$\cdots$	$\cdots$					$\cdots$				
T	Type λ_T	$G_T(0)$	$G_T(1)$	$G_T(2)$	$\cdots$	$G_T(K)$	0	N_T	$2N_T$	$\cdots$	KN_T

Recurrence Formula to Calculate the Maximum Value $G_{\max}$

The key nature of designing a dynamic programming algorithm is to find the subproblems of the overall problem and write the correct recurrence formula. Here we define $OPT(t,k,w)$, $\forall t \in [1,T]$, $\forall k \in [0,K]$, and $\forall w \in [0,W]$ as the optimal outcome that includes the decisions from the Tth row to the tth row, with two conditions: (1) the kth option in the tth row is chosen, and (2) the occupied weight is no more than w. Because the algorithm starts from the Tth row, we first provide the calculation of $OPT(T,k,w)$, $\forall k \in [0,K]$, and $\forall w \in [0,W]$, given as

$$OPT(T,k,w) = \begin{cases} G_T(k), & \text{if } w \geq kN_t, \\ -\infty, & \text{if } w < kN_t, \end{cases} \tag{3.27}$$

where $-\infty$ implies that $OPT(T,k,w)$ is impossible to achieve due to the lack of weight capacity. This expression is straightforward because it only includes the Tth row in Table 3.2. From $OPT(T,k,w)$, we can calculate $OPT(t,k,w)$ for all $t \in [1,T-1]$, $k \in [0,K]$ and $w \in [0,W]$ using the following recurrence formula:

$$OPT(t,k,w) = \begin{cases} \max\limits_{l=k,\ldots,K} \Big[G_t(k) + OPT(t+1,l,w-kN_t)\Big], & \text{if } w \geq kN_t, \\ -\infty, & \text{if } w < kN_t. \end{cases} \tag{3.28}$$

Here is the meaning of this formula: If we want to choose k in the tth row, then the option that made in the $(t+1)$th row must be within $[k,K]$ due to the constraint of $w_1 \leq \cdots \leq w_T$. In addition, choosing k in the tth row with total weight limit of w indicates that there is only $w - kN_t$ left for the other rows from $t+1$ to T. If $w - kN_t < 0$, the outcome is $-\infty$ because choosing k in the tth row is impossible.

Let $G_{\max}$ denote $\max\limits_{\{w_t\}} \sum_{t=1}^{T} G_t(w_t)$. Then, we have the following expression:

$$G_{\max} = \max_{k=0\ldots K} \Big[OPT(1,k,W)\Big]. \tag{3.29}$$

Thus, we have to calculate $OPT(1,k,W)$ for all $k \in [0,K]$, by iteratively using (3.28).

Method to Find the Parameters $\{w_t\}$ *that Achieve* $G_{\max}$

Note that the earlier calculation only considers the value of the optimal result $G_{\max}$. To record the exact values of $\{w_t\}$ chosen for this optimal result by the algorithm, we have to add another data structure, which is given as $D(t,k,w)$. We let $D(t,k,w) = l$ if $OPT(t,k,w)$ chooses l to maximize its value in the upper line of (3.28), which is given by

$$D(t,k,w) = \begin{cases} \arg\max\limits_{l=k,\ldots,K} \Big[G_t(k) + OPT(t+1,l,w-kN_t)\Big], & \text{if } w \geq kN_t, \\ 0, & \text{if } w < kN_t. \end{cases} \tag{3.30}$$

After acquiring $G_{\max}$ in (3.29), we can use $D(t,k,w)$ to inversely find the optimal values of $\{w_t\}$ along the "path" of the optimal solution. Specifically, we have

$$\begin{cases} \hat{w}_1 = \arg \max_{k=0\ldots K} \Big[OPT(1,k,W)\Big], \\ \hat{w}_t = D(t-1, \hat{w}_{t-1}, W - \sum_{i=1}^{t-2} \hat{w}_i N_i), \quad \forall t = 2,\ldots,T, \end{cases} \tag{3.31}$$

where we define $\sum_{i=1}^{t-2} \hat{w}_i N_i$ as 0 if $t-2=0$ for the sake of simplicity.

Solution Overview

By now, we have presented the key part of our solution: the dynamic programming algorithm to solve the optimization problem in (3.26). Equation (3.25) (i.e., our final goal) can be directly solved by setting different values of W in (3.26) and comparing the corresponding results with the consideration of $C(W)$.

An overview of our entire solution is given in Algorithm 3.1. It can be observed that the computational complexity of calculating $OPT(t,k,w)$ for all $k \in [0,K]$, $w \in [0,W]$, and $t \in [1,T]$ is $O(TK^2W)$. Therefore, the overall complexity is $O(MTK^2W)$, which can also be written as $O(TM^4)$ because $W \le M$ and $K \le W$. Although M^4 seems to be non-negligible, there are usually only hundreds of available channels of a MBS to be allocated in practice.[5]

Algorithm 3.1 Optimal contract for UAV offloading.

Input: Type information $\{\lambda_1,\ldots,\lambda_T\}$, $\{N_1,\ldots,N_T\}$, and the number of total channels M.

Output: Optimal pricing strategy $\{\hat{p}_1,\ldots,\hat{p}_T\}$, optimal quality assignment $\{\hat{w}_1,\ldots,\hat{w}_T\}$.

```
1  begin
2      Calculate G_t(k) for all t ∈ [1,T] and k ∈ [0,M] by (3.23);
3      Calculate C(m) for all m ∈ [0,M] by (3.8);
4      Initialize R̂ = 0, w_t = 0 for all t ∈ [1,T], and p_t = 0 for all t ∈ [1,T];
5      for W is from 0 to M do
6          Let K = W, to be the upper bound for each w_i;
7          Calculate OPT(T,k,w) for ∀k ∈ [0,K] and ∀w ∈ [0,W] by (3.27);
8          Calculate OPT(t,k,w) for ∀k ∈ [0,K], ∀w ∈ [0,W] and ∀t ∈ [1,T−1]
             by (3.28);
9          Acquire G_max from {OPT(1,w,t)} according to (3.29);
10         if G_max − C(W) > R̂ then
11             Update the overall maximum revenue R̂ = G_max − C(W);
12             Update ŵ_t for all t ∈ [1,T] according to (3.30) and (3.31);
13             Update p̂_t for all t ∈ [1,T] based on {ŵ_t} according to (3.19);
14         end
15     end
16 end
```

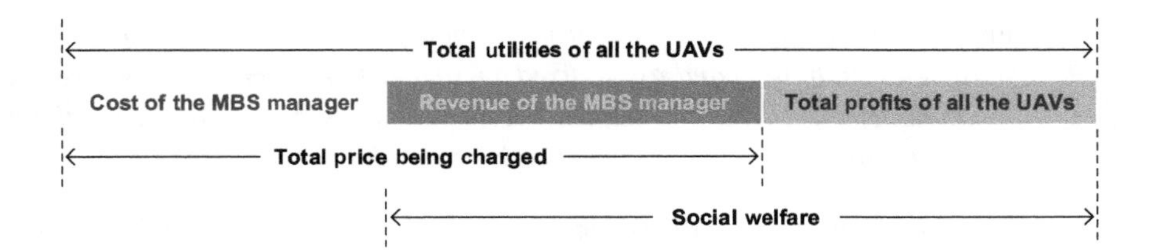

Figure 3.3 Relation of the social welfare, the revenue of the MBS manager, and the total profit of the UAVs operators

Socially Optimal Contract

To better discuss the effectiveness of the MBS optimal contract, in the following, we briefly discuss another contract that aims to maximize social welfare. Before that, we briefly explain the true meaning of social welfare. In our context, social welfare indicates the sum of the revenue of the MBS and the total profits of the UAVs (as shown in Fig. 3.3), which also means the increase of the number of users that can be served by the overall system.[6] Therefore, social welfare can be seen as the parameter to indicate the effectiveness of the UAV offloading system.

The objective of socially optimal contract is given by

$$\hat{S} = \max_{\{w(\lambda)\},\{p(\lambda)\}} \sum_{\lambda \in \Lambda} \Big(N_\lambda \cdot U(\lambda, w(\lambda))\Big) - C\Big(\sum_{\lambda \in \Lambda} N_\lambda \cdot w(\lambda)\Big), \tag{3.32}$$

where the first term is the total utility of the UAVs, the second term is the cost of the MBS, and we omit the constraints because they are the same as those in (3.12). This optimization problem has a structure similar to (3.12) and can be solved by the proposed dynamic programming algorithm with only minor changes. To calculate the optimal $\{w(\lambda)\}$ and $\{p(\lambda)\}$, we need to replace $G_t(k)$ with $N_t U(t,k)$ in line 2 of Table 3.2. In addition, we use $U_{\max}$ to replace $G_{\max}$ to represent the maximum overall utility of the UAVs. Lastly, the equation in line 11 of Table 3.2 should be replaced by $\hat{S} = U_{\max} - C(W)$ to represent the maximum social welfare. For convenience, in the rest part of this section, we call the solutions of (3.12) and (3.32) the MBS optimal contract and the socially optimal contract, respectively. In addition, the relation of social welfare and MBS's revenue is illustrated in Fig. 3.3.

3.1.3 Theoretical Analysis and Discussions

In this section, we briefly discuss the impact of the height of the UAVs, H. Because H influences the optimal revenue of the MBS $\hat{R}$, through the types of the UAVs $\{\lambda_t\}$, we first discuss the impact of H on $\{\lambda_t\}$ and then discuss the impact of $\{\lambda_t\}$ on $\hat{R}$.

Impact of the Height on the UAV Types

We first define $\sigma_n = \varepsilon_n / S_n$ as the average density of active users in the effective coverage region of UAV$_n$. Based on this concept, we provide the following proposition:

PROPOSITION 3.5 *With fixed transmission power P_{UAV} and P_{MBS}, fixed terrain parameters a, b, η_{Los} and η_{NLoS}, fixed average active user density σ_n, fixed horizontal locations of the UAVs, and unified height $H \in [0, +\infty)$ of the UAVs, there exists a height $\hat{H}_n$ that can maximize the effective offloading area of UAV_n.*

Proof We use ϕ (in radian) instead of θ (in degrees) to denote the elevation angle, where $\phi = \theta \cdot \pi/180°$. For a user with horizontal distance r to the UAV, the average pathloss is given by $\bar{L}_{UAV}(\phi,r) = L_{LoS}(d)P_{LoS}(\theta) + L_{NLoS}(d)\big[1 - P_{LoS}(\theta)\big]$. With minor deduction, we have

$$\bar{L}_{UAV}(\phi,r) = L_{NLoS}(d) - \eta \cdot P_{LoS}(\theta), \tag{3.33}$$

where $\eta = \eta_{NLoS} - \eta_{LoS} < \eta_{NLoS}$, $d = \frac{r}{\cos\phi}$, $\theta = \frac{180°}{\pi}\phi$. By denoting $L_{NLoS}(d)$ as L_1 and denoting $\eta P_{LoS}(\theta)$ as L_2 for simplicity, we can provide the following assertions based on (3.1): As ϕ increases from 0 to $\pi/2$, L_1 increases monotonously from $L_{NLoS}(r)$ to infinity, while L_2 monotonously increases within a subinterval of $(0, \eta)$. Therefore, $0 < \bar{L}_{UAV}(0,r) < L_{NLoS}(r)$, and $\bar{L}_{UAV}(\phi,r) \to +\infty$ as $\phi \to \pi/2$. In addition, $\bar{L}_{UAV}(\phi,r)$ has the lower bound $\big[L_{NLoS}(r) - \eta_{NLoS}\big]$ in the whole definition domain $[0, \pi/2]$. By considering the partial derivative of $\bar{L}_{UAV}(\phi,r)$ with respect to ϕ, we have

$$\frac{\partial \bar{L}_{UAV}(\phi,r)}{\partial \phi} = \frac{\partial L_1}{\partial \phi} - \frac{\partial L_2}{\partial \phi} = \frac{20}{\ln 10}\tan\phi - \frac{180°\pi^{-1}ab\eta\exp[-b(\theta-a)]}{\{1+a\exp[-b(\theta-a)]\}^2}, \tag{3.34}$$

where we have $\partial L_1/\partial\phi = 0$ as $\phi = 0$, $\partial L_1/\partial\phi \to +\infty$ as $\phi \to +\infty$, and $\partial L_2/\partial\phi > 0$ as $\forall\phi \in [0, \pi/2)$. (Also note that (3.34) is the same as the one in [49].) Therefore, we can conclude that $\bar{L}_{UAV}(\phi,r)$ decreases near $\phi = 0$ and rapidly increases to $+\infty$ near $\phi = \pi/2$.

By now we have confirmed three things: (1) $\bar{L}_{UAV}(\phi,r)$ decreases near $\pi = 0$, (2) $\bar{L}_{UAV}(\phi,r)$ increases to infinity as $\phi \to \pi/2$, and (3) $\bar{L}_{UAV}(\phi,r)$ has a lower bound in $[0, \pi/2)$. Therefore, there is at least one minimal value $\phi \in (0, \pi/2)$ that is smaller than $\bar{L}_{UAV}(0,r)$, which makes the existence of a minimum value $\phi \in (0, \pi/2)$. Figure 3.4 provides an exemplary illustration of $\bar{L}_{UAV}(\phi,r)$ with different r values.

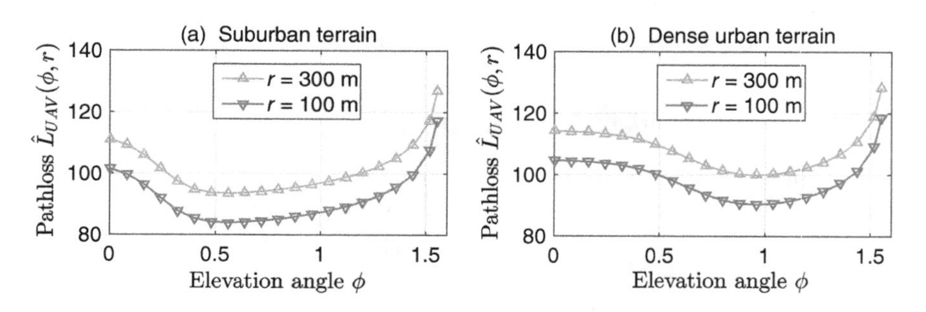

Figure 3.4 $\hat{L}_{UAV}(\phi,r)$ with different r values: (a) Pathloss in a typical suburban terrain, where parameters $a = 5$, $b = 0.2$, $\eta_{LoS} = 0.1$, and $\eta_{NLoS} = 21$. (b) Pathloss in a typical dense urban terrain, where parameters $a = 14$, $b = 0.12$, $\eta_{LoS} = 1.6$, and $\eta_{NLoS} = 23$

The effective offloading region of the UAV, however, is based on the SNR of each possible location. Rigorous mathematical analysis would be highly difficult, so we only include a simple discussion here. Because we have assumed that the UAVs have the same height and fixed horizontal locations, we can first conclude that, if a user is horizontally nearest to UAV_n, then the SNR from UAV_n is always the largest among all the UAVs no matter how large H is. Therefore, the user partition among UAVs are independent of H, and we only have to care about whether the SNR from UAV_n (γ_{UAV_n}) is greater than the SNR from the MBS (γ_{MBS}). For any given location, the scope of H that satisfies $\gamma_{UAV_n} > \gamma_{MBS}$ can be either an empty interval or one or more disjoint intervals (called the *effective height interval* of this user), depending on the number and the values of the minimal points of $\bar{L}_{UAV}(\phi, r)$.

At the height of H, the effective offloading area of UAV_n (given by S_n) depends on whether the value of H resides in the effective height interval of each possible location on the ground. The theoretical deduction of the optimal height that maximizes S_n is intractable. However, the existence of such an optimal height can be guaranteed because the effective height intervals are either empty or within $[0, +\infty)$. □

Because finding the optimal height is an intractable problem, the numerical method to obtain it can be done by numerically trying different values of H in our algorithm and seeing which value achieves the highest revenue for the MBS operator, as shown in Fig. 3.8.

Note that this conclusion is different from existing studies (such as [62]) because we define the effective coverage region of a UAV as the area that has a higher receive SNR from this UAV compared with the receive SNR from the MBS. From this proposition, we know that in the process of H varying from 0 to $+\infty$, different UAVs are able to achieve their maximum effective offloading areas at different heights. However, if all the UAVs are horizontally symmetrically distributed around the MBS (as shown in Fig. 3.8 in Section 3.1.4), their optimal heights will be the same because the UAVs have symmetrical positions. Therefore, there is a globally optimal height $\hat{H}$ that can maximize S_n for all $n \in \{1, 2, \ldots N\}$. Due to the fact that the types of the UAVs is given by $\lambda_n = \sigma_n S_n$, we can also achieve the largest type for each UAV.

Impact of the UAV Types on the Optimal Revenue

For any two random sets of types $\{\lambda_1, \ldots, \lambda_{T_1}\}$ and $\{\lambda'_1, \ldots, \lambda'_{T_2}\}$, there is no obvious relation between the outcomes of the corresponding two MBS optimal contracts. However, some properties can be explored when we add some constraints, as given in the following proposition:

PROPOSITION 3.6 *Given a fixed number of types T, two sets of types $\{\lambda_t\}$, $\{\lambda'_t\}$, and the constraint $\lambda_t \leq \lambda'_t$, $\forall t \in [1, T]$, we have $\hat{R} \leq \hat{R}'$, where $\hat{R}$ is the MBS's revenue of a MBS optimal contract with inputs $\{\lambda_t\}$, and $\hat{R}'$ is the MBS's revenue of a MBS optimal contract with inputs $\{\lambda'_t\}$.*

Proof For the MBS optimal contract based on $\{\lambda_t\}$, the bandwidth allocation is denoted as $\{w_t\}$, and the corresponding cost of the MBS is denoted as $C\left(\sum w_t\right)$.

If we change the types from $\{\lambda_t\}$ to $\{\lambda'_t\}$ and assume that the bandwidth allocation remains $\{w_t\}$, the cost of the MBS will still be $C\left(\sum w_t\right)$. Because $\lambda_t \leq \lambda'_t$, we have $U(\lambda_t, w) < U(\lambda'_t, w)$ according to Proposition 3.1. Besides, based on (3.19), we can deduce that p_t will be greater, for any $t = 1, \ldots, T$. Therefore, as the sum of prices increases, and the revenue of the MBS will increase from $\hat{R}$ to $\hat{R}_w$. Note that this discussion is based on the assumption that $\{w_t\}$ remains the same, which is probably not an optimal bandwidth allocation for $\{\lambda'_t\}$. If we run the algorithm in Section 3.1.2, the final revenue $\hat{R}'$ is based on another bandwidth allocation $\{w'_t\}$ being greater than $\hat{R}_w$. Therefore, we have $\hat{R} \leq \hat{R}_w \leq \hat{R}'$. □

With Propositions 3.5 and 3.6, we can directly obtain the conclusion that there exists a highest value of the MBS's revenue by manipulating the height of the UAVs, as long as the UAVs are horizontally symmetrically distributed around the MBS, as shown in Section 3.1.4.

3.1.4 Performance Evaluation

In this section, we simulate and compare the outcomes of the MBS optimal contract and the socially optimal contract under different settings.

Simulation Setups

We set M within [100, 300], which is sufficient to generally evaluate a real system such as LTE. The terrain parameters are set as $a = 11.95$ and $b = 0.136$, indicating a typical urban environment. We also set the transmission power as $P_{UAV} < P_{MBS}$, due to the typical consideration of UAVs that they have limited battery capacities. Parameter setting details can be found in Table 3.3.

Table 3.3 Simulation parameters

Terrain parameters a and b	11.95 and 0.136
Additional pathloss parameters η_{LoS} and η_{NLoS}	2 dB and 20 dB
Transmission power P_{MBS} and P_{UAV}	10 W and 50 mW
Downlink transmission frequency f	3 GHz
Height of UAVs H	Between 200 m and 1,000 m
Average active user density σ_n (also as μ_n/S_n) (km^{-2})	Between 10 and 20
Number of UAV types T	Between 1 and 20
Number of each type of UAVs $\{N_t\}$	Between 1 and 10
Average active user number of UAVs $\{\lambda_t\}$	Between 1 and 20
Average active user number of MBS λ_{BS}	Between 10 and 200
Number of total channels of MBS M	Between 100 and 300
Total energy of UAV$_n$ E_n	Between 2,000 mAh and 8,000 mAh
Stabilization energy consumption of UAV$_n$ during one time slot e_n	200 mAh
Moving energy consumption of UAV$_n$ between time slots q_n	Between 1 mAh/m and 5 mAh/m

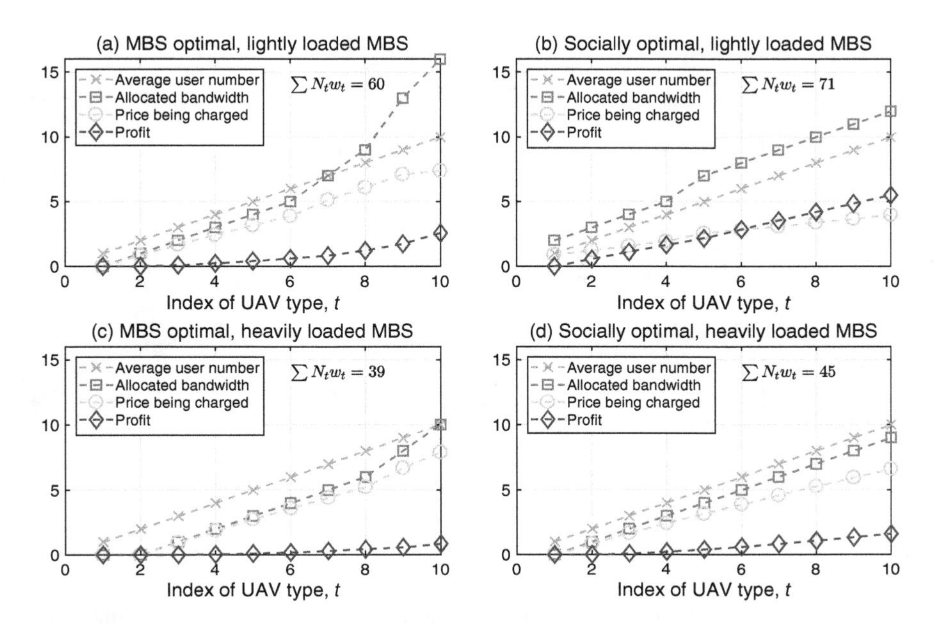

Figure 3.5 Structure of the optimal contracts where $T = 10$, $\{N_t\} = (1, 1, \ldots, 1)$, $\{\lambda_t\} = (1, 2, \ldots, 10)$, and $M = 200$, with $\lambda_{BS} = 120$ for (a) and (b), and $\lambda_{BS} = 160$ for (c) and (d). In addition, (a) and (c) show MBS optimal contracts, whereas (b) and (d) show socially optimal contracts

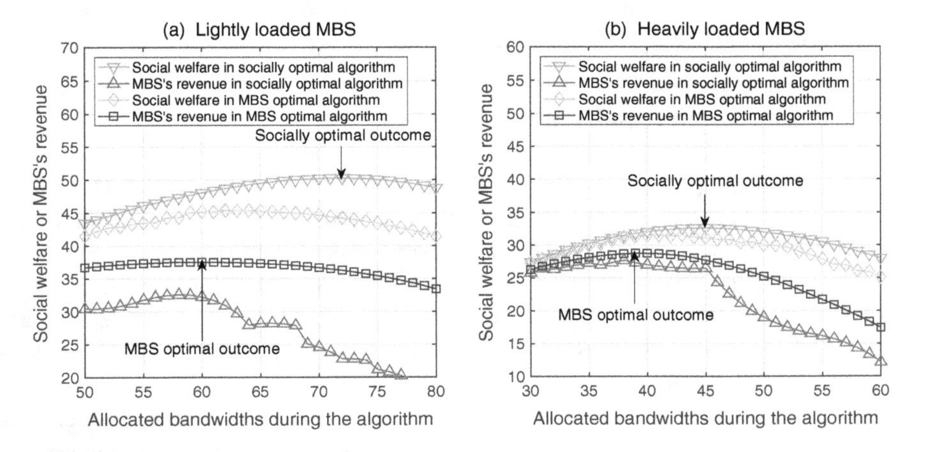

Figure 3.6 Change in welfare and MBS's revenue during the socially optimal algorithm and MBS optimal algorithm, where $T = 10$, $\{N_t\} = (1, 1, \ldots, 1)$, $\{\lambda_t\} = (1, 2, \ldots, 10)$, $M = 200$, with $\lambda_{BS} = 120$ for (a) and $\lambda_{BS} = 160$ for (b)

In the following simulations, we first study the UAV offloading system based on the given UAV types (i.e., fixed active user number for each UAV), from which we can acquire basic comprehension of the MSB optimal contract and the socially optimal contract, shown in Figs. 3.5, 3.6, and 3.7. Then, we further study a more practical

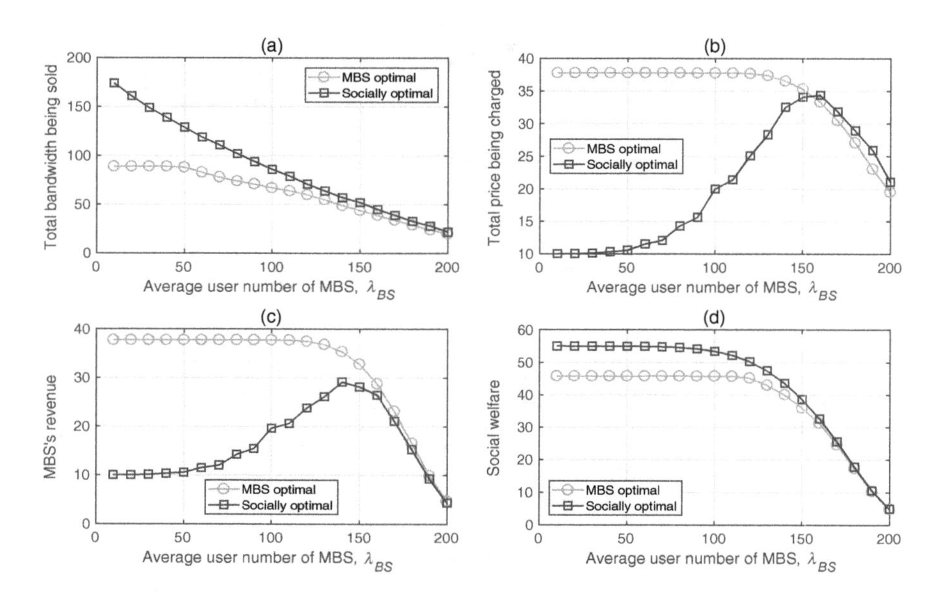

Figure 3.7 Impact of λ_{BS} on (a) bandwidth, (b) price, (c) MBS revenue, and (d) social welfare, where $T = 10$, $\{N_t\} = (1, 1, \ldots, 1)$, $\{\lambda_t\} = (1, 2, \ldots, 10)$, and $M = 200$

scenario where the height of the UAVs determines their types, as shown in Fig. 3.8. Lastly, we present the results of multiple time slots where the mobility and the energy constraints of a UAV influence its operator's long-term profit, as shown in Fig. 3.9.

Simulation Results and Discussions

We first illustrate the typical structure of the contract designed according to our algorithm, as given in Fig. 3.5, where we set $T = 10$, $\{N_t\} = (1, 1, \ldots, 1)$, $\{\lambda_t\} = (1, 2, \ldots, 10)$, and $M = 200$. All four subplots show the patterns of $\{w_t\}$, $\{p_t\}$, and $\{U(t, w_t) - p_t\}$ with respect to different type λ_t. To be specific, Fig. 3.5a and 3.5b show the results of a lightly loaded MBS ($\lambda_{BS} = 120$), and Fig. 3.5c and 3.5d show the results of a heavily loaded MBS ($\lambda_{BS} = 160$). In addition, Fig. 3.5a and 3.5c are the outcomes of MBS optimal contracts, whereas Fig. 3.5b and 3.5d are the outcomes of socially optimal contracts. In any one of these subplots, we can see that a higher type of UAV is allocated with more channels but also a higher price. It can also be observed that a higher type yields higher profit compared with a lower type; that is, $U(i, w_i) - p_i \leq U(j, w_j) - p_j$ as long as $i < j$. In Fig. 3.5a, it is noticeable that for λ_8, λ_9, and λ_{10}-types the allocated channels exceed their respective average user numbers. Such a phenomenon is reasonable because a UAV needs more channels w than its average active user number λ to deal with the situation of burst access. And due to the IP property, higher types consider additional channels more valuable than lower types. Therefore, only λ_8, λ_9, and λ_{10}-types are allocated with excessive channels. By comparing Fig. 3.5a with Fig. 3.5b, or Fig. 3.5c with Fig. 3.5d, we find that a socially optimal contract allocates more channels than a MBS optimal contract,

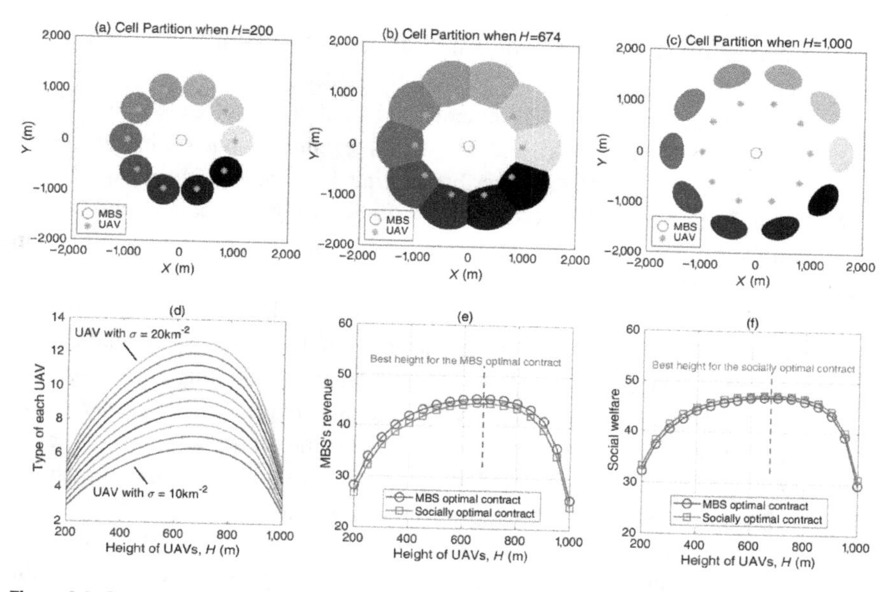

Figure 3.8 Impact of the height of UAVs. Subplots (a), (b), and (c) show the top views of the cell partition with different height settings. The white areas represent MBS's effective service regions, and gray areas represent the UAVs' effective offloading regions. Subplot (d) provides the impact on the type of each UAV with different active user density. Subplots (e) and (f) illustrate the impacts of the height of UAVs on MBS's revenue and social welfare, respectively

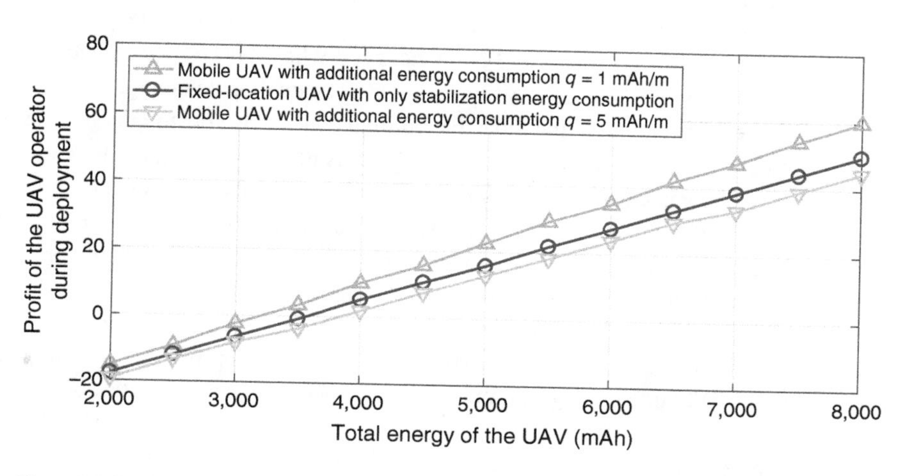

Figure 3.9 Impact of the mobility and battery capacity of the UAV, with $H = 200$ m, $M = 200$, $\lambda_{BS} = 120$

where we have 60 versus 71 in Fig. 3.5a and 3.5b, and 39 versus 45 in Fig. 3.5c and 3.5d. Thus, a socially optimal contract can be considered more "generous" than a MBS optimal contract. By comparing Fig. 3.5a with Fig. 3.5c, or Fig. 3.5b with Fig. 3.5d, we can also find a difference in the numbers of totally allocated channels.

This is because the cost of a heavily loaded MBS allocating the same number of channels is greater than that of a lightly loaded MBS.

To better explain the aforementioned bandwidth differences, we provide Fig. 3.6 to show how social welfare and the MBS's revenue change during the algorithm with W setting from 0 to M (as described in line 5 in Table 3.2). In Fig. 3.6a, the line with cricles shows the change in social welfare during the socially optimal algorithm. The highest point of this line curve represents the corresponding socially optimal contract, which makes $W = 71$, as in Fig. 3.5b. The line with boxes shows the corresponding change of the MBS's revenue during the socially optimal algorithm. For the MBS optimal algorithm, the resulting line curve of the the MBS's revenue lies above the one from the socially optimal algorithm, whereas the resulting curve of social welfare lies below the one from the socially optimal algorithm. Because the two groups of curves do not coincide, we can deduce that the structure of the solutions of the two algorithms are not identical. For a fixed W, the MBS optimal algorithm somehow changes the allocation of channels among different types to increase the MBS's revenue, which results in a reduction in social welfare. And the bandwidth allocation of the MBS optimal contract is $W = 60$, as in Fig. 3.5a. In Fig. 3.6b, we also show the situation of heavily loaded MBS, where the relation of these curves is similar as well as the reason that causes this.

Figure 3.7 illustrates the impacts of the load of the MBS, λ_{BS}, on the different part of the utility of the whole system, as presented in Fig. 3.3. From Fig. 3.7a we can see that the difference of allocated channels between the MBS optimal contract and the socially optimal contract becomes smaller as the load of MBS gets heavier. This is due to the fact that the cost of MBS rises fast when it is heavily loaded, and neither the MBS optimal or the socially optimal contract can allocate enough channels as desired. Figure 3.7b shows us that the MBS optimal contract is able to guarantee that high total prices are being charged because the MBS is not heavily loaded. In addition, the total price being charged according to the socially optimal contract is not monotonous and may rapidly change. For the case $\lambda_{BS} > 150$, although the total price being charged in the MBS optimal contract is lower than that in the socially optimal contract, the final revenue of the MBS is still higher in the MBS optimal contract, as shown in Fig. 3.7c. This is because the MBS optimal contract has less total bandwidth being sold, which reduces the cost of the MBS. The social welfare given in Fig. 3.7d, implies that, for both MBS and socially optimal contracts, a heavier loaded MBS could bring a lower overall system efficiency.

Next, we study the impact of the height of the UAVs, as presented in Fig. 3.8, where $M = 200$, $\lambda_{BS} = 150$. The considered 10 UAVs are located 1,000 m horizontally from the MBS and symmetrically distributed. The average active user density of the effective offloading region of UAV$_n$ (i.e., σ_n) is set from 10 km^{-2} to 20 km^{-2}. From the top three subplots in Fig. 3.8, we can see that the offloading regions of these UAVs first expand and then shrink when the height of the UAVs monotonously increases. The maximum offloading areas can be achieved at $H = 674$, where the UAVs can cover the largest number of active users, as given in Fig. 3.8d. In addition, the MBS's revenue can be maximized when offloading areas become the largest, as discussed in Section

3.1.3. It can also be observed in Fig. 3.8f that the profile of the social welfare in the MBS optimal contract is close to that of the social welfare in the socially optimal contract. In addition, the best height for the socially optimal contract ($H = 676$) is close to the best height for the MBS optimal contract ($H = 674$). Therefore, we can infer that the height H designated by the selfish MBS manager will generally maintain a high social welfare. In other words, the performance of the overall system will not be significantly impaired.

Lastly, we take a look at the influence of a UAV's mobility and energy constraint. We generate the initial distribution of users according to a Poisson point process (PPP) and then acquire the distribution of users in the next time slot according to random walk with a maximum moving distance of 10 m. Ten UAV operators are added into the system with a disjoint target region of users. Each UAV has a fixed cost of deploying, given by $C_n = 40$, $n = 1, 2, \ldots, 10$. We focus on only one of these UAVs, which can adjust its horizontal location between time slots with a maximum moving distance of 20 m, to greedily maximize its number of covered users (based on the algorithm in [63]). Figure 3.9 shows the result, comparing a fixed and mobile UAV. Here we further set the mobility cost to 1 mAh/m and 5 mAh/m to illustrate the difference between a low-cost movement and a high-cost movement. Because an adjustable UAV is able to cover as many users as possible in each time slot, the UAV's profit is expected to be higher. However, the energy consumption of mobility may also reduce the number of time slots of the deployment. Therefore, a low additional energy consumption ($q = 1$ mAh/m) of mobility could result in a better outcome for the UAV operator, whereas a high additional energy consumption ($q = 1$ mAh/m) of mobility could make it worse to adjust the position of UAV. Moreover, it can also be observed that the profit of the UAV operator has an approximate linear relation with the total energy of its UAV because a higher battery capacity can increase the time of deployment. To guarantee the total profit to be positive in the long term, the UAV operator should use a high-capacity battery for UAV offloading.

3.1.5 Conclusion

In this section, we focused on the scenario where the UAVs were deployed in a cellar network to better serve local mobile users. Considering the selfish MBS manager and the selfish UAV operators, we modeled the utilities and the costs of spectrum trading among them and formulated the problem of designing the optimal contract for the MBS manager. To deduce the optimal contract, we first derived the optimal pricing strategy based on a fixed quality assignment and then analyzed and transformed the optimal quality assignment problem, which can be solved by the proposed dynamic programming algorithm in polynomial time. In the simulations, by comparing with the socially optimal contract, we found that the MBS optimal contract allocated fewer channels to the UAVs to guarantee a lower level of costs. In addition, the best height of the UAVs for the selfish MBS manager can maintain a high performance of the overall system. Moreover, the UAV's mobility is able to increase the long-term profit of the UAV operator, but a high-capacity battery is also necessary.

3.2 UAVs Serving as Relays

Recently, UAVs have become especially helpful in the situations with widely scattered users, large obstacles such as hills or buildings that deteriorate the quality of links, and communication challenges caused by natural disasters [64]. Wireless communication assisted by a UAV (i.e., UAV relay) has been widely discussed. In UAV-aided relay networks, UAVs are deployed to provide wireless connectivity between two or more distant users or user groups without reliable direct communication links [48]. In [65], an energy efficiency maximization algorithm is proposed for a UAV relay with a circular trajectory. In [66], the authors study throughput maximization of a rectilinear trajectory UAV relay network. However, most of the works only consider the UAV location as a fixed point or on a fixed trajectory. In practice, UAVs can move freely in 3D space to achieve a better performance, but the trajectory design and power control on this condition have not been well studied.

In this section, we consider a half-duplex uplink UAV relay network with a UAV, a BS, and a mobile device (MD). The UAV works as an amplify-and-forward (AF) relay, which can adjust its transmit power and flying trajectory. We formulate the trajectory design and power control as a nonconvex outage probability minimization problem. The problem is decoupled into trajectory design and power control subproblems. We address these two subproblems using a gradient descent method and extremum principles, respectively. Finally, an approximate solution is obtained to approach the minimum outage probability.

The rest of this section is organized as follows. In Section 3.2.1, we describe the system model and formulate the outage probability minimization problem. In Section 3.2.2, a joint trajectory design and power control algorithm is given to solve this problem. Simulation results are presented in Section 3.2.3, and finally we summarize this section in Section 3.2.4.

3.2.1 System Model and Problem Formulation

As shown in Fig. 3.10, we consider an uplink scenario in a cellular network with one BS and one MD that is beyond the coverage of the BS [25]. A UAV works as an AF relay[7] to provide communication service for the MD. During the transmission, the UAV adjusts its location to improve the service quality. We assume that the transmission process contains N time slots. In two consecutive time slots, the MD transmits signals to the UAV in the first time slot, and the UAV amplifies and forwards the received signals from the MD to the BS in the second time slot. We denote the locations of the BS and MD by B and M, respectively.

Let L be the distance between the BS and MD. In time slot t, the distance between the MD and UAV and the distance between the UAV and BS are given by d_M^t and d_B^t, respectively. The flying distance of the UAV from time slot i to time slot j is $d_{i,j}$. We assume that the UAV can fly for a maximum distance of v in each time slot, where $v \ll L$. The transmit power of the MD in time slot t is given by P_M^t, and the transmit

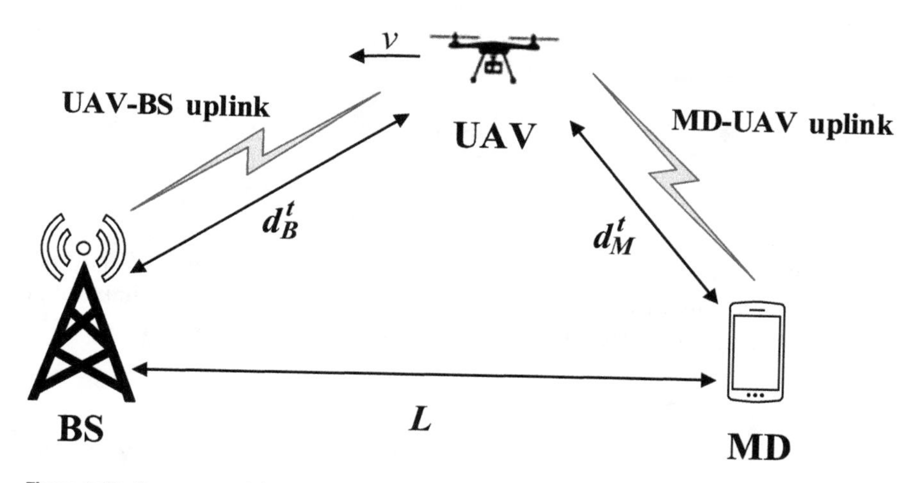

Figure 3.10 System model for communication with UAV relay

power of the UAV in time slot $(t+1)$ is given by P_U^{t+1}. The total transmit power in time slot t and $(t+1)$ is constrained[8] (i.e., $P_M^t + P_U^{t+1} \leq P_{\max}$).

In time slot t, the received power at the UAV from the MD is expressed as [69]

$$P_{M,U}^t = P_M^t \left(d_M^t\right)^{-\alpha} \left\|h_M^t\right\|^2. \tag{3.35}$$

In time slot $(t+1)$, the received power at the BS is shown as

$$P_{U,B}^{t+1} = P_U^{t+1}\left(d_B^{t+1}\right)^{-\alpha}\left\|h_B^{t+1}\right\|^2, \tag{3.36}$$

where α is the pathloss exponent, and h_M^t, h_B^{t+1} are independent small-scale channel fading coefficients with zero mean and unit variance. The noise at each node satisfies the Gaussian distribution with zero mean and N_0 as variance. The received signal of the UAV relay is expressed as

$$Y_{UAV}^t = \sqrt{P_M^t \left(d_M^t\right)^{-\alpha}} h_M^t X_M^t + n_U^t, \tag{3.37}$$

where X_M^t is the signal of unit energy from the MD, and n_U^t is the noise received at the UAV relay. The amplification coefficient of the UAV relay is given by

$$G_t = \sqrt{P_U^{t+1} \Big/ \left(P_M^t \left(d_M^t\right)^{-\alpha} \left\|h_M^t\right\|^2 + N_0\right)}. \tag{3.38}$$

After being amplified by the UAV relay, the received signal at the BS can be expressed as

$$\begin{aligned} Y_{BS}^{t+1} &= G_t \sqrt{P_M^t \left(d_M^t\right)^{-\alpha} \left(d_B^{t+1}\right)^{-\alpha}} h_M^t h_B^{t+1} X_M^t \\ &+ n_U^t G_t \sqrt{\left(d_B^{t+1}\right)^{-\alpha}} h_B^{t+1} + n_B^{t+1}, \end{aligned} \tag{3.39}$$

where n_B^{t+1} is the noise received at the BS. According to (3.39), the SNR of the uplink network is given by

$$\gamma_t = \frac{P_M^t \left(d_M^t\right)^{-\alpha} \left\|h_M^t\right\|^2 G_t^2 \left(d_B^{t+1}\right)^{-\alpha} \left\|h_B^{t+1}\right\|^2}{N_0 G_t^2 \left(d_B^{t+1}\right)^{-\alpha} \left\|h_B^{t+1}\right\|^2 + N_0}. \tag{3.40}$$

The outage probability is defined as the probability that the SNR falls below a predetermined threshold γ_{th}. Thus, the outage probability of the uplink can be derived by integrating the probability density function (PDF) of γ_t, shown as

$$P_{out}^t = P[\gamma_t \leq \gamma_{th}] = \int_0^{\gamma_{th}} f(\gamma_t) d\gamma_t. \tag{3.41}$$

THEOREM 3.4 *The approximate solution of the outage probability in time slot t and* $(t+1)$ *is*

$$\begin{aligned} P_{out}^t &= 1 - \exp\left(-\frac{N_0 \gamma_{th}}{P_M^t \left(d_M^t\right)^{-\alpha}}\right) \times (1 + 2V^2 \ln V), \\ V &= \sqrt{(N_0 \gamma_{th}) \Big/ \left(P_U^{t+1} \left(d_B^{t+1}\right)^{-\alpha}\right)}. \end{aligned} \tag{3.42}$$

Proof According to (3.40), we rewrite SNR γ_t as

$$\gamma_t = \left(a P_{M,U}^t P_{U,B}^{t+1}\right) \Big/ \left(a P_{U,B}^{t+1} N_0 + N_0\right), \tag{3.43}$$

where $a = G_t^2 / P_U^{t+1}$. Variables $P_{M,U}^t$ and $P_{U,B}^{t+1}$ obey exponential distribution for their physical significance. The outage probability in (3.41) can be simplified as

$$P_{out}^t = P\left(P_{M,U}^t \leq S + W\right), \tag{3.44}$$

where $S = N_0 \gamma_{th}$ and $W = N_0 \gamma_{th} / a P_{U,B}^{t+1}$. Let $\Phi = P_M^t (d_M^t)^{-\alpha}$, and the outage probability can be rewritten by

$$\begin{aligned} P_{out}^t &= E_{S+W}\left\{P\left(P_{M,U}^t \leq s + w | s + w\right)\right\} \\ &= E_{S+W}\left\{\int_0^{s+w} (1/\Phi) \exp(-x/\Phi) dx\right\} \\ &= E_{S+W}\{1 - \exp(-(s+w)/\Phi)\}. \end{aligned} \tag{3.45}$$

Because S and W are independent variables, we further obtain

$$P_{out}^t = 1 - E_S\{\exp(-S/\Phi)\} E_W\{\exp(-W/\Phi)\}. \tag{3.46}$$

Because variable S is a constant, E_S can be expressed by

$$E_S\{\exp(-S/\Phi)\} = \exp\left(-\left(N_0 \gamma_{th}\right)/\Phi\right). \tag{3.47}$$

We can also derive the PDF of W from the PDF of $P_{U,B}^{t+1}$, which is given by

$$\begin{aligned} f_W(w) &= \frac{d}{dw} P(W \leq w) = \frac{d}{dw} P\left(\frac{N_0 \gamma_{th}}{ay} \leq w\right) \\ &= \frac{N_0 \gamma_{th}}{aw^2} \frac{1}{\Psi} \exp\left(-\frac{N_0 \gamma_{th}}{a \Psi w}\right), \end{aligned} \tag{3.48}$$

where $\Psi = P_U^{t+1}\left(d_B^{t+1}\right)^{-\alpha}$. Thus, E_W can be expressed as

$$E_W\left\{\exp\left(-\frac{W}{\Phi}\right)\right\} = \frac{1}{\Psi}\int_0^{+\infty} \exp\left(-\frac{w}{\Phi}\right) \exp\left(-\frac{N_0 \gamma_{th}}{a \Phi w}\right) \frac{N_0 \gamma_{th}}{aw^2} dw. \tag{3.49}$$

By substituting (3.47) and (3.49) into (3.46), we have

$$P_{out}^t = 1 - \frac{1}{\Psi} \exp\left(-\frac{N_0 \gamma_{th}}{\Phi}\right) \int_0^{+\infty} \exp\left(-\frac{w}{\Phi}\right) \exp\left(-\frac{N_0 \gamma_{th}}{a \Psi w}\right) \frac{N_0 \gamma_{th}}{aw^2} dw. \tag{3.50}$$

According to the results in [70], (3.50) can be rewritten as

$$\begin{aligned} P_{out}^t &= 1 - 2V \exp\left(-\frac{N_0 \gamma_{th}}{\Phi}\right) K_{-1}(2V), \\ V &= \sqrt{\left(N_0 \gamma_{th}(\Phi + N_0)\right) / (\Phi \Psi)}, \end{aligned} \tag{3.51}$$

where $K_{-1}(x)$ is the negative first-order modified Bessel function of the second kind. Because $\Phi \gg N_0$, V in (3.51) can be simplified as $V = \sqrt{\frac{N_0 \gamma_{th}}{\Phi}}$. Note that the modified Bessel function of the second kind has the property $K_{-1}(x) = K_1(x)$. Thus, we expand $K_1(x)$ according to [70] and have

$$K_1(x) \simeq 1/x + x/2 \times \ln(x/2). \tag{3.52}$$

By substituting (3.52) into (3.51), the analytical approximate solution of the outage probability is shown as (3.42). □

Our objective is to minimize the outage probability by optimizing both the UAV trajectory and the power of the UAV and MD. Equation (3.42) shows that the outage probability in time slot t is only affected by the power and location parameters in time slot t and $(t+1)$. Therefore, we simplify the optimization objective as the outage probability in a single time slot, and the problem can be formulated by

$$\min_{P_M^t, P_U^{t+1}, d_M^t, d_B^{t+1}} \quad P_{out}^t, \tag{3.53a}$$

$$s.t. \; P_U^{t+1} + P_M^t \leq P_{\max}, \tag{3.53b}$$

$$P_U^{t+1} \geq 0, P_M^t \geq 0, \tag{3.53c}$$

$$d_{t,t+1} \leq v, \tag{3.53d}$$

where (3.53b) and (3.53c) are the power constraints for UAV and MD, and (3.53d) shows the UAV mobility constraint.

3.2.2 Power and Trajectory Optimization

Equation (3.42) shows that the joint power and trajectory optimization problem (3.53) is nonconvex. In this section, we tackle the problem through alternating minimization, where trajectory design and power control are optimized iteratively.

The algorithm is illustrated in Algorithm 3.2. In each iteration, we design the trajectory given the power control results obtained by the last iteration and then solve the power control subproblem given the UAV trajectory. In iteration k, let $S^k = \sum_{t=1}^{N} P_{out}^t$. The algorithm converges when $S^k - S^{k-1} < \epsilon$, where ϵ is a predefined error tolerance threshold.

Trajectory Design

Given the power control variables P_M^t and P_U^{t+1}, (3.53) can be expressed as

$$\min_{d_M^t, d_B^{t+1}} P_{out}^t, \tag{3.54a}$$

$$s.t.\ d_{t,t+1} \leq v. \tag{3.54b}$$

Equation (3.54) is also nonconvex. To achieve a local minimum outage probability, the UAV will fly in the direction with the maximum outage probability descent velocity (i.e., $-\nabla P_{out}^t$). Let $(0,0,0)$ and $(L,0,0)$ be the locations of the BS and the MD, respectively. In time slot t, we denote the location of the UAV by $\boldsymbol{l}_t = (x_t, y_t, z_t)$, and the trajectory by $\boldsymbol{\Delta l}_t = (\Delta x_t, \Delta y_t, \Delta z_t)$, with $|\boldsymbol{\Delta l}_t| \ll |\boldsymbol{l}_t|$. When the high-order terms are neglected, the trajectory direction of the UAV (i.e., the gradient of the outage probability function) is expressed as

$$-\nabla P_{out}^t = ((R-M)x_t + ML)\hat{\boldsymbol{\imath}} + (R-M)y_t\hat{\boldsymbol{\jmath}} + (R-M)z_t\hat{\boldsymbol{k}}, \tag{3.55}$$

Algorithm 3.2 Power and trajectory optimization algorithm

Input: Total transmit power $P_{\max}$, and the maximum speed v.
Output: Transmit powers $P_M^t, P_U^{t+1}\}$, and location parameters d_M^t, d_B^{t+1}.

```
1  begin
2      Initialize k = 0, S^0 = 0, P_M^t = P_U^{t+1} = P_max/2, ∀t = 1,3,...,N;
3      repeat
4          k = k + 1;
5          for t is from 1 to N do
6              Solve trajectory design subproblem (3.54) for slot t;
7          end
8          for t is from 1 to N do
9              Solve power control subproblem (3.57) for slot t;
10         end
11     until S^k − S^{k−1} ≤ ε;
12 end
```

where

$$\begin{aligned} M &= \frac{N_0\gamma_{th}}{P_M^t}\left((x_t-L)^2+y_t^2+z_t^2\right)^{\alpha/2-1}(1+Q\ln Q),\\ R &= \frac{N_0\gamma_{th}}{P_U^{t+1}}(x_t^2+y_t^2+z_t^2)^{\alpha/2-1}(1+\ln Q),\\ Q &= \frac{N_0\gamma_{th}}{P_U^{t+1}}(x_t^2+y_t^2+z_t^2)^{\alpha/2}; \end{aligned} \tag{3.56}$$

and $\hat{\boldsymbol{i}}$, $\hat{\boldsymbol{j}}$, and $\hat{\boldsymbol{k}}$ are the unit vectors of x, y, and z axis, respectively.

Because $v \ll L$, we set the step size of the gradient descent process as v for each time slot. We also set a minimum outage probability threshold δ, with $\delta \to 0^-$. When $-\nabla P_{out}^t \geq \delta$, we consider the minimum outage probability to be achieved, and the UAV stops moving. In the trajectory design for time slot $(t+1)$, the location of UAV is updated as $\boldsymbol{l}_{t+1} = \boldsymbol{l}_t + \boldsymbol{\Delta l}_t$.

Remark 3.1 When N is sufficiently large, the outage probability will tend to be stable at the minimum value.

THEOREM 3.5 *When the outage probability is minimized, the BS, MD, and UAV are collinear, and $d_B^t < d_M^t$ is satisfied.*

Proof When $-\nabla P_{out}^t = 0$, the outage probability is minimized. In (3.55), it can be easily found that the root of $-\nabla P_{out}^t = 0$ contains $y_t = 0$ and $z_t = 0$, which means the BS, MD, and UAV are collinear. The solution of x_t satisfies $\frac{L}{x_t} = 1 - \frac{N}{M}$, which cannot be solved easily. When we substitute $x_t = L/2$, we have $\frac{L}{x_t} < 1 - \frac{N}{M}$. It can be proved that the right side of the inequation is monotonically increasing with x_t at $x_t = L/2$, while the left side is monotonically decreasing. Therefore, the solution of x_t exists in $0 < x_t < L/2$, showing that the UAV is closer to the BS than to the MD. □

Power Control

Given the UAV trajectory $\boldsymbol{l}_t$, (3.53) can be rewritten as

$$\min_{P_M^t,\, P_U^{t+1}} P_{out}^t, \tag{3.57a}$$

$$\text{s.t. } P_U^{t+1} + P_M^t \leq P_{\max}, \tag{3.57b}$$

$$P_U^{t+1} \geq 0,\, P_M^t \geq 0. \tag{3.57c}$$

THEOREM 3.6 *The minimized outage probability is obtained when $P_U^{t+1} + P_M^t = P_{\max}$ is satisfied.*

Proof We note that both $P_{M,U}^t$ and $P_{U,B}^{t+1}$ are negatively related to P_{out}^t. Because received power is positively related with transmit power, the increment of transmit power decreases the outage probability. Therefore, minimum outage probability requires maximized total transmit power (i.e., $P_U^{t+1} + P_M^t = P_{\max}$). □

We substitute $P_M^t = P_{\max} - P_U^{t+1}$ into (3.57), and P_{out}^t is a function of P_U^{t+1}. The outage probability function in (3.57) is convex with respect to P_U^{t+1}. Therefore, when $\frac{d(P_{out})}{dP_U^{t+1}} = 0$, the optimal power control is realized. The closed-form solution of power control is given in the following theorem.

THEOREM 3.7 *The proposed power control is mainly determined by* d_M^t/d_B^{t+1}. *When* $d_M^t \ll d_B^{t+1}$ *or* $d_M^t \gg d_B^{t+1}$, $P_U^{t+1} \to 0$, *and the transmit power of the MD is* $P_M^t \to P_{\max}$.

Proof We substitute (3.42) into $\frac{d(P_{out})}{dP_U^{t+1}} = 0$ and assume that $P_{M,U}^t \gg N_0$. We then have

$$\ln\left(\frac{P_U^{t+1}}{N_0\gamma_{th}\left(d_B^{t+1}\right)^{\alpha}}\right) = \frac{\left(P_U^{t+1}\right)^2\left(d_M^t\right)^{\alpha}}{\left(P_{\max}-P_U^{t+1}\right)^2\left(d_B^{t+1}\right)^{\alpha}} + 1. \tag{3.58}$$

With $\theta = \frac{P_U^{t+1}}{P_{\max}}$ and $u = \frac{\theta}{1-\theta}$, (3.58) can be simplified as

$$u^2 = \ln(B/A) + \ln\theta, \tag{3.59}$$

where $A = \exp\frac{(d_M^t)^{\alpha}}{\left(d_B^{t+1}\right)^{\alpha}}$, and $B = \frac{P_{\max}}{eN_0\gamma_{th}\left(d_B^{t+1}\right)^{\alpha}}$. Using Taylor expansion, $\ln(\theta) = \theta - 1 + o(\theta) \simeq -\frac{1}{u+1}$, and u is the root of the following equation:

$$u^3 + u^2 - u\ln(B/A) + 1 - \ln(B/A) = 0. \tag{3.60}$$

It is solved that

$$u = \left(-q/2 + \sqrt{q^2/4 + p^3/27}\right)^{1/3} + \left(-q/2 + \sqrt{q^2/4 - p^3/27}\right)^{1/3}, \tag{3.61}$$

where $p = -\ln(B/A) - 1/3$, and $q = 2 + 9\ln(B/A) + 27(1-\ln(B/A))^2/27$. The approximate solution for the transmit power is $P_M^t = \frac{P_{\max}}{u}$, and $P_U^{t+1} = P_{\max} - P_M^t$. The power control solution is determined by the ratio of A and B, which is mostly affected by d_M^t and d_B^{t+1}. The key factor of power control is d_M^t/d_B^{t+1} because A is an exponential function of d_M^t and d_B^{t+1}.

When $d_M^t \ll d_B^{t+1}$, we have $A \simeq 1$, and $B \gg A$: therefore, $|\ln\frac{B}{A}| \gg 1$. When $d_M^t \gg d_B^{t+1}$, it is shown that $A \gg B$ because A is an exponential function of d_M^t, and we also have $|\ln\frac{B}{A}| \gg 1$. In both cases, (3.60) can be simplified as $u = 1$. The power control is given as $P_M^t = P_{\max}$, and $P_U^{t+1} = 0$. This means that the relay will be redundant if it is too close to the source or destination. □

3.2.3 Performance Evaluation

In this part, we evaluate the performance of Algorithm 3.2. The selection of the simulation parameters is based on the existing works and 3GPP specifications [71]. We consider $N = 1{,}500$ time slots, and the distance between the MD and BS $L = 500$. The maximum initial MD-UAV and UAV-BS distances (d_M^1 and d_U^1) are set as $\frac{3}{5}L$. We set the maximum moving distance for UAV in each time slot as $v = 0.1$ and set $\delta = -10^{-2}$, $\epsilon = -10^{-2}$. The maximum total transmit power $P_{\max}$ is given as 26 dBm, and the noise variance N_0 is given as -96 dBm. The SNR threshold γ_{th} is 0 dB, and the pathloss exponent α is 4. All curves are generated by averaging over 10^5 instances.

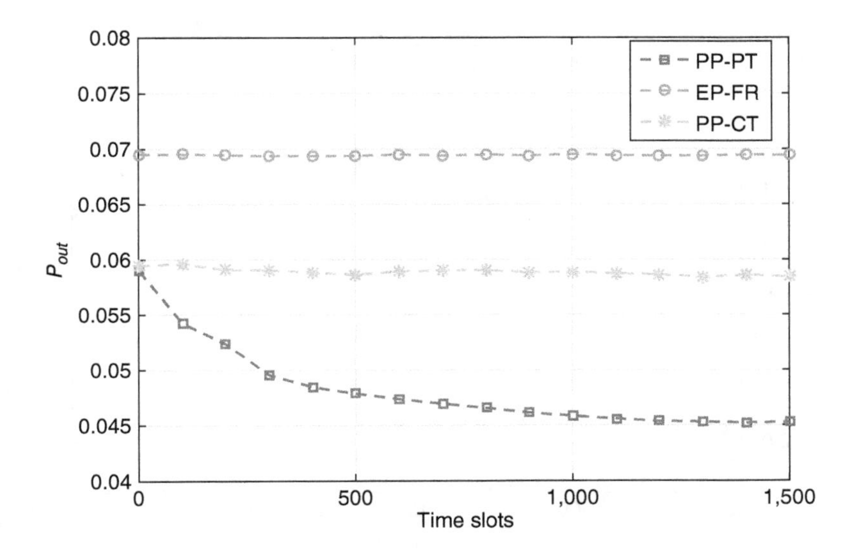

Figure 3.11 Time slot versus outage probability

We compare two schemes to the proposed power control and trajectory design scheme (PP-PT):

- **Proposed power control with a circle trajectory scheme (PP-CT)**: The trajectory is a circle with a center that is $(L/2, 0, 0)$ and a radius of 100. The initial location of the UAV is a random point on this circle, and the moving distance is v for a time slot. The power control is the same as our proposed algorithm.
- **Equal power allocation with a fixed relay scheme (EP-FR)**: The location of the UAV is fixed in different time slots, and it obeys uniform distribution in a circle area with $(L/2, 0, 0)$ being the center, and $L/2$ being the radius. The MD and UAV use the same transmit power in different time slots.

Figure 3.11 depicts the average outage probability using the time axis. The outage probability obtained by PP-PT decreases with time at the beginning. After about 1,000 time slots, the outage probability decreases about 23% and becomes stable at the minimum level, which is consistent with Remark 3.1. The average outage probability of the PP-CT scheme is similar to that of PP-PT at the beginning, but it does not decrease with time because the trajectory is fixed. The outage probability obtained by EP-FR is about 18% higher than the scheme with PP-CT.

Figure 3.12 illustrates the distance between the MD and BS versus the minimum achievable outage probability. The number of transmission time slots N is sufficiently large for the UAV to achieve the minimum outage probability with our proposed solution. The minimum outage probability of exhaustive search is obtained by enumerating the outage probability of over 1,000 power control strategies and over 5,000 UAV location possibilities. It is shown that the minimum outage probability increases monotonically with the increment of L. When the maximum total transmit power is

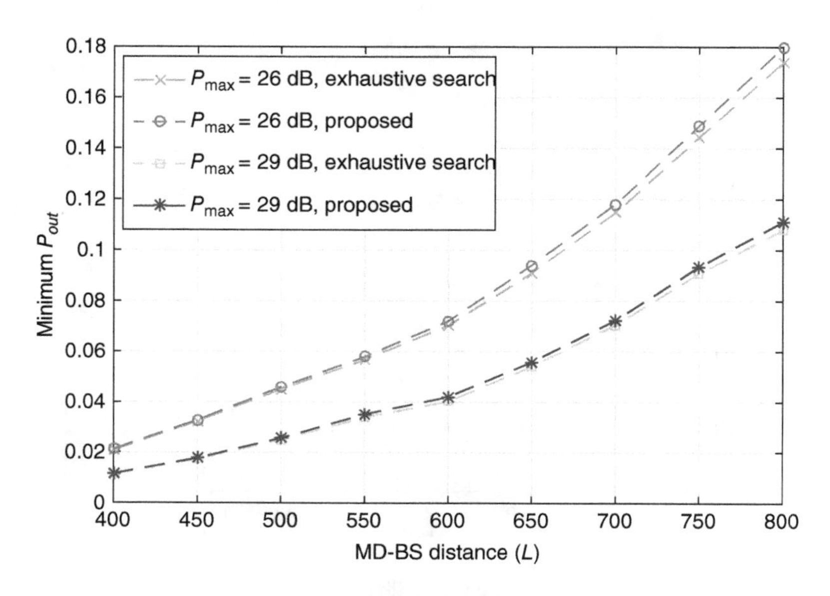

Figure 3.12 MD-BS distance (L) versus minimum outage probability

raised from 26 dBm to 29 dBm, the outage probability will decrease about 40%. The difference between the minimum outage probability obtained by the proposed solution and exhaustive search is less than 5% in all of our simulations.

3.2.4 Conclusion

In this section, we considered an uplink network model with a UAV working as an AF relay. An analytical expression of the outage probability has been derived. With power and UAV speed constraints, we have given a joint solution for the trajectory design and power control. The outage probability of the proposed solution converged to the minimum level and significantly outperformed the fixed power and circle trajectory schemes. The minimum outage probability obtained by the proposed algorithm is close to the exhaustive search minimum outage probability, with a difference less than 5%.

3.3 Summary

This chapter investigated the applications of UAVs serving as flying infrastructures. First, we utilized the UAVs as BSs for data offloading. To tackle the selfishness of an MBS manager and UAV operators, an optimal contract was designed for UAV placement. Second, we used UAVs as relays and designed a power and trajectory optimization algorithm to minimize the outage probability.

4 UAVs Serving as Flying Users

In addition to applications of flying infrastructures, UAVs are also utilized for sensing purposes due to their advantages of on-demand flexible deployment, larger service coverage compared with conventional fixed sensor nodes, and ability to hover. In particular, UAVs equipped with cameras or sensors have come to execute critical real-time sensing tasks, such as smart agriculture, security monitoring, forest fire detection, and traffic surveillance. Due to the limited computation capability of UAVs, real-time sensory data needs to be transmitted to the BS for real-time data processing. In this regard, the cellular networks are needed support the data transmission for UAVs. We refer to this approach as **cellular-assisted UAV sensing**. To support real-time sensing streaming, it is desirable to design joint sensing and communication protocols, develop novel beamforming and estimation algorithms, and study efficient distributed resource optimization methods.

In this chapter, we first present the architecture of cellular-assisted UAV sensing, which are called *cellular Internet of UAVs*, in Section 4.1 and investigate the trajectory design problem in Section 4.2. To improve the communication quality in the cellular internet of UAVs, we propose UAV-to-everything (U2X) communications in Section 4.3. We present an application of the cellular internet of UAVs for AQI monitoring in Section 4.4 and a chapter summary in Section 4.5.

4.1 Cellular Internet of UAVs

The wide availability of UAVs is generating considerable interest in civilian applications [72], as illustrated in Fig. 4.1. Recently, UAVs equipped with cameras or sensors have been widely used to execute a variety of sensing missions. In particular, UAVs are regarded as one of the best candidates to execute critical sensing missions such as emergency search and rescue and traffic monitoring due to their ease of deployment, high autonomy, and ability to hover. Moreover, their low cost and high flexibility also make them useful for sensing applications with a huge volume of data (e.g., video recording and landscape photography). To turn these technological visions into reality, seamless connectivity over wireless networking is necessary. Existing works focus on the ad hoc networking for UAV sensing [18, 73]. In [18], the authors surveyed UAV ad hoc networks and discussed the routing and handover among UAVs.

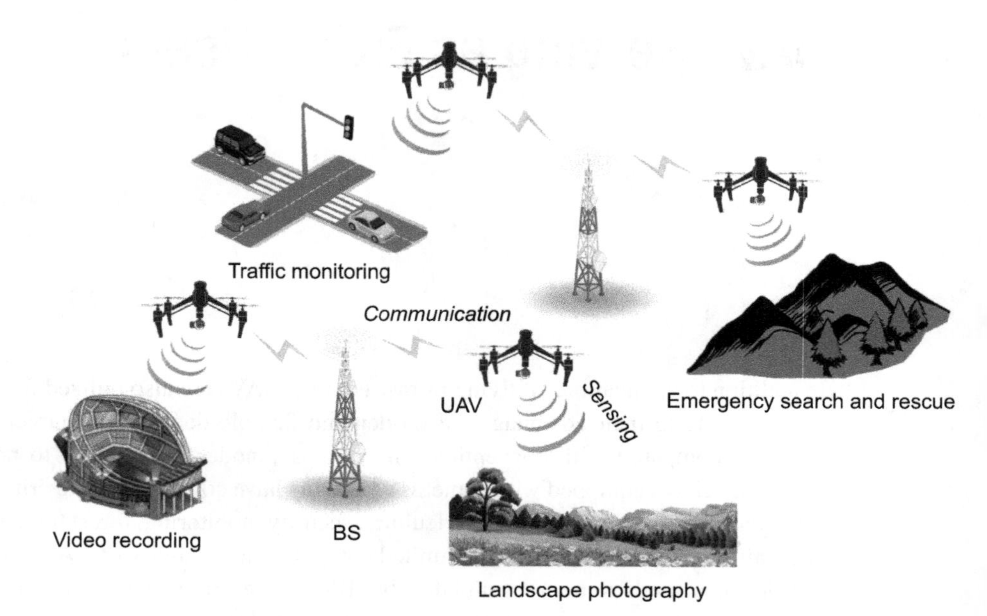

Figure 4.1 UAV sensing applications

In [73], an adaptive communication protocol was proposed to meet the demands of autonomy in airborne ad hoc networks for sensing applications. However, these ad hoc UAV networks are operated over unlicensed bands using carrier sense multiple access with collision avoid (CSMA/CA) based protocols, which cannot guarantee QoS requirements. In addition, multihop routing among UAVs may lead to higher delay. Therefore, a more reliable network is necessary.

Recently, the Third Generation Partnership Project (3GPP) has approved a study item on enhanced support to seamlessly integrate UAVs into future cellular networks [74]. Notably, it has produced accurate modeling of UAV-to-ground channels and defined various UAV scenarios together with their respective features. Therefore, terrestrial cellular networks, such as LTE and 5G, are considered a promising enabler for UAV sensing applications, which we refer to as the cellular internet of UAVs [26]. In the cellular internet of UAVs, the sensory data needs to be transmitted to the BS for timely processing. Because the sensing and transmission processes are coupled together and need to be designed jointly, we propose a sense-and-send protocol to coordinate multiple UAVs to perform the sensing tasks. However, because different applications have different requirements on the UAVs, we have two versions of the sense-and-send protocol, centralized and decentralized, to meet the different requirements of the sensing applications.

The rest of this section is organized as follows. In Section 4.1.1, we give an overview of the cellular internet of UAVs. In Sections 4.1.2 and 4.1.3, we elaborate on centralized and decentralized sense-and-send protocols.

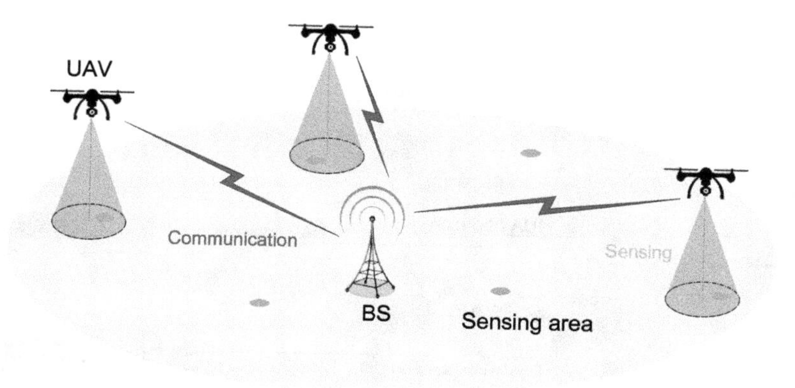

Figure 4.2 System architecture of cellular internet of UAVs

4.1.1 System Overview

As illustrated in Fig. 4.2, in the cellular internet of UAVs, each UAV is required to execute several sensing tasks independently and upload the sensory data by cellular communications concurrently. Note that the UAVs will not process the sensory data due to their limited computation capabilities. The collected data will be transmitted to the BS for further processing. The decision-making of the UAVs that have the capability to process the sensory data can be studied in the future work. To avoid severe interference among different UAVs, they will use orthogonal subchannels for data transmission.

To realize joint sensing and communication, in the following we propose centralized and decentralized sense-and-send protocols to coordinate multiple UAVs.

4.1.2 Centralized Sense-and-Send Protocol

In this part, we introduce a centralized sense-and-send protocol, which contains two phases: UAV sensing and UAV communications. Because the sensing and communications are coupled by the trajectory of the UAVs, it is necessary to consider them jointly.

UAV Sensing

To guarantee the quality of sensing, the probability of a UAV sensing failure needs to be less than a tolerance threshold. We define the feasible sensing area as the one where the requirement for sensing failure is satisfied. According to the probability model in [75], the probability of successful sensing is related to the distance between the location of the UAV and the center of the sensing area. Note that the UAV has a minimum flying height constraint, and thus the feasible sensing area for a task is a spherical crown.

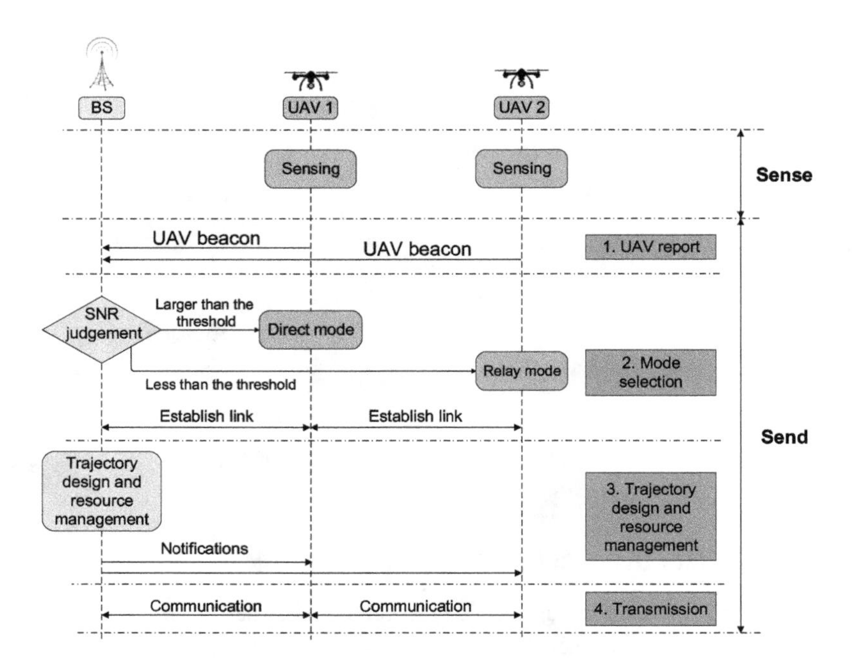

Figure 4.3 Centralized sense-and-send protocol

For a sensing task, the UAV hovers for a period over the sensing point and collects data. Meanwhile, the UAV also transmits the sensory data to the BS. Note that better sensing quality requires the UAV to get close to the sensing area, and the UAV can obtain better QoS for communications when it is closer to the BS. Therefore, the sensing and the communications impact each other.

UAV Communications

As shown in Fig. 4.3, the communication phase consists of four steps: UAV report, mode selection, trajectory design and resource management, and transmission. In the following, we will elaborate on these steps.

- **UAV report:** At the beginning of the communication phase, each UAV needs to send a beacon, which contains its ID and location, to the BS over the control channel in a time division manner.
- **Mode selection:** After receiving the beacon, the BS can determine the communication mode for each UAV according to the received signal-to-noise ratio (SNR). For example, if the received SNR is higher than the predefined threshold, the UAV can communicate with the BS directly, and otherwise, the UAV may need to transmit through another UAV.
- **Trajectory design and resource management:** The BS performs trajectory design and resource management. Afterward, the results are sent to the UAVs over the control channel.

- **Transmission:** These UAVs start to transmit the sensory data to the corresponding BS or another UAV over the channel allocated by the BS. The transmission step lasts until the end of this time slot.

4.1.3 Decentralized Sense-and-Send Protocol

In the following, we introduce the decentralized sense-and-send protocol. We first introduce the sense-and-send cycle, which consists of the beaconing phase, the sensing phase, and the transmission phase. After that, we describe the uplink subchannel allocation mechanism of the BS.

Sense-and-Send Cycle

The UAVs perform the sensing tasks in a synchronized iterative manner. Specifically, the process is divided into cycles, which are indexed by k. In each cycle, each UAV senses its task and then sends the collected sensory data to the BS. In order to synchronize the transmissions of the UAVs, we further divide each cycle into T_c *frames*, which are the basic time units for the subchannel allocation. The duration of the time unit frame is set to be the duration of the transmission and acknowledgement of the sensory data frame.[1]

The cycle consists of three separated phases: *beaconing phase*, *sensing phase*, and *transmission phase*, which contain T_b, T_s, and T_u frames, respectively. The duration of the beaconing phase and sensing phase are considered to be fixed and determined by the time necessary for transmitting beacons and collecting sensory data. On the other hand, the duration of the transmission phase is decided by the BS and is related to the network conditions, such as the number of UAVs in the network. Moreover, as illustrated in Fig. 4.4, we consider the sensing and transmission phases separate to avoid the possible interference.[2]

In the beaconing phase, each UAV sends its location to the BS in its beacon through the control channel, which can be obtained by the UAV from the on-board GPS location. Collecting the beacons sent by the UAVs, the BS then broadcasts to inform the UAVs of the general network settings as well as the locations of all the UAVs. By this means, UAVs can obtain the locations of other UAVs at the beginning of each cycle. Based on the acquired information, each UAV then decides its trajectory in the cycle and informs the BS using another beacon.

In the sensing phase, each UAV senses the task for T_s frames continuously, during which it collects the sensory data. In each frame of the transmission phase, the UAVs attempt to transmit the collected sensory data to the BS if subchannels are allocated to them by the BS. Specifically, there are four possible situations for each UAV, which are shown in the Fig. 4.4 and can be described as follows:

- **No subchannel allocated**: In this case, no uplink subchannel is allocated to UAV i by the BS. Therefore, the UAV cannot transmit its collected sensory data to the BS. It will wait for the BS to assign a subchannel to it in order to transmit the sensory data.

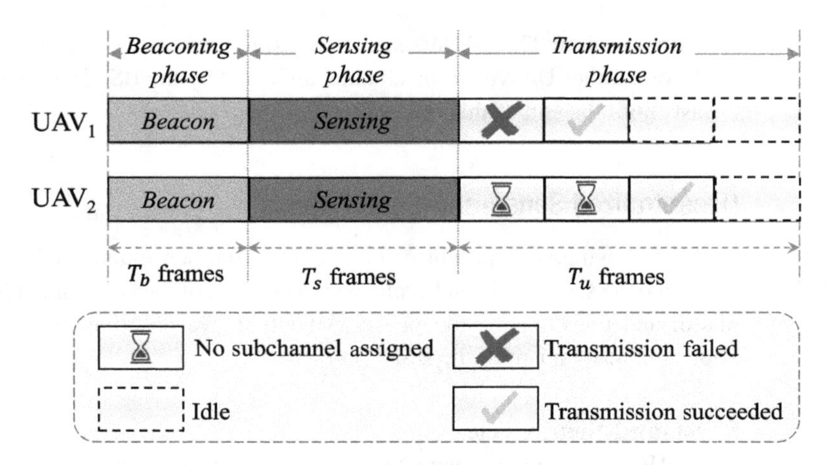

Figure 4.4 Decentralized sense-and-send protocol

- **Transmission failed**: In this case, an uplink subchannel is allocated to UAV i by the BS. However, the transmission is unsuccessful due to the low SNR at the BS, and thus, UAV i attempts to send the sensory data again to the BS in the next frame.
- **Transmission succeeded**: In this case, an uplink subchannel is allocated to UAV i, and the UAV succeeds in sending its collected sensory data to the BS.
- **Idle**: In this case, UAV i has successfully sent its sensory data in the former frames and will remain idle for the rest of the cycle until the beginning of the next cycle.

Although in this section we have assumed that the transmission of sensory data occupies a single frame, it can be extended to the case where the sensory data transmission takes n frames. In that case, the channel scheduling unit becomes n frames instead of a single frame.

Uplink Subchannel Allocation Mechanism

Because the uplink subchannel resources are usually scarce, in each frame of the transmission phase, there may exist more UAVs requesting to transmit sensory data than the number of available uplink subchannels. To deal with this problem, the BS adopts the following subchannel allocation mechanism to allocate the uplink subchannels to the UAVs.

In each frame, the BS allocates the C available uplink subchannels to the UAVs with uplink requirements in order to maximize the sum of successful transmission probabilities of the UAVs. Based on the matching algorithm in [76], the BS allocates the C available subchannels to the first C UAVs with the highest successful transmission probabilities. The successful transmission probabilities of UAVs can be calculated by the BS based on (4.5), using the information on the trajectories of the UAVs collected in the beaconing phase. Moreover, we denote the transmission states of the UAVs in the kth cycle as the vector $\boldsymbol{I}^{(k)}(t)$, in which $\boldsymbol{I}^{(k)}(t) = (I_1^{(k)}(t), \ldots, I_N^{(k)}(t))$. Here,

$I_i^{(k)}(t) = 0$ if UAV i has not succeeded in transmitting its sensory data to the BS at the beginning of the tth frame; otherwise, $I_i^{(k)}(t) = 1$. Based on these notations, the uplink subchannel allocation can be expressed by the channel allocation vector $\boldsymbol{v}^{(k)}(t) = (v_1^{(k)}(t), \ldots, v_N^{(k)}(t))$, in which the elements are determined by

$$v_i^{(k)}(t) = \begin{cases} 1, & Pr_{\mathrm{Tx},i}^{(k)}(t) I_i^{(k)}(t) \geq (\boldsymbol{Pr}_{\mathrm{Tx}}^{k}(t)\boldsymbol{I}^{(k)}(t))_C, \\ 0, & otherwise. \end{cases} \quad (4.1)$$

Here, $v_i^{(k)}(t)$ is the *channel allocation indicator* for UAV i; that is, $v_i^{(k)}(t) = 1$ only if an uplink subchannel is allocated to UAV i in the tth frame, $Pr_{\mathrm{Tx},i}^{(k)}(t)$ denotes the successful transmission probability of UAV i in the tth frame of the kth cycle, and $(\boldsymbol{Pr}_{\mathrm{Tx}}^{(k)}(t)\boldsymbol{I}^{(k)}(t))_C$ denotes the Cth largest successful transmission probabilities among the UAVs that have not succeeded in sending the sensory data before the tth frame.

Because the trajectory of UAV i determines UAV i's distance to the BS, it influences the successful transmission probability. And because the UAVs are allocated subchannels only when they have the C-largest transmission probabilities, the UAVs have the incentive to compete with each other by selecting trajectories where their successful transmission probabilities are among the highest C ones. Consequently, the UAVs need to design their trajectories by considering not only their distance to the BS and the task but also the trajectories of other UAVs.

4.2 Trajectory Design

In order to enable the real-time sensing applications, cellular UAV transmission is considered a promising solution where the uplink QoS is guaranteed compared to that in ad-hoc sensing networks. However, it remains a challenge for UAVs to determine their trajectories in the cellular internet of UAVs. When a UAV is far from the task, it risk obtaining invalid sensing data, whereas if it is far from the BS, the low uplink transmission quality may lead to difficulties in transmitting the sensory data back to the BS. Therefore, the UAVs need to take both the sensing accuracy and the uplink transmission quality into consideration in designing their trajectories. Moreover, it is even more challenging when the UAVs belong to different entities and are noncooperative. Because the spectrum resource is scarce, the UAVs have the incentive to compete for the limited uplink channel resources. Therefore, each UAV should also consider other UAVs that are competing for the spectrum dynamically when it determines its trajectory. Therefore, a decentralized trajectory design approach is necessary for the UAV trajectory design problem in which the locations of both the task and the BS, as well as the behaviors of the other UAVs, should be taken into consideration by each UAV.

To tackle these problems, we consider the scenario where multiple UAVs in a cellular network perform different real-time sensing tasks. We first analyzed the performance using a nested Markov chain based on the decentralized sense-

and-send protocol. Under this condition, the UAV trajectory design problem can be seen as a Markov decision problem, which makes reinforcement learning the suitable and promising approach to solve the problem. To be specific, we formulate the UAV trajectory design problem under a reinforcement-learning framework and propose an enhanced multi-UAV Q-learning algorithm to solve the problem efficiently.

The rest of this section is organized as follows. In Section 4.2.1, the system model is described. We first analyze the performance of the proposed sense-and-send protocol in Section 4.2.2 and derive the probability of successful valid sensory data transmission using the nested Markov chains. Following that, the reinforcement-learning framework and the enhanced multi-UAV Q-learning algorithm are given in Section 4.2.3, together with the analyses of complexity, convergence, and scalability. The simulation results are presented in Section 4.2.4.

4.2.1 System Description

As illustrated in Fig. 4.5, we consider a single-cell orthogonal frequency-division multiple access (OFDMA) cellular internet of UAVs that consists of N UAVs to perform real-time sensing tasks [77]. We set the horizontal location of the BS as the origin, and the location of the BS and the UAVs can be specified by 3D Cartesian coordinates; that is, the location of the ith UAV can be denoted as $s_i = (x_i, y_i, h_i)$, and the location of the BS can be denoted as $S_0 = (0, 0, H_0)$, with H_0 being its height. The location of the UAV i's real-time sensing task is denoted as $S_i = (X_i, Y_i, 0)$. To perform the real-time sensing task, each UAV continuously senses the condition of its task and sends the collected sensory data back to the BS immediately. Therefore, the sensing process and transmission process jointly determine the UAVs' performance on their real-time sensing tasks. The sensing and transmission models of the UAVs are described in the following.

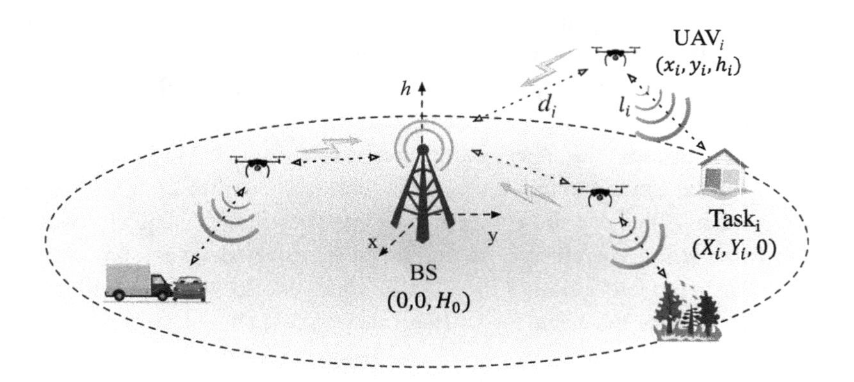

Figure 4.5 System model of cellular Internet of UAVs in which UAVs perform real-time sensing tasks

UAV Sensing

To evaluate the sensing quality of the UAV, we utilize the probabilistic sensing model. Supposing that UAV i senses task i for a second, the probability that the UAV will sense the condition of its task successfully (i.e., the successful sensing probability) can be expressed as

$$\Pr_{s,i} = e^{-\lambda l_i}, \tag{4.2}$$

in which λ is the parameter evaluating the sensing performance, and l_i denotes the distance from UAV i to task i.

It is worth noticing that UAV i cannot determine whether the sensing is successful or not from its collected sensory data, due to its limited on-board data-processing ability. Therefore, UAV i needs to send the sensory data to the BS, and the BS will decide whether the sensory data is valid or not. Nevertheless, UAV i can evaluate its sensing performance by calculating the successful sensing probability based on (4.2).

UAV Transmission

In the UAV transmission, the UAVs transmit the sensory data to the BS over orthogonal subchannels to avoid mutual interference. To be specific, we adopt the 3GPP channel model to evaluate the urban macro cellular support for UAVs [74, 78].

Denoting the transmit power of UAVs as P_u, we can express the received SNR at the BS of UAV i as

$$\gamma_i = \frac{P_u \| H_i \|}{N_0 10^{\mathrm{PL}_{a,i}/10}}, \tag{4.3}$$

in which $\mathrm{PL}_{a,i}$ denotes the air-to-ground pathloss, N_0 denotes the power of noise at the receiver of the BS, and H_i is the small-scale fading coefficient. Specifically, the pathloss $\mathrm{PL}_{a,i}$ and small-scale fading H_i are calculated in two cases separately: when line-of-sight component exists (LoS case), and when no LoS components exist (NLoS case). The probability of the UAV i-BS channel containing a LoS component can be calculated as

$$\Pr_{\mathrm{LoS},i} = \begin{cases} 1, & r_i \le r_c, \\ \frac{r_c}{r_i} + e^{-r_i/p_0 + r_c/p_0}, & r_i > r_c, \end{cases} \tag{4.4}$$

in which $r_i = \sqrt{x_i^2 + y_i^2}$, $p_0 = 233.98 \log_{10}(h_i) - 0.95$, and $r_c = \max\{294.05 \log_{10}(h_i) - 432.94, 18\}$.

When the channel contains a LoS component, the pathloss from UAV i to the BS can be calculated as $\mathrm{PL}_{\mathrm{LoS},i} = 30.9 + (22.25 - 0.5 \log_{10}(h_i)) \log_{10}(d_i) + 20 \log_{10}(f_c)$, where f_c is the carrier frequency and d_i is the distance between the BS and UAV i. In the LoS case, the small-scale fading H_i obeys the Rice distribution with scale parameter $\Omega = 1$ and shape parameter $K[\mathrm{dB}] = 4.217 \log_{10}(h_i) + 5.787$. On the other hand, when the channel contains no LoS components, the pathloss from UAV i to the BS can be calculated as $\mathrm{PL}_{\mathrm{NLoS},i} = 32.4 + (43.2 - 7.6 \log_{10}(h_i)) \times \log_{10}(d_i) + 20 \log_{10}(f_c)$, and the small-scale fading H_i obeys the Rayleigh distribution with zero means and unit variance.

To achieve a successful transmission, the SNR at the BS needs to be higher than the decoding threshold γ_{th}. Therefore, each UAV can evaluate its successful transmission probability by calculating the probability that the SNR at BS is larger than γ_{th}. The successful transmission probability $\mathrm{Pr}_{\mathrm{Tx},i}$ for UAV i can be calculated as

$$\mathrm{Pr}_{\mathrm{Tx},i} = \mathrm{Pr}_{\mathrm{Los},i}(1-F_{ri}(\chi_{\mathrm{LoS},i}))+(1-\mathrm{Pr}_{\mathrm{LoS},i})(1-F_{ra}(\chi_{\mathrm{NLoS},i})),$$

where $\chi_{\mathrm{NLoS},i} = N_0 10^{0.1\mathrm{PL}_{\mathrm{NLoS},i}} \gamma_{th}/P_u$, and $\chi_{\mathrm{LoS},i} = N_0 10^{0.1\mathrm{PL}_{\mathrm{LoS},i}} \gamma_{th}/P_u$. Here, $F_{ri}(x) = 1 - Q_1(\sqrt{2K}, x\sqrt{2(K+1)})$ is the cumulative distribution function (CDF) of the Rice distribution with $\Omega = 1$ [79], and $F_{ra}(x) = 1 - e^{-x^2/2}$ is the CDF of the Rayleigh distribution with unit variance. Here $Q_1(x)$ denotes the Marcum Q-function of order 1 [80].

4.2.2 Markov Chain Modeling

To facilitate the coordination among UAVs, we utilize the decentralized sense-and-send protocol introduced in Section 4.1.3. In the following, we will analyze the performance by calculating the probability of successful valid sensory data transmission, which plays an important role in solving the UAV trajectory design problem. We first specify the state transitions of the UAVs by using nested bilevel Markov chains. The outer Markov chain depicts the state transitions in the UAV sensing process, and the inner Markov chain depicts the state transitions in the UAV transmission process, which will be elaborated on in the following sections.

Outer Markov Chain of UAV Sensing

In the outer Markov chain, the state transition takes place among different cycles. As shown in Fig. 4.6, for each UAV, it has two states in each cycle: state $\mathcal{H}_f$ to denote that the sensing was failed, and state $\mathcal{H}_s$ to denote that the sensing was successful. Supposing the successful sensing probability of UAV i in the kth cycle is $p_{s,i}^{(k)}$, UAV i transits to the $\mathcal{H}_s$ state with probability $p_{s,i}^{(k)}$ and transits to the $\mathcal{H}_f$ state with probability $(1 - p_{s,i}^{(k)})$ after the kth cycle. The value at the right side of the transition probability denotes the number of valid sensory data that have been transmitted successfully to the BS in the cycle.

We denote the probability of UAV i successfully transmiting the sensory data to the BS as $p_{u,i}^{(k)}$. Therefore, UAV i successfully transmits valid sensory data to the BS with the probability $p_{s,i}^{(k)} p_{u,i}^{(k)}$. On the other hand, with probability $p_{s,i}^{(k)}(1 - p_{u,i}^{(k)})$, no valid sensory data is transmitted to the BS, though the sensing is successful in the kth cycle. The probability $p_{u,i}^{(k)}$ can be analyzed by the inner Markov chain of the UAV transmission in the next subsection, and $p_{s,i}^{(k)}$ can be calculated as follows.

Because the change of the UAVs' locations during each frame is small, we assume that the location of each UAV is fixed within each frame. Therefore, the location of UAV i in the kth cycle can be expressed as a function of the frame index t; that is, $s_i^{(k)}(t) = (x_i^{(k)}(t), y_i^{(k)}(t), h_i^{(k)}(t))$, $t \in [1, T_c]$. Similarly, the distance between UAV i and its task can be expressed as $l_i^{(k)}(t)$, and the distance between the UAV and the BS

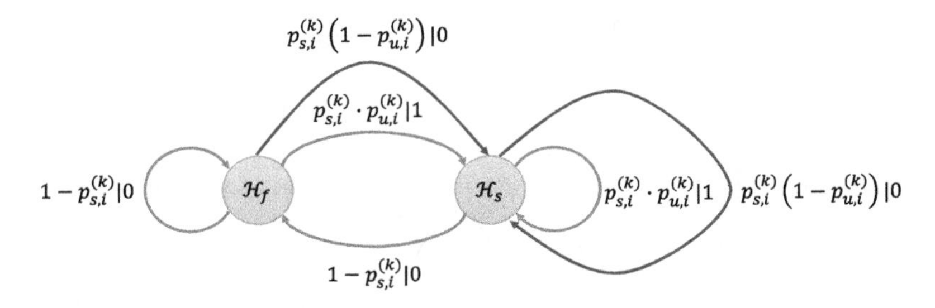

Figure 4.6 Outer Markov chain of UAV sensing

can be expressed as $d_i^{(k)}(t)$. Moreover, we assume that the UAVs move with uniform speed and direction in each cycle after the beginning of the sensing phase. Based on these assumptions, at the tth frame of the kth cycle, the location of UAV i can be expressed as

$$\boldsymbol{s}_i^{(k)}(t)=\boldsymbol{s}_i^{(k)}(T_b)+\frac{t}{T_c}\left(\boldsymbol{s}_i^{(k+1)}(1)-\boldsymbol{s}_i^{(k)}(T_b)\right),\ t\in[T_b,T_c]. \tag{4.5}$$

Therefore, the successful sensing probability of UAV i in the cycle can be calculated as

$$p_{s,i}^{(k)}=\prod_{t=T_b+1}^{T_s+T_b}\left(\Pr_{s,i}^{(k)}(t)\right)^{t_f}=\prod_{t=T_b+1}^{T_s+T_b}e^{-\lambda t_f l_i^{(k)}(t)}, \tag{4.6}$$

in which t_f denotes the duration of a frame, and $l_i^{(k)}(t)=\|\boldsymbol{s}_i^{(k)}(t)-\boldsymbol{S}_i\|$.

Inner Markov Chain of UAV Transmission

For simplicity, we consider a general cycle and omit the superscript cycle index k. Because the general state transition diagram is complicated, we illustrate the inner Markov chain by giving an example where the number of available uplink subchannels is $C=1$, the number of UAVs is $N=3$, and the number of uplink transmission frames is $T_u=3$.

Taking UAV 1 as an example, the state transition diagram is illustrated in Fig. 4.7. The state of the UAVs in frame t can be represented by the transmission state vector $\boldsymbol{I}(t)$, as defined in Section 4.1.3. Initially, $t=T_b+T_s+1$, and the transmission state is $\boldsymbol{I}(T_b+T_s+1)=\{0,0,0\}$, which indicates that UAVs 1, 2, and 3 have not succeeded in uplink transmission at the beginning of the transmission phase, and all of them are competing for the uplink subchannels. In the next frame, the transmission state will transit to the *successful Tx* state for UAV 1, if the sensory data of UAV 1 has been successfully transmitted to the BS. The probability for this transition equals $\Pr_{Tx,1}(T_b+T_s+1)\nu_1(T_b+T_s+1)$. (i.e., the probability for successful uplink transmission if a subchannel is allocated to UAV 1); otherwise, it equals zero.

However, if UAV 1 does not succeed in uplink transmission, the transmission transits into other states, which is determined by whether other UAVs have succeeded in uplink transmission (e.g., it transits to $\boldsymbol{I}(T_b+T_s+2)=(0,0,1)$ if UAV 3 succeeds in the first transmission frame). Note that when other UAVs succeed in transmitting

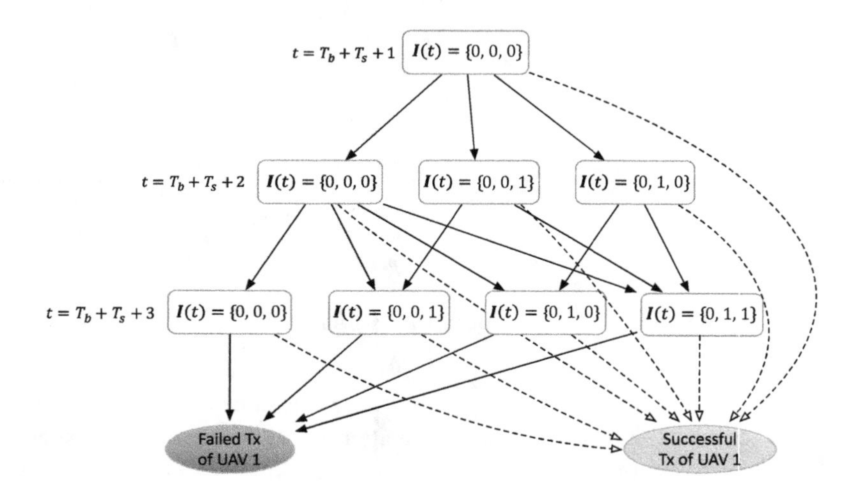

Figure 4.7 Inner Markov chain of UAV 1's transmission given $C = 1$, $N = 3$, and $T_u = 3$

sensory data in the previous frames, UAV 1 will face fewer competitors in the following frames, and thus, it will have a larger probability of transmitting successfully. Finally, when $t = T_c$ (i.e., the last transmission frame in the cycle), UAV 1 will enter the *failed Tx* state if it does not transmit the sensory data successfully, which means that the sensory data in this cycle fails to be uploaded. Therefore, obtaining the $p_{u,i}$ in the outer Markov chain is equivalent to calculating the absorbing probability of a successful Tx state in the inner Markov chain.

From the previous example, it can be observed that the following general recursive equation holds for UAV i when $t \in [T_b + T_s + 1, T_c]$:

$$\begin{aligned}\Pr_{u,i}\{t|\boldsymbol{I}(t)\} &= \Pr_{\mathrm{Tx},i}(t)\nu_i(t) \\ &+ \sum_{\substack{\boldsymbol{I}(t+1),\\ I_i(t+1)=0}} \Pr\{\boldsymbol{I}(t+1)|\boldsymbol{I}(t)\}\Pr_{u,i}\{t+1|\boldsymbol{I}(t+1)\},\end{aligned}$$

in which $\Pr\{\boldsymbol{I}(t+1)|\boldsymbol{I}(t)\}$ denotes the probability for the transmission state vector of the $(t+1)$th frame to be $\boldsymbol{I}(t+1)$, on the condition that the transmission state vector of the tth frame is $\boldsymbol{I}(t)$, and $\Pr_{u,i}\{t|\boldsymbol{I}(t)\}$ denotes the probability for UAV i to transmit sensory data successfully after the tth frame in the current cycle, given the transmission state $\boldsymbol{I}(t)$.

Because the successful uplink transmission probabilities of the UAVs are independent, we have $\Pr\{\boldsymbol{I}(t+1)|\boldsymbol{I}(t)\} = \prod_{i=1}^{N} \Pr\{I_i(t+1)|I_i(t)\}$, in which $\Pr\{I_i(t+1)|\, I_i(t)\}$ can be calculated as

$$\begin{cases}\Pr\{I_i(t+1) = 0|I_i(t) = 0\} = 1 - \Pr_{\mathrm{Tx},i}(t), \\ \Pr\{I_i(t+1) = 1|I_i(t) = 0\} = \Pr_{\mathrm{Tx},i}(t), \\ \Pr\{I_t(t+1) = 0|I_i(t) = 1\} = 0, \\ \Pr\{I_t(t+1) = 1|I_i(t) = 1\} = 1.\end{cases} \tag{4.7}$$

Here, the first two equations hold because the successful transmission probability of UAV i in the tth frame is $\mathrm{Pr}_{\mathrm{Tx},i}(t)$. The third and forth equations indicate that the UAVs remain idle in the rest of frames once they have successfully sent their sensory data to the BS.

Based on (4.7), the recursive algorithm can be proposed to solve $\mathrm{Pr}_{u,i}\{t|\boldsymbol{I}(t)\}$, as presented in Algorithm 4.1. Therefore, the successful transmission probability can be obtained by $p_{u,i} = \mathrm{Pr}_{u,i}\{T_b + T_s + 1|\boldsymbol{I}(T_b + T_s + 1)\}$.

In summary, the probability of successful valid sensory data transmission for UAV i in the kth cycle can be calculated as

$$p_{\mathrm{sTx},i}^{(k)} = p_{s,i}^{(k)} p_{u,i}^{(k)}. \tag{4.8}$$

Analysis on the Spectral Efficiency

In this section, we evaluate the spectral efficiency using the average number of valid sensory data transmissions per second, which is denoted as N_{vd}. The value of N_{vd} is influenced by many factors, such as the distance between the BS and the tasks, the number of available subchannels, the number of UAVs in the network, and the duration of the transmission phase.

In addition, we analyze the influence of the duration of transmission phase T_u on N_{vd} in a simplified case in which we assume all the UAVs are equivalent (i.e., they have the same probabilities for successful uplink transmission in a frame), the same probabilities for successful sensing, and the same probabilities to be allocated subchannels. Based on these assumptions, the following proposition can be derived.

PROPOSITION 4.1 (Optimal duration of transmission phase) *When the N UAVs are equivalent and have the probability for successful sensing p_s and the probability for successful uplink transmission p_u, then N_{vd} first increases and then decreases with the increment of T_u, and the optimal T_u^* can be calculated as*

$$T_u^* = \frac{N}{C\ln(1-p_u)}\left(1 + W_{-1}\left(-\frac{(1-p_u)^{\frac{CT_u}{N}}}{e}\right)\right) - T_b - T_s, \tag{4.9}$$

in which $W_{-1}(\cdot)$ denotes the lower branch of Lambert-W function in [81].

Proof Denoting the UAVs' probability for successful uplink transmission as p_u and their probability for successful sensing as p_s, we can calculate the average number of valid sensory data transmissions per second by

$$N_{vd} = N \cdot \frac{p_s\left(1-(1-p_u)^{\frac{CT_u}{N}}\right)}{(T_b + T_s + T_u)t_f},$$

in which t_f is the duration of single frame in seconds.

The partial derivative of N_{vd} with respect to T_u can be calculated as

$$\frac{\partial N_{vd}}{\partial T_u} = \frac{p_s F(T_u)}{t_f(T_b + T_s + T_u)^2},$$

in which $F(T_u) = p_f^{\frac{CT_u}{N}}(N - C(T_b + T_s + T_u)\ln p_f) - N$, and $p_f = 1 - p_u$. Taking the partial derivative of $F(T_u)$ with regard to T_u, we can derive that $\partial F(T_u)/\partial T_u = -C^2 p_f^{CT_u/N}(T_s + T_b + T_u)\ln p_f/N < 0$. Besides, when $T_u \to \infty$, $F(T_u) \to -N$ and $N_{vd} \to 0$, and when $T_u = 0$, $N_{vd} = 0$. Therefore, $\partial F(T_u)/\partial T_u < 0$ indicates that there is a unique maximum point for N_{vd} when $T_u \in (0, \infty)$.

The maximum of N_{vd} is reached when $F(T_u^*) = 0$, in which T_u^* can be obtained by

$$T_u^* = \frac{N}{C\ln p_f}\left(1 + W_{-1}\left(-\frac{p_f^{\frac{CT_u}{N}}}{e}\right)\right) - T_b - T_s, \tag{4.10}$$

where $W_{-1}(\cdot)$ denotes the lower branch of the Lambert-W function [81]. □

This proposition sheds light on the relation between the spectral efficiency and the duration of transmission phase in the general cases. In the cases where the UAVs are not equivalent, the spectral efficiency also first increases and then decreases with the duration of transmission phase. This is because when $T_u = 0$, $N_{vd} = 0$, and $T_u \to \infty$, $N_{vd} \to 0$.

4.2.3 Reinforcement Learning for Trajectory Design

In this section, we first formulate the decentralized trajectory design problem of UAVs and then analyze the problem in the reinforcement-learning framework. After that, we describe the single-agent and multiagent reinforcement-learning algorithms under the framework and propose an enhanced multi-UAV Q-learning algorithm to solve the UAV trajectory design problem efficiently.

UAV Trajectory Design Problem

Before the formulation of the trajectory design problem, we first set up the model to describe the UAVs' trajectories. In this section, we focus on the cylindrical region with the maximum height $h_{\max}$ and the radius of the cross section $R_{\max}$, which satisfies $R_{\max} = \max\{R_i | R_i = \sqrt{X_i^2 + Y_i^2}, \forall i \in [1, N]\}$, because it is inefficient for the UAVs to move beyond the farthest task. Moreover, we assume that the space is divided into a finite set of discrete spatial points $\mathscr{S}_p$, which are arranged in a square lattice pattern as shown in Fig. 4.8. Therefore, the trajectory of UAV i starting from the kth cycle can be represented as the sequence of spatial points $\mathscr{S}_i^{(k)} = \{\mathbf{s}_i^{(k)}, \mathbf{s}_i^{(k+1)}, \ldots\}$ that the UAV locates at the beginning of each cycle.

To determine the trajectories, UAVs select their next spatial point at the beginning of each cycle. To be specific, UAV i locates at the spatial point $\mathbf{s}_i^{(k)}$ at the beginning of the kth cycle and decides which spatial point $\mathbf{s}_i^{(k+1)}$ it will move to at the beginning of the $(k+1)$th cycle. After the UAV has selected its next spatial point, it will move to the point with a uniform speed and direction within this cycle. Moreover, the available spatial points that UAV i can reach is within the maximum distance it can fly in a cycle, which is denoted as D. We set the distance between two adjacent spatial points to be

Algorithm 4.1 Algorithm for successful transmission probability in a cycle

Input: Frame index (t); transmission state vector $(\boldsymbol{I}(t))$; length of beaconing phase (T_b); length of sensing phase (T_s); length of transmission phase (T_u); location of UAVs $(\boldsymbol{s}(t))$; number of subchannels (C).

Output: $\Pr_{u,i}\{t|\boldsymbol{I}(t)\}$, $i = 1,\ldots,N$.

1 **begin**
2 **if** $t = T_b + T_s + 1$ **then**
3 $\Pr_{u,i}\{t|\boldsymbol{I}(t)\} := 0, i = 1,\ldots,N;$
4 **else**
5 **if** $t > T_c$ **then**
6 **return** $\Pr_{u,i}\{t|\boldsymbol{I}(t)\} = 0,\ i = 1,\ldots,N;$
7 **end**
8 **end**
9 Calculate the successful transmission probabilities $\Pr_{\mathrm{Tx},i}(t)$, $i = 1,\ldots,N$ based on (4.5);
10 Determine the subchannel allocation indicator $\boldsymbol{\nu}(t)$ based on (4.1);
11 **for** $i \in [1,N]$ *and* $I_i(t) = 0$ **do**
12 $\Pr_{u,i}\{t|\boldsymbol{I}(t)\} := \Pr_{Tx,i}(t)\nu_i(t);$
13 **end**
14 **for** *all* $\boldsymbol{I}(t+1)$ *with* $\Pr\{\boldsymbol{I}(t+1)|\boldsymbol{I}(t)\} > 0$ **do**
15 Solve $\Pr_{u,i}\{t+1|\boldsymbol{I}(t+1)\}$ by calling Algorithm 4.1, in which $t := t+1$ and $\boldsymbol{I}(t) := \boldsymbol{I}(t+1)$ and other parameters hold;
16 $\Pr_{u,i}\{t|\boldsymbol{I}(t)\} := \Pr_{u,i}\{t|\boldsymbol{I}(t)\} + \Pr\{\boldsymbol{I}(t+1)|\boldsymbol{I}(t)\}\Pr_{u,i}\{t+1|\boldsymbol{I}(t+1)\};$
17 **end**
18 **return** $\Pr_{u,i}\{t|\boldsymbol{I}(t)\}$, $i = 1,\ldots,N;$
19 **end**

$\Delta = D/\sqrt{3}$, and thus, the available spatial point UAV i can fly to in the $k+1$ cycle is within a cube centered at $(x_i^{(k)}, y_i^{(k)}, h_i^{(k)})$ with side length equal to 2Δ, as illustrated in Fig. 4.8.

In Fig. 4.8, there are at most 27 available spatial points that can be selected by the UAVs in each cycle. We denote the set of all the vectors from the center to the available spatial points as the available action set of the UAVs, which is denoted as $\mathscr{A}$. Moreover, it is worth noting that when the UAV is at the marginal location (e.g., flying at the minimum height), there are fewer available actions to be selected. To handle the differences among the available action sets at different spatial points, we denote the available action set at the spatial point $\boldsymbol{s}$ as $\mathscr{A}(\boldsymbol{s})$.

In this section, we consider the utility of each UAV to be the total number of successful valid sensory data transmissions for its task. Therefore, the UAVs have incentive to maximize the total successful valid sensory data transmissions by designing their trajectories. In addition, we assume that the UAVs have discounting valuation on the successfully transmitted valid sensory data. Specifically, for the UAVs in the

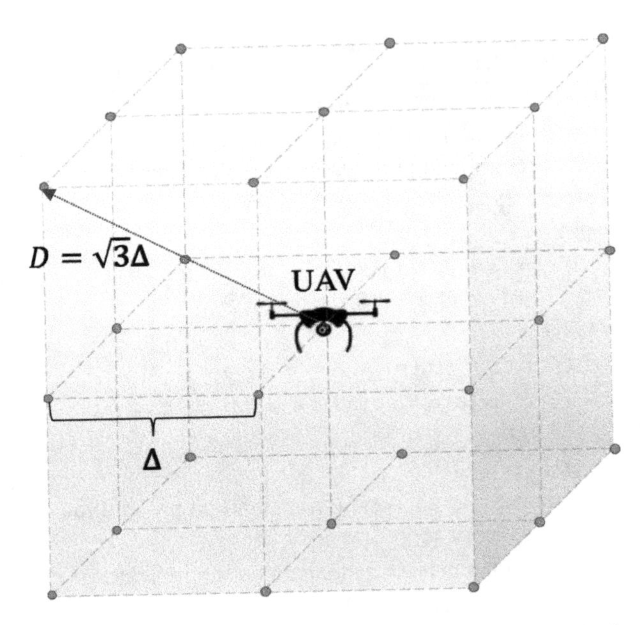

Figure 4.8 Set of available spatial points that the UAV can reach at the beginning of the next cycle

kth cycle, the successfully valid sensory data transmitted in the k'th cycle is worth only $\rho^{|k'-k|}$ ($\rho \in [0, 1)$) of the successful valid sensory data transmitted in the current cycle, due to the timeliness requirements of real-time sensing tasks. Therefore, at the beginning of kth cycle, the utility of UAV i is defined as the *total discounted rewards* in the future and can be denoted as

$$U_i^{(k)} = \sum_{n=0}^{\infty} \rho^n R_i^{(k+n)}, \tag{4.11}$$

in which $R_i^{(k)}$ denotes the reward of UAV i in the kth cycle, and $R_i^{(k)} = 1$ if valid sensory data is successfully transmitted to the BS by UAV i in the kth cycle; otherwise, $R_i^{(k)} = 0$.

Based on such assumptions, the UAV trajectory design problem can be formulated as

$$\max_{\mathscr{S}_i^{(k)}} \quad U_i^{(k)} = \sum_{n=0}^{\infty} \rho^n R_i^{(k+n)}, \tag{4.12}$$

$$s.t. \quad \mathbf{s}_i^{(k'+1)} - \mathbf{s}_i^{(k')} \in \mathscr{A}\left(\mathbf{s}_i^{(k')}\right),\ k' \in [k, \infty). \tag{4.12a}$$

Reinforcement-Learning Framework

Generally, the UAV trajectory design problem (4.12) is hard to solve because the rewards of the UAVs in the future cycles are influenced by the trajectories of all UAVs, which are determined in a decentralized manner and hard to model. Fortunately, reinforcement learning is able to deal with the problem of agent programming in an environment with deficient understanding, which removes the burden of developing accurate models and solving the optimization with respect to those models [82]. For this reason, we adopt reinforcement learning to solve the UAV trajectory design problem.

To begin with, we formulate a reinforcement-learning framework for the problem. With the help of [83], the reinforcement-learning framework can be given as follows, in which the superscript cycle index k is omitted for description simplicity.

DEFINITION 4.1 *A reinforcement learning framework for UAV trajectory design problem is described by a tuple* $<\mathscr{S}_1,\ldots,\mathscr{S}_N,\mathscr{A}_1,\ldots,\mathscr{A}_N,\mathscr{T},p_{R,1},\ldots,p_{R,N},\rho>$, *where*

- $\mathscr{S}_1,\ldots,\mathscr{S}_N$ *are finite state spaces of all the possible locations of the UAVs at the beginning of each cycle, and the state space of UAV* i *equals to the finite spatial space (i.e.,* $\mathscr{S}_i=\mathscr{S}_p$, $\forall i\in[1,N]$*).*
- $\mathscr{A}_1,\ldots,\mathscr{A}_N$ *are the corresponding finite sets of actions available to each agent. The set* $\mathscr{A}_i$ *consists of all the available actions of UAV* i *(i.e.,* $\mathscr{A}_i=\mathscr{A}$, $\forall i\in[1,N]$*).*
- $\mathscr{T}$: $\prod_{i=1}^{N}\mathscr{S}_i\times\prod_{i=1}^{N}\mathscr{A}_i\rightarrow(\mathscr{S}_p)^N$ *is the state transition function. It maps the location profile and the action profile of the UAVs in a certain cycle to the location profile of the UAVs in the next cycle.*
- $p_{R,i}:\prod_{j=1}^{N}\mathscr{S}_j\times\prod_{j=1}^{N}\mathscr{A}_j\rightarrow\Pi(0,1)$, $i=1,\ldots,N$ *represents the reward function for UAV* i*. That is, it maps the location profile and action profile of all the UAVs in the current cycle to the probability for UAV* i *to get a unit reward from a successful valid sensory data transmission.*
- $\rho\in[0,1)$ *is the discount factor, which indicates UAVs' evaluation of the rewards it obtained in the future or in the past.*

In the framework, the rewards of the UAVs are informed by the BS. Specifically, we assume that the BS informs each UAV whether it has transmitted valid sensory data in a certain cycle using the BS beacon at the beginning of the next cycle. For each UAV, it obtains one reward if the BS informs it that the valid sensory data has been received successfully. Therefore, the probability of UAV i obtaining one reward is equal to the probability of it transmitting valid sensory data successfully in the cycle (i.e., $p_{R,i}=p_{\mathrm{sTx},i}$). Because the probability of successful valid sensory data transmission is influenced by both the successful sensing probability and the successful transmission probability, the UAV's trajectory learning process is associated with the sensing and transmission processes through the obtained reward in each cycle.

Under the reinforcement-learning framework for the UAV trajectory design, the following two kinds of reinforcement-learning algorithms can be adopted, which are single-agent Q-learning and multiagent Q-learning algorithms.

Single-Agent Q-learning Algorithm

One of the most basic reinforcement-learning algorithms is the single-agent Q-learning algorithm [84]. It is a form of model-free reinforcement learning and provides a simple way for the agent to learn how to act optimally. The agent selects actions by following its policy in each state, which is denoted as $\pi(\boldsymbol{s})$, $\boldsymbol{s} \in \mathscr{S}$ and is a mapping from states to actions. The essence of the algorithm is to find the Q value of each state and action pair, which is defined as the accumulated reward received when taking the action in the state and then following the policy thereafter. In its simplest form, the agent maintains a table containing its current estimated Q values, which is denoted as $Q(\boldsymbol{s},\boldsymbol{a})$ with $\boldsymbol{a}$ indicating the action. It observes the current state $\boldsymbol{s}$ and selects the action $\boldsymbol{a}$ that maximizes $Q(\boldsymbol{s},\boldsymbol{a})$ with some exploration strategies. Q-learning has been studied extensively in single-agent tasks where only one agent is acting alone in a stationary environment.

In the UAV trajectory design problem, multiple UAVs take actions at the same time. When each UAV adopts the single-agent Q-learning algorithm, it assumes that the other agents are part of the environment. In the UAV trajectory design problem, the single-agent Q-learning algorithm can be adopted as follows. For UAV i, the policy of UAV i to select action is

$$\pi_i(\boldsymbol{s}_i) = \arg \max_{\boldsymbol{a}_i' \in \mathscr{A}(\boldsymbol{s}_i)} Q_i(\boldsymbol{s}_i,\boldsymbol{a}_i). \tag{4.13}$$

Upon receiving a reward R_i after the end of the cycle and observing the next state $\boldsymbol{s}_i'$, it updates its table of Q values according to the following rule:

$$Q_i(\boldsymbol{s}_i,\boldsymbol{a}_i) \leftarrow Q_i(\boldsymbol{s}_i,\boldsymbol{a}_i) + \alpha \left(R_i + \rho \max_{\boldsymbol{a}_i' \in \mathscr{A}(\boldsymbol{s}_i)} Q_i(\boldsymbol{s}_i,\boldsymbol{a}_i) \right), \tag{4.14}$$

in which $\alpha \in (0,1)$ denotes the learning rate. With the help of [85], the single-agent Q-learning algorithm for UAV trajectory design problem can be given in Algorithm 4.2.

Multi-Agent Q-learning Algorithm

Although the single-agent Q-learning algorithm has many favorable properties such as a small state space and easy implementation, it lacks of consideration of the states and the strategic behaviors of other agents. Therefore, we adopt a multiagent Q-learning algorithm called opponent modeling Q-learning to solve the UAV trajectory design problem, which enables the agent to adapt to other agents' behaviors.

Opponent modeling Q-learning is an effective multiagent reinforcement-learning algorithm [86, 87], in which explicit models of the other agents are learned as stationary distributions over their actions. These distributions, combined with learned joint state-action values from standard temporal difference, are used to select an action in each cycle. Specifically, at the beginning of each cycle, UAV i selects the action a_i that maximizes the expected discounted reward according to the observed frequency distribution of other agents' action in the current state $\boldsymbol{s}$. That is, its policy at $\boldsymbol{s}$ is

$$\pi_i(\boldsymbol{s}) = \arg\max_{\boldsymbol{a}_i'} \sum_{\boldsymbol{a}_{-i}'} \frac{\Phi(\boldsymbol{s},\boldsymbol{a}_{-i}')}{n(\boldsymbol{s})} Q_i(\boldsymbol{s},(\boldsymbol{a}_i',\boldsymbol{a}_{-i}')), \tag{4.15}$$

in which the state $\boldsymbol{s} = (s_1,\ldots,s_N)$ is the location profile of all the UAVs, $\Phi(\boldsymbol{s},\boldsymbol{a}_{-i}')$ denotes the number of times the agents other than agent i select action profile $\boldsymbol{a}_{-i}'$ in the state $\boldsymbol{s}$, and $n(\boldsymbol{s})$ is the total number of times the state $\boldsymbol{s}$ has been visited.

Algorithm 4.2 Single-agent Q-learning algorithm for the UAV trajectory design problem of UAV i

1 **begin**
2 Initialize $Q_i(\boldsymbol{s}_i, \boldsymbol{a}_i) := 0, \forall \boldsymbol{s}_i \in \mathscr{S}_p, \boldsymbol{a}_i \in \mathscr{A}_i(\boldsymbol{s}_i), \pi_i(\boldsymbol{s}_i) := \text{Rand}(\{\mathscr{A}_i(\boldsymbol{s}_i)\})$;
3 **for** $k = 1$ *to max-number-of-cycles* **do**
4 With probability $1 - \epsilon^{(k)}$, choose action $\boldsymbol{a}_i$ from the policy at the state $\pi_i(\boldsymbol{s}_i)$, and with probability $\epsilon^{(k)}$, randomly choose an available action for exploration;
5 Perform the action $\boldsymbol{a}_i$ in the kth cycle;
6 Observe the transited state $\boldsymbol{s}_i'$ and the reward R_i;
7 Select $\boldsymbol{a}_i'$ in the transited state $\boldsymbol{s}_i'$ according to $\pi_i(\boldsymbol{s}_i')$;
8 Update the Q-value for the former state-action pair, i.e., $Q_i(\boldsymbol{s}_i, \boldsymbol{a}_i) := Q_i(\boldsymbol{s}_i, \boldsymbol{a}_i) + \alpha^{(k)}(R_i + \rho Q(\boldsymbol{s}_i', \boldsymbol{a}_i') - Q_i(\boldsymbol{s}_i, \boldsymbol{a}_i))$;
9 Update the policy at state $\boldsymbol{s}_i$ as $\pi_i(\boldsymbol{s}_i, \boldsymbol{a}_i') := 1$, where $\boldsymbol{a}_i = \arg\max_{\boldsymbol{m} \in \mathscr{A}_i(\boldsymbol{s}_i)} Q_i(\boldsymbol{s}_i, \boldsymbol{m})$;
10 Update the state $\boldsymbol{s}_i := \boldsymbol{s}_i'$;
11 **end**
12 **end**

After the agent i observes the transited state $\boldsymbol{s}'$, the action profile $\boldsymbol{a}_{-i}$, and the reward in the previous cycle, it updates the Q value as follows:

$$Q_i(\boldsymbol{s}, (\boldsymbol{a}_i, \boldsymbol{a}_{-i})) = (1 - \alpha) Q_i(\boldsymbol{s}, (\boldsymbol{a}_i, \boldsymbol{a}_{-i})) + \alpha(R_i + \rho V_i(\boldsymbol{s}')), \tag{4.16}$$

in which $V_i(\boldsymbol{s}') = \max_{\boldsymbol{a}_i'} \sum_{\boldsymbol{a}_{-i}'} \frac{\Phi(\boldsymbol{s}', \boldsymbol{a}_{-i}')}{n(\boldsymbol{s}')} Q(\boldsymbol{s}, (\boldsymbol{a}_i', \boldsymbol{a}_{-i}'))$ indicating that agent i selects the action in the transited state $\boldsymbol{s}'$ to maximize the expected discounted reward based on the empirical action profile distribution then. With the help of [87], the multiagent Q-learning algorithm for UAV trajectory design can be given in Algorithm 4.3.

Enhanced Multi-UAV Q-Learning Algorithm for UAV Trajectory Design

In the opponent modeling multiagent reinforcement-learning algorithm, UAVs need to tackle too many state-action pairs, resulting in a slow convergence speed. Therefore, we enhance the opponent modeling Q-learning algorithm in the UAV trajectory design problem by reducing the available action set and adopting an model-based reward representation. These two enhancing approaches are elaborated as follows, and the proposed enhanced multi-UAV Q-learning algorithm is given in Algorithm 4.4.

Available Action Set Reduction

It can be observed that although the UAVs can reach all the spatial points in the finite location space $\mathscr{S}_p$, it makes no sense for the UAVs to move away from the vertical planes passing the BS and their tasks (i.e., the BS-task planes), which decreases the successful sensing probability as well as the successful transmitting probability. Therefore, we confine the available action set of the UAV to the actions that do not

Algorithm 4.3 Opponent modeling Q-learning algorithm for the UAV trajectory design problem of UAV i.

1 **begin**
2 Initialize $Q_i(\boldsymbol{s},(\boldsymbol{a}_i,\boldsymbol{a}_{-i})) := 0, \forall \boldsymbol{s} \in \prod_i^N \mathscr{S}_i$, $\boldsymbol{a}_i \in \mathscr{A}_i(\boldsymbol{s}_i)$, $\boldsymbol{a}_{-i} \in \prod_{j \neq i}^N \mathscr{A}_j$, $\pi_i(\boldsymbol{s}_i) := \text{Rand}(\{\mathscr{A}_i(\boldsymbol{s}_i)\})$;
3 **for** $k = 1$ *to max-number-of-cycles* **do**
4 With probability $1 - \epsilon^{(k)}$, choose action $\boldsymbol{a}_i$ according to the policy $\pi_i(\boldsymbol{s})$, or with probability $\epsilon^{(k)}$, randomly choose an available action for exploration;
5 Perform the action $\boldsymbol{a}_i$ in the kth cycle;
6 Observe the transited state $\boldsymbol{s}'$ and the reward R_i;
7 Select action $\boldsymbol{a}_i'$ in the transited state $\boldsymbol{s}'$ according to the strategy in state $\boldsymbol{s}'$ according to (4.15);
8 Update the Q value for the former state-action pair according to (4.16);
9 Update the policy at state $\boldsymbol{s}$ to the action that maximizes the expected discounted reward according to (4.15);
10 Update the state $\boldsymbol{s} := \boldsymbol{s}'$;
11 **end**
12 **end**

increase the horizontal distance between it and the BS-task plane, which is illustrated by the arrows in Fig. 4.9.

Ideally, the UAVs should be in the BS-task plane and only move within the plane. However, because the spatial space is discrete, the UAV cannot only move within the BS-task plane and needs to deviate from the plane in order to reach different locations on or near the plane. Therefore, we mitigate the constraint by allowing the UAVs to move to the spatial points from which the distance to their BS-task planes is within Δ, as the dots illustrate in Fig. 4.9. The reduced available action set of UAV i at state $\boldsymbol{s}_i = (x_i, y_i, h_i)$ can be defined as follows.

DEFINITION 4.2 (Reduced available action set of UAV) *Supposing UAV i at the state $\boldsymbol{s}_i = (x_i, y_i, h_i)$, the location of its task at $\boldsymbol{S}_i = (X_i, Y_i, 0)$, and the location of BS at $\boldsymbol{S}_0 = (0, 0, H_0)$, the action $\boldsymbol{a} = (a_x, a_y, a_h)$ in the reduced available action set $\mathscr{A}_i^+(\boldsymbol{s}_i)$ satisfies the following conditions:*

1. $\text{Dist}(\boldsymbol{s}_i + \boldsymbol{a}; \boldsymbol{S}_i, \boldsymbol{S}_0) \leq \text{Dist}(\boldsymbol{s}_i; \boldsymbol{S}_i, \boldsymbol{S}_0)$ *or* $\text{Dist}(\boldsymbol{s}_i + \boldsymbol{a}; \boldsymbol{S}_i, \boldsymbol{S}_0) \leq \Delta$;
2. $x_i + a_x \in [\min(x_i, X_i, 0), \max(x_i, X_i, 0)]$,
   $y_i + a_y \in [\min(y_i, Y_i, 0), \max(y_i, Y_i, 0)]$, *and*
   $h_i + a_h \in [h_{\min}, h_{\max}]$.

Here, $\text{Dist}(\boldsymbol{s}_i; \boldsymbol{S}_i, \boldsymbol{S}_0)$ *denotes the horizontal distance between s_i to the vertical plane passing through $\boldsymbol{S}_i$ and $\boldsymbol{S}_0$.*

Algorithm 4.4 Enhanced multi-UAV Q-learning algorithm for trajectory design problem of UAV i

1 **begin**
2 **for** $k = 1$ *to max-number-of-cycles* **do**
3 Obtain the available action set $\mathscr{A}_j^+(s_j)$, $\forall j \in [1, N]$ for the current state $\boldsymbol{s}$ according to Definition 4.2;
4 **if** $\boldsymbol{s}$ *has not been visited before* **then**
5 Initialize $Q_i(\boldsymbol{s}, \boldsymbol{a}) := p_{\mathrm{sTx},i}(\boldsymbol{s}, \boldsymbol{a})$, $\forall \boldsymbol{s} \in \prod_i^N \mathscr{S}_i$, $\boldsymbol{a} \in \prod_{j=1}^N \mathscr{A}_j^+(\boldsymbol{s}_j)$, $\pi_i(\boldsymbol{s}_i) := \text{Rand}(\{\mathscr{A}_i^+(\boldsymbol{s}_i)\})$;
6 **end**
7 With probability $1 - \epsilon^{(k)}$, choose action a_i from the strategy at the state $\pi_i(\boldsymbol{s})$, or with probability $\epsilon^{(k)}$, randomly choose an available action for exploration;
8 Perform action $\boldsymbol{a}_i$ in the kth cycle;
9 Observe the transited state $\boldsymbol{s}'$ and the action profile $\boldsymbol{a}$ in the previous state;
10 Select action $\boldsymbol{a}_i'$ in the transited state $\boldsymbol{s}'$ according to policy $\pi_i(\boldsymbol{s}')$ defined in (4.15);
11 Calculate the probability of successful valid sensory data transmission in the previous cycle $p_{\mathrm{sTx},i}(\boldsymbol{s}, \boldsymbol{a})$;
12 Update the Q function for the former state-action pair according to (4.16), with $R_i := p_{\mathrm{sTx},i}(\boldsymbol{s}, \boldsymbol{a})$;
13 Update the policy at state $\boldsymbol{s}$ to the action that maximizes the expected discounted rewards according to (4.15);
14 Update the state $\boldsymbol{s} := \boldsymbol{s}'$;
15 **end**
16 **end**

In Definition 4.2, condition 1 limits the actions to those leading the UAV to the spatial points near the BS-task plane, and condition 2 stops the UAV from moving away from the cross region between the location of its task and the BS.

Moreover, instead of initializing the Q values for all the possible state-action pairs at the beginning, we propose that the UAVs initialize the Q values only for the reached state and the actions within the reduced available action set. In this way, the state sets of the UAVs are reduced to much smaller sets, which makes the learning more efficient.

Model-Based Reward Representation

In both the single-agent Q-learning and the opponent modeling Q-learning algorithms, the UAVs update their Q values based on the information provided by the BS, which indicates the validity of the latest transmitted sensory data. Nevertheless, because the UAVs can only observe the reward to be either one or zero, the Q values converge slowly, and the performance of the algorithms is likely to be poor.

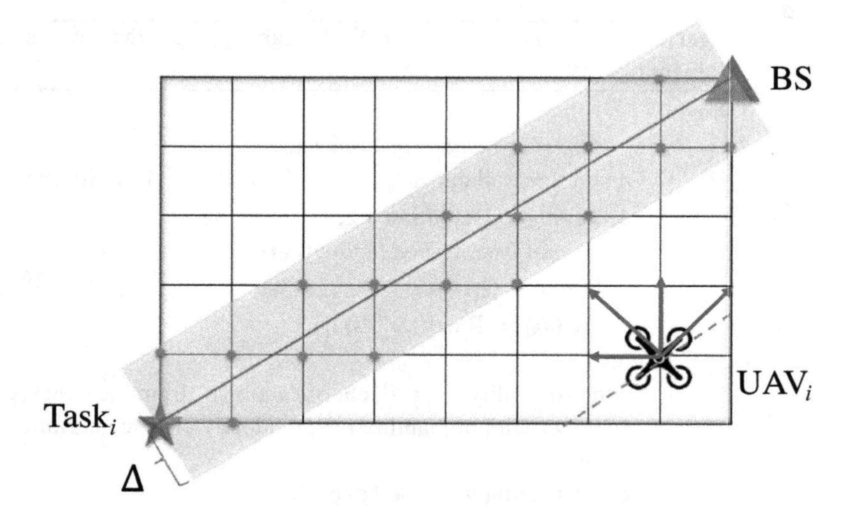

Figure 4.9 Constrained available action set of UAV i

Therefore, we propose that the UAVs update their Q values based on the probabilities of a successful valid sensory data transmission. Specifically, UAV i calculates the probability $p_{\text{sTx},i}$ after observing the state-action profile $(s,(a_i,\boldsymbol{a}_{-i}))$ in the latest cycle according to (4.8) and takes it as the reward R_i for the kth cycle.

Moreover, to make the learning algorithm converge more quickly, in the initialization of the enhanced multi-UAV Q-learning algorithm, we propose that UAV i initializes its $Q_i(s,(a_i,\boldsymbol{a}_{-i}))$ with the calculated $p_{sTx,i}$ for the state-action pair. In this way, the update of the Q-values is more accurate, and the learning algorithm is expected to have a higher convergence speed.

Remark 4.1 (Signaling in UAVs' learning algorithms) In the previously mentioned reinforcement-learning algorithms, UAVs need to know the location profile in the beginning of each cycle and the rewards in the last cycle associated with the actions they took. This information gathering can be done in beaconing phase of the cycle, as described in Section 4.1.3, in which the BS can include the rewards of UAVs in the last cycle in the broadcasted beacon.

Analysis of Reinforcement-Learning Algorithms

In the final part of this section, we analyze the convergence, the complexity, and the scalability of the proposed reinforcement-learning algorithms.

Convergence Analysis

For the convergence of the reinforcement-learning algorithms, it was proved in [88] that under certain conditions, a single agent Q-learning algorithm is guaranteed to converge to the optimal Q^*. Therefore, the policy π of the agent converges to the optimal policy π^*. This can be summarized in the following Theorem 4.3.

THEOREM 4.3 (Convergence of Q-learning algorithm) *The Q-learning algorithm given by*

$$Q^{(k+1)}\left(s^{(k)},a^{(k)}\right) = (1-\alpha^{(k)})Q^{(k)}\left(s^{(k)},a^{(k)}\right) + \alpha^{(k)}\left[R(s^{(k)},a^{(k)}) + \gamma \max_{a'} Q(s^{(k+1)},a')\right] \quad (4.17)$$

converges to the optimal Q^* values if the following conditions are satisfied:

1. *The state and action spaces are finite.*
2. $\sum_k \alpha^{(k)} = \infty$ *and* $\sum_k (\alpha^{(k)})^2 < \infty$.
3. *The variance of $R(s,a)$ is bounded.*

Therefore, in the multiagent reinforcement-learning cases, if other agents play or converge to stationary strategies, the single-agent reinforcement-learning algorithm also converges to the optimal policy.

However, it is generally hard to prove convergence with other agents that are learning simultaneously. This is because when the agent is learning the Q value of its actions in the presence of other agents it faces a nonstationary environment and the convergence of Q values is not guaranteed. The theoretical convergence of the Q learning in multi-agent cases is guaranteed only in a few situations, such as in iterated dominance solvable games and team games [85]. Like single-agent Q-learning algorithm, the convergence of opponent modeling Q learning is not generally guaranteed, except for in the setting of iterated dominance solvable games and team matrix game [87].

To handle this problem, we adopt $\alpha^{(k)} = 1/k^{2/3}$ in [89], which satisfies the conditions for convergent in single-agent Q learning, and analyze the convergence of the reinforcement learning in the multiagent case through simulation results, which will be provided in Section 4.2.4.

Complexity Analysis

For the single-agent Q-learning algorithm, the computational complexity in each iteration is $\mathcal{O}(1)$ because the UAV does not consider the other UAVs in the learning process. For the multiagent Q-learning algorithm, the computational complexity in each iteration is $\mathcal{O}(2^N)$ due to the calculation of the expected discounted reward in (4.15).

As for the proposed enhanced multi-UAV Q-learning algorithm, each UAV needs to calculate the probability of a successful valid data transmission based on Algorithm 4.1. It can be seen that the recursive Algorithm 4.1 runs for at most 2^{CT_u} times, and each iteration has the complexity of $\mathcal{O}(N)$, which makes its overall complexity $\mathcal{O}(N)$. Therefore, the complexity of the proposed enhanced algorithm is still $\mathcal{O}(2^N)$ due to the expectation of the joint action space.

Although the computational complexity of the enhanced multi-UAV Q-learning algorithm in each iteration is in the same order as the opponent modeling Q-learning algorithm, it reduces the computational complexity and speeds up the convergence by the following means.

- Due to the available action set reduction, the available action set of each UAV is at least reduced to one-half of its original size. This makes the joint action space 2^N times smaller.
- The reduced available action set leads to a much smaller state space for each UAV. For example, for UAV i and its task at $(X_i, Y_i, 0)$, the original size of its state space can be estimated as $\pi R_{\max}^2 (h_{\max} - h_{\min})/\Delta^3$, and the size of its state space after available action set reduction is $2(X_i + Y_i)(h_{\max} - h_{\min})/\Delta^2$, which is $2\Delta/(\pi R_{\max})$ of the original one.
- The proposed algorithm adopts a model-based reward representation, which makes the Q-value updating more precise and reduces the number of iterations needed to estimate the accurate Q values of the state-action pairs.

Scalability Analysis

With the growth of the number of UAVs, the state spaces of the UAVs in the multiagent Q-learning algorithm and the enhanced multi-UAV Q-learning algorithm grow exponentially. Besides, it can be seen that the enhanced multi-UAV Q-learning algorithm still has exponential computational complexity in each iteration, and thus, it is not suitable for large-scale UAV networks.

To adapt the algorithms to large-scale UAV networks, reinforcement-learning methods need to be combined with function approximation approaches in order to estimate Q values efficiently. The function approximation approaches take examples from a desired function (Q function in the case of reinforcement learning) and generalize from them to construct an approximation of the entire function. In this regard, it can be used to estimate the Q values of the state-action pairs in the entire state space efficiently when the state space is large.

4.2.4 Performance Evaluation

In order to evaluate the performance of the proposed reinforcement-learning algorithms for the UAV trajectory design problem, simulation results are presented in this section. Specifically, we use MATLAB to build a frame-level simulation of the UAV sense-and-send protocol based on the system model described in Section 4.2.1 and the parameters in Table 4.1. The learning ratio in the algorithm is set to be $\alpha^{(k)} = 1/k^{2/3}$ in order to satisfy the converge condition in Theorem 4.3, and the exploration ratio is set to be $\epsilon^{(k)} = 0.8e^{-0.03k}$, which approaches 0 when $k \to \infty$.

Figure 4.10 shows UAV 1's probability of successful valid sensory data transmission versus UAV 1's height and its distance to the BS, given that task 1 is located at $(500, 0, 0)$, and the locations of UAV 2 and UAV 3 are fixed at $(-125, 125, 75), (-125, -125, 75)$, respectively. It can be seen that the optimal point at which UAV 1 has the maximum probability of successful valid sensory data transmission is located in the region between the BS and task 1. This is because when the UAV approaches the BS, its successful sensing probability drops, and when the UAV approaches the task, its successful transmission probability suffers. In addition, it is shown that the

Table 4.1 Simulation parameters

Parameter	Value
BS height H	25 m
Number of UAVs N	3
Noise power N_0	−85 dBm
BS decoding threshold γ_{th}	10 dB
UAV sensing parameter λ	10^{-3}/s
UAV transmit power P_u	10 dBm
Duration of frame t_f	0.1 s
Distance between adjacent spatial points Δ	25 m
UAVs' minimum flying height $h_{\min}$	50 m
UAVs' maximum flying height $h_{\max}$	150 m
Discount ratio ρ	0.9
Duration of beaconing phase in frames T_b	3
Duration of sensing phase in frames T_s	5
Duration of transmission phase in frames T_u	5

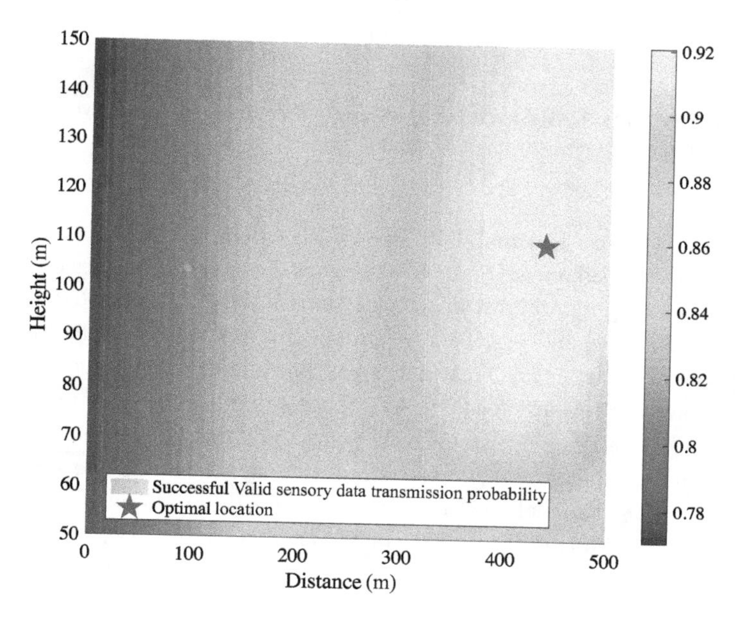

Figure 4.10 Successful valid sensory data transmission probability versus the location in the task-BS surface

optimal point for UAV 1 to sense and send is *above*, rather than *on* the BS-task line, where UAV 1 can be closer to both the BS and its task. This is because in the transmission model in Section 4.2.1, with the increment of the height of the UAV, the LoS probability increases, and thus, the successful uplink transmission probability of the UAV increases.

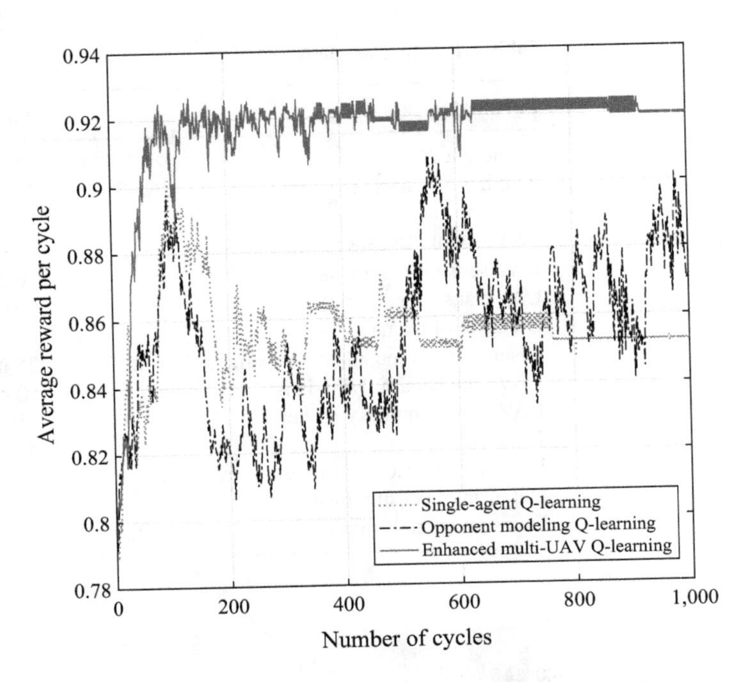

Figure 4.11 UAVs' average reward per cycle versus number of cycles of different reinforcement-learning algorithms

Figures 4.11 and 4.12 show the average reward per cycle and the average total discounted reward of the UAVs versus the number of cycles in different reinforcement-learning algorithms, in which tasks 1, 2, and 3 are located at $(500, 0, 0)$, $(-250\sqrt{2}, 250\sqrt{2}, 0)$ and $(-250\sqrt{2}, -250\sqrt{2}, 0)$, respectively. It can be seen that compared to the single-agent Q-learning algorithm, the proposed algorithm converges to higher average rewards for the UAVs. This is because the UAV in the enhanced multi-UAV Q-learning algorithm takes the states of all the UAVs into consideration, which makes the estimation of Q values more accurate. It can also be seen that compared to the opponent modeling Q-learning algorithm, the proposed algorithm converges faster due to the available action set reduction and the model-based reward representation.

Moreover, in Fig. 4.13, we can observe that for different distances between the tasks and the BS, the proposed algorithm converges to higher average total discounted rewards for UAVs after 1,000 cycles compared to the other algorithms. It can be seen that the average total discounted reward in the three algorithms decreases with the increment of the distance between the BS and the tasks. Nevertheless, the decrement in the proposed algorithm is less than those in the other algorithms. This indicates that the proposed algorithm is more robust to the variance of the tasks' locations.

Figure 4.14 shows the average number of successful valid sensory data transmissions versus the duration of the transmission phase T_u in the proposed algorithm under different conditions of the distance between the tasks and the BS. It can be seen that the average number of successful valid sensory data transmissions per second first

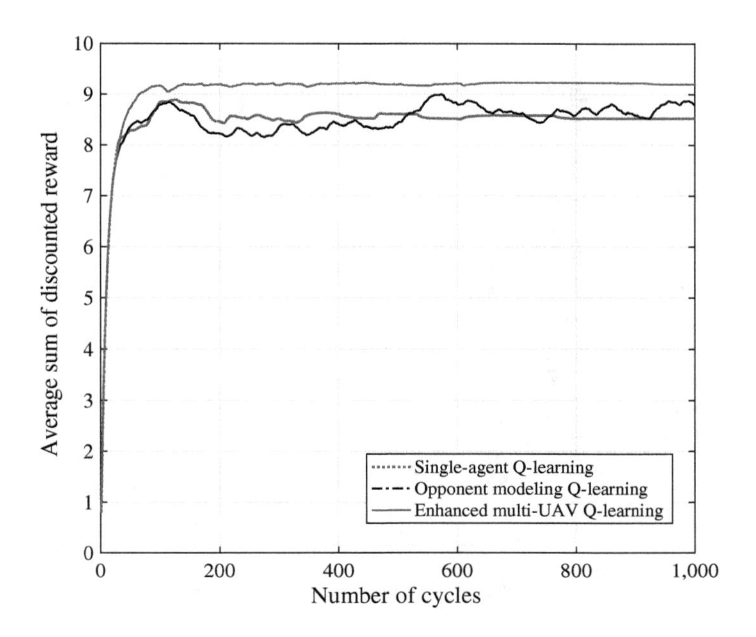

Figure 4.12 UAVs' average discounted reward versus number of cycles of different reinforcement-learning algorithms

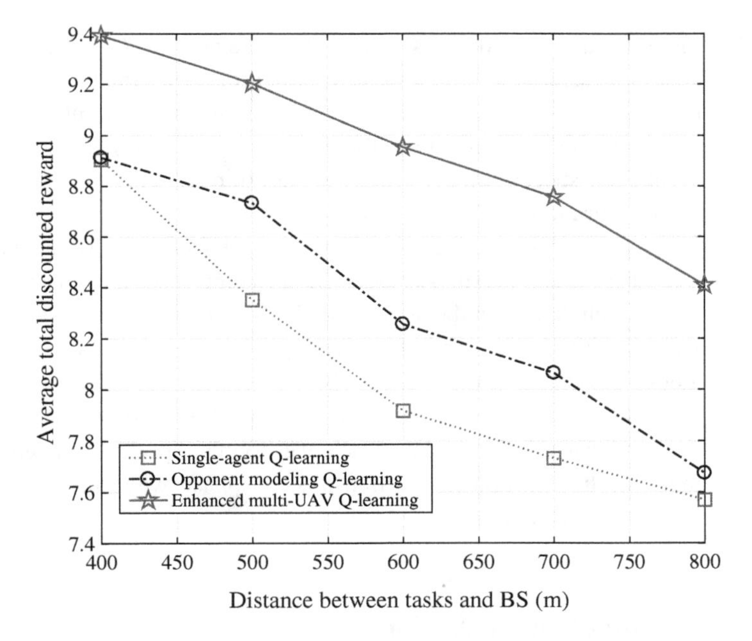

Figure 4.13 UAVs' average discounted reward versus distance between tasks and BS in different reinforcement-learning algorithms

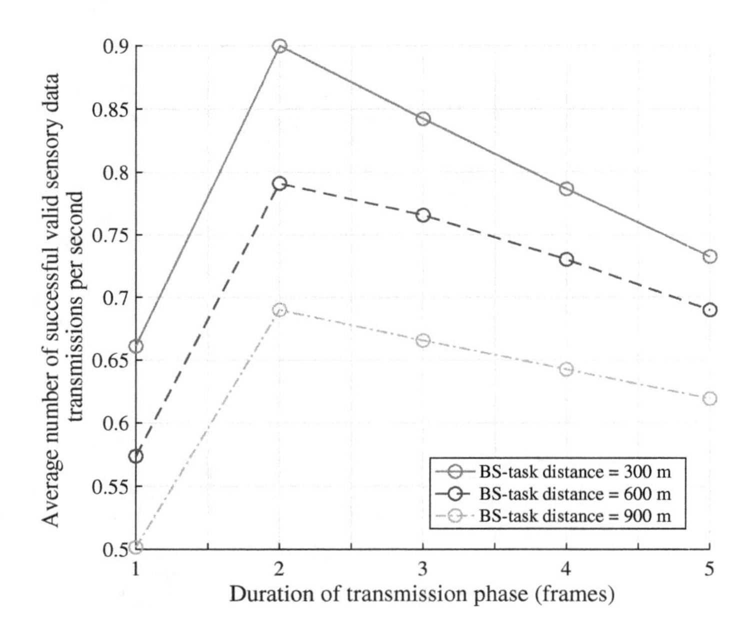

Figure 4.14 Average number of successful valid sensory data transmissions per second versus duration of transmission phase T_u under different task distance conditions

increases and then decreases with the increment of T_u. This is because when T_u is small the successful uplink transmission probability increases rapidly with the increment of T_u. However, when T_u is large, the successful uplink transmission probability is already high and increases slightly when T_u becomes larger. Therefore, the average number of successful valid sensory data transmissions per second drops due to the cycles' duration.

Figure 4.15 shows the average number of successful valid sensory data transmissions per second versus T_u under the different conditions of the number of UAVs. It can be seen that when the number of UAVs increases the average number of successful valid sensory data transmissions per second decreases. This is because the competition among the UAVs for the limited subchannels becomes more intense. When the number of UAVs increases, the optimal duration of the transmission phase becomes longer. This indicates that in order to achieve optimal spectral efficiency, the BS needs to increase the duration of the transmission phase when the number of UAVs in the network increases.

4.3 UAV-to-X Communications

The cellular internet of UAVs multiplexes the spectrum resources and infrastructure of the terrestrial cellular user equipment (UE) and consumes the communication services supported by the powerful hardware foundation in the 6G era. However, due

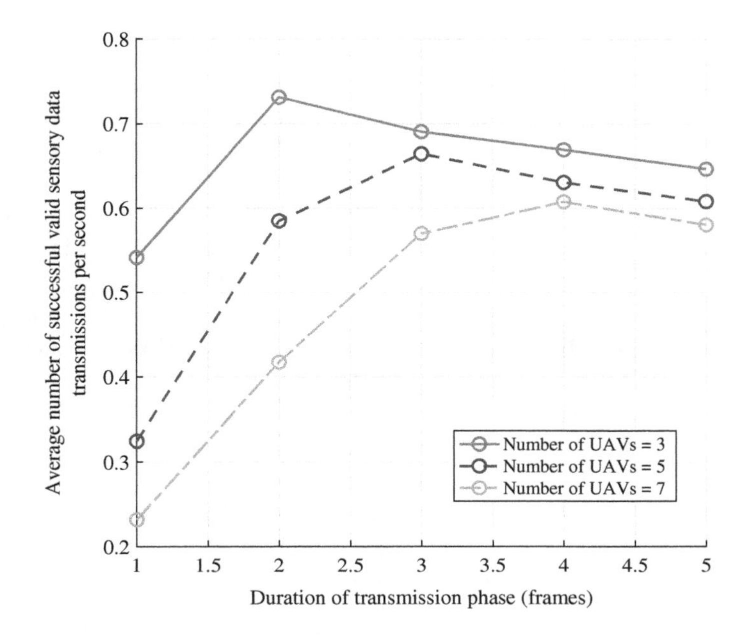

Figure 4.15 Average number of successful valid sensory data transmissions per second versus duration of transmission phase T_u for different numbers of UAVs, where the distance between the BS and the tasks equals 800 m

to the diverse requirements for UAV communications, the UAVs cannot achieve a high data rate by applying the terrestrial cellular networks directly[90]. Unlike conventional terrestrial UE that can obtain a high transmission rate by communicating with a nearby BS, the sensing UAVs have full dimension transmission destinations, including the BS, terrestrial UE, and other adjacent UAVs [91]. Therefore, a more flexible framework is necessary for the UAVs that gives them freedom to choose the optimal transmission links [28]. Moreover, different from terrestrial UE, the UAVs are likely to move to the cell edge for the sake of sensing performance. In this case, it is difficult to guarantee a high transmission rate with the conventional cellular framework, where all the sensory data are required to be uploaded to the BS over cellular links.

To tackle these challenges, we propose the concept of full-dimension UAV-to-everything (U2X) communications, which contains three different modes:

- **UAV-to-network (U2N) mode**: In U2N mode,[3] a UAV uploads its collected data to the BS directly overlaying the cellular uplink spectrum resource. This U2N communication provides a high data rate and low latency transmission when the UAV is close to the BS. It also enables multiple UAVs to communicate with the BS over different subchannels simultaneously, which reduces the potential transmission interference. This type of communication is essential for applications that require the BS to collect massive data, such as crowdsourcing. Furthermore, the existing infrastructure guarantees a high downlink rate for terrestrial cellular

UE, and enable the UE to access highly reliable sensory data collected by the BS from the UAVs.

- **UAV-to-UAV (U2U) mode**: In U2U mode,[4] a UAV communicates with another UAV underlaying the cellular and U2N communications, which reduces the latency and provides high QoS for the communications between adjacent UAVs. Such communications also allow a UAV to broadcast data to every direction that has adjacent UAVs, thus providing a physical mechanism for efficient information diffusion. Moreover, the underlaying transmission improves the spectrum efficiency of the network, which is necessary for the network with massive UE.
- **UAV-to-device (U2D) mode**: In U2D mode, a UAV communicates with destination UE directly underlaying the cellular and U2N communications. Such communications allow the UE to receive the data from a UAV directly bypassing the BS. This type of communication is especially efficient when a UAV is close to its transmission destination UE. It not only saves the spectrum and energy resources for UAV-to-BS and BS-to-UE links but also reduces the transmission latency. It is promising for low latency applications without a huge amount of computation, such as live video streaming.

The U2X communications enable the UAVs to adopt different transmission modes according to the specific requirements of their corresponding sensing applications and provide a feasible architecture for the utilization of UAV sensing in the 6G network [92].

In this section, we give a case study to introduce the key technologies in U2X communications, and some other examples can be found in [90, 93]. First, because the U2U transmissions underlie the spectrum resources of the U2N and cellular user (CU) transmissions, the U2N and CU transmissions may have interference from the U2U transmissions when sharing the same subchannel. Correspondingly, the U2U transmissions experience interference from the U2N, CU, and U2U links on the same subchannel. Moreover, different channel models are utilized for the U2N, U2U, and CU transmissions due to the different characteristics of air-to-ground, air-to-air, and ground-to-ground communications. Therefore, an *efficient spectrum allocation algorithm* is studied to manage the mutual interference. Second, to complete the data collection of the sensing tasks given time requirements, the *UAV speed optimization* problem is formulated and solved. Third, to avoid data loss and provide a relatively high data rate for the UAVs with a low SNR for the link to the BS, an *efficient communication method* is proposed.

The rest of this section is organized as follows. In Section 4.3.1, we present the system model. In Section 4.3.2, we formulate the uplink sum-rate maximization problem by optimizing the subchannel allocation and UAV speed jointly. The iterative subchannel allocation and UAV speed optimization algorithm (ISASOA) is proposed in Section 4.3.3, followed by the corresponding analysis. Simulation results are presented in Section 4.3.4, and finally we summarize the section in Section 4.3.5.

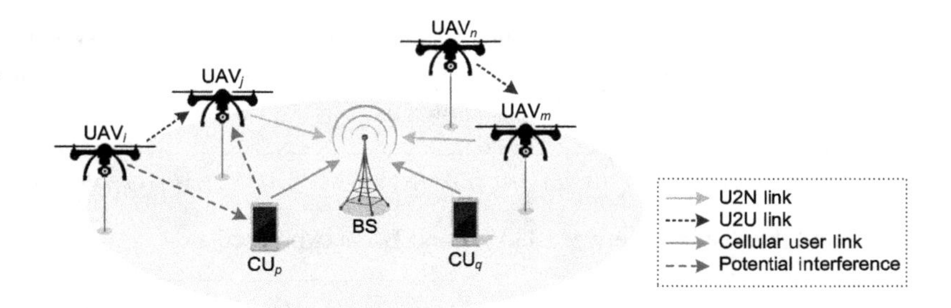

Figure 4.16 System model of U2X communications

4.3.1 System Model

In this part, we first describe the working scenario and then introduce the data transmission of this network. Finally, we present the channel models for U2N, U2U, and CU transmissions.

Scenario Description

We consider a single cell cellular network as shown in Fig. 4.16, which consists of one BS, M CUs, denoted by $\mathscr{M} = \{1,2,\ldots,M\}$, and N UAVs, denoted by $\mathscr{N} = \{1,2,\ldots,N\}$ [28]. The UAVs collect various required data with their sensors in each time slot, and the data will be transmitted to the BS for further processing. In each time slot, the UAVs first perform UAV sensing and then perform data transmission. The length of time for UAV sensing and data transmission are given in each time slot. We assume that each UAV flies along a predetermined trajectory during the sensing and transmission process. The speeds of the UAVs in each time slot are not given, but all the UAVs are required to arrive at the endpoints of their trajectories within a number of time slots for timely sensing and transmission. To provide a high data transmission rate for all the UAVs, we distinguish between the UAVs with different QoS requirements for the U2N links using two transmission modes: U2N transmission and U2U transmission. The UAVs in the U2N mode transmit over the cellular ones, whereas the UAVs in the U2N mode can transmit under the U2N and CU transmissions by reusing the spectrum resources occupied by the U2N and CU transmissions.

We denote the location of UAV i in time slot t by $\boldsymbol{l}_i(t) = (x_i(t), y_i(t), h_i(t))$ and the location of the BS by $(0,0,H)$. Each UAV moves along a predetermined trajectory. Let $v_i(t)$ be the speed of UAV i in time slot t. The location of UAV i in time slot $t+1$ is given as $\boldsymbol{l}_i(t+1) = \boldsymbol{l}_i(t) + v_i(t) \cdot \boldsymbol{\omega}_i(t)$, where $\boldsymbol{\omega}_i(t)$ is the trajectory direction of UAV i in time slot t. Due to its mechanical limitations, the speed of a UAV is no more than $v_{\max}$.[5] Let L_i be the length of UAV i's trajectory. With proper transmission rate requirements, the UAVs are capable of uploading the sensory data to the BS with low latency. Therefore, the task completion time of a UAV can be defined as the time that it takes to move along the trajectory, which is determined by its speed in each time slot.

For timely data collection, the task completion time of each UAV is required to be no more than T time slots; that is, $\sum_{t=1}^{T} v_i(t) \geq L_i, \forall i \in \mathcal{N}$. In time slot t, the distance between UAV i and UAV j is shown as

$$d_{i,j}(t) = \sqrt{(x_i(t) - x_j(t))^2 + (y_i(t) - y_j(t))^2 + (h_i(t) - h_j(t))^2}, \tag{4.18}$$

and the distance between UAV i and BS is expressed as

$$d_{i,BS}(t) = \sqrt{(x_i(t))^2 + (y_i(t))^2 + (h_i(t) - H)^2}. \tag{4.19}$$

The location of CU i is given as (x_i^c, y_i^c, h_i^c). We assume that the locations of the CUs are fixed in different time slots because the mobility of the CUs are much lower than that of the UAVs. Therefore, the distance between CU i and UAV j can be denoted by

$$d_{i,j}^c(t) = \sqrt{(x_i^c(t) - x_j(t))^2 + (y_i^c(t) - y_j(t))^2 + (h_i^c(t) - h_j(t))^2}, \tag{4.20}$$

and the distance between UAV i and BS can be shown as

$$d_{i,BS}^c(t) = \sqrt{x_i^c(t)^2 + y_i^c(t)^2 + (h_i^c(t) - H)^2}. \tag{4.21}$$

Data Transmission

There are two UAV transmission modes in this network: U2N and U2U modes. A UAV may transmit in either the U2N or U2U mode in one time slot. The criterion of adopting U2N or U2U mode is given as follows:

- **U2N mode**: A UAV with a high SNR for the link to the BS performs U2N transmission in the network. It uploads its collected data to the BS directly over the assigned subchannel.
- **U2U mode**: A UAV with a low SNR for the link to the BS performs U2U communication transmitting the collected data to a UAV in U2N transmission mode.

The transmission mode is selected according to the following criteria. When receiving the beacons of the UAVs, the BS categorizes the UAVs into U2N and U2U modes according to the received SNR. A SNR threshold γ_{th} is given to distinguish the UAVs that transmit in the U2N and U2U modes.[6] The UAVs with the SNR for U2N links that are larger than γ_{th} are considered to perform U2N transmission, and the UAVs with the SNR for the U2N links that are lower than γ_{th} are considered to perform U2U transmission.

Let $\mathcal{N}_h(t) = \{1, 2, \ldots, N_h(t)\}$ and $\mathcal{N}_l(t) = \{1, 2, \ldots, N_l(t)\}$ be the set of UAVs in U2N and U2U modes in time slot t, respectively, with $\mathcal{N} = \mathcal{N}_h(t) \cup \mathcal{N}_l(t)$. The UAVs in $\mathcal{N}_h(t)$ send the data to the BS by U2N transmissions. For the UAVs in $\mathcal{N}_l(t)$, the SNR of the direct communication links is low, and thus it is difficult for these UAVs to provide high data rates to support timely data upload via U2N modes. Therefore, the UAVs send the collected data to the neighboring UAVs with the high SNR for the U2N link via U2U transmissions, and the data is sent to the BS later by the relaying UAVs.

The transmission bandwidth of this network is divided into K orthogonal subchannels, denoted by $\mathcal{K} = \{1, 2, \ldots, K\}$. It is worthwhile to mention that a single UAV can perform U2N transmission and U2U reception over different subchannels simultaneously. For the sake of transmission quality, we assume that a subchannel can serve at most one U2N or CU link but multiple U2U links in one time slot. In addition, to guarantee fairness among the users, we also assume that each transmission link can be allocated to no more than $\chi_{\max}$ subchannels. In time slot t, we define a binary U2N and CU subchannel pairing matrix $\mathbf{\Phi}(t) = [\phi_{i,k}(t)]_{(N_h+M)\times K}$, and a binary U2U subchannel pairing matrix $\mathbf{\Psi}(t) = [\psi_{i,k}(t)]_{N_l \times K}$, to describe the resource allocation for CU, U2N, and U2U transmissions. For $i \le N_h$, $\phi_{i,k}(t) = 1$ when subchannel k is assigned to UAV i for U2N transmission; otherwise, $\phi_{i,k}(t) = 0$. For $i > N_h$, $\phi_{i,k}(t) = 1$ when subchannel k is assigned to CU $i - N$ for CU transmission; otherwise, $\phi_{i,k}(t) = 0$. Likewise, the value of $\psi_{i,k}(t) = 1$ when subchannel k is assigned to UAV i in the U2U mode; otherwise, $\psi_{i,k}(t) = 0$.

We denote $\xi_{i,j}(t) = 1$ when UAV i performs U2U transmission with UAV j in time slot t, and $\xi_{i,j}(t) = 0$ otherwise. In order to avoid high communication latency for the UAVs, the data rate of each U2U communication link should be no less R_0; that is, $\sum_{k=1}^{K} \psi_{i,k}(t) R_{i,j}^{k}(t) \ge R_0, \forall i, j \in \mathcal{N}, \xi_{i,j} = 1$.

Channel Model

In this subsection, we introduce the channel model in this network. The channel models of the U2N, CU, and U2U transmissions are different due to the different characteristics in LoS probability and elevation angle, which will be introduced as follows.

U2N Channel Model

We use the probabilistic propagation model for the U2N transmission. In time slot t, the LoS and NLoS pathloss from UAV i to the BS is given by

$$PL_{LoS,i}(t) = L_{FS,i}(t) + 20\log(d_{i,BS}(t)) + \eta_{LoS}, \tag{4.22}$$

$$PL_{NLoS,i}(t) = L_{FS,i}(t) + 20\log(d_{i,BS}(t)) + \eta_{NLoS}, \tag{4.23}$$

where $L_{FS,i}(t)$ is the free space pathloss given by $L_{FS,i}(t) = 20\log(f) + 20\log(\frac{4\pi}{c})$, and f is the system carrier frequency. The additional attenuation factors are η_{LoS} and η_{NLoS} due to the LoS and NLoS connections. Considering the antennas on UAVs and the BS placed vertically, the probability of LoS connection is given by

$$P_{LoS,i}(t) = \frac{1}{1 + a\exp(-b(\theta_i(t) - a))}, \tag{4.24}$$

where a and b are constants that depend on the environment, and the elevation angle $\theta_i(t) = \sin^{-1}((h_i(t) - H)/d_{i,BS}(t))$. The average pathloss in decibels can then be expressed as

$$PL_{avg,i}(t) = P_{LoS,i}(t) \times PL_{Los,i}(t) + P_{NLoS,i}(t) \times PL_{NLoS,i}(t), \tag{4.25}$$

where $P_{NLoS}(t) = 1 - P_{LoS}(t)$. The average received power of BS from UAV i over its paired subchannel k is given by

$$P^k_{i,BS}(t) = \frac{P_U}{10^{PL_{avg,i}(t)/10}}, \tag{4.26}$$

where P_U is the transmit power of a UAV or CU over one subchannel. Because each subchannel can be assigned to at most one U2N or CU link, the interference to the U2N links only comes from the U2U links due to spectrum sharing. When UAV i in the U2N mode over subchannel k, the U2U interference is expressed as

$$I_{k,U2U}(t) = \sum_{j=1}^{N_I} \psi_{j,k}(t) P^k_{j,BS}(t). \tag{4.27}$$

Therefore, the signal to interference plus noise ratio (SINR) of the BS over subchannel k is given by

$$\gamma^k_{i,BS}(t) = \frac{P^k_{i,BS}(t)}{\sigma^2 + I_{k,U2U}(t)}, \tag{4.28}$$

where σ^2 is the variance of the additive white Gaussian noise (AWGN) with zero mean. The data rate that BS receives from UAV i over subchannel k is shown as

$$R^k_{i,BS}(t) = \log_2(1 + \gamma^k_{i,BS}(t)). \tag{4.29}$$

CU Channel Model

We utilize the macrocell pathloss model as proposed in [94]. For CU i, the pathloss in decibels can be expressed by

$$PL^k_{i,C}(t) = -55.9 + 38\log(d^c_{i,BS}(t)) + (24.5 + 1.5f/925)\log(f). \tag{4.30}$$

When CU i transmits signals to the BS, the received power is expressed as

$$P^k_{i,C}(t) = \frac{P_U}{10^{PL^k_{i,C}(t)/10}}. \tag{4.31}$$

We denote the set of UAVs that share subchannel k with CU i by $U_i = \{m|\psi_{m,k}(t) = 1, \forall m \in \mathcal{N}_e\}$, and the received power at the BS over subchannel k is shown as

$$y^k_{i,j}(t) = \sqrt{P^k_{i,C}(t)} + \sum_{m \in U_i} \sqrt{P^k_{m,BS}(t)} + n^k_j(t), \tag{4.32}$$

where $n^k_j(t)$ is the AWGN with zero mean and σ^2 variance. Therefore, the received signal at the BS over subchannel k can be given by

$$\gamma^k_{i,BS}(t) = \frac{P^k_{i,C}(t)}{\sigma^2 + I_{k,U2U}(t)}, \tag{4.33}$$

where $I_{k,U2U}(t)=\sum_{j=1}^{N}\psi_{j,k}(t)P_{j,BS}^{k}(t)$ is the U2U interference. The data rate for CU i over subchannel k is expressed as

$$R_{i,BS}^{k}(t)=\log_2(1+\gamma_{i,BS}^{k}(t)). \tag{4.34}$$

U2U Channel Model

For U2U communication, the free-space channel model is utilized. When UAV i transmits signals to UAV j over subchannel k, the received power at UAV j from UAV i is expressed as

$$P_{i,j}^{k}(t)=P_U G(d_{i,j}(t))^{-\alpha}, \tag{4.35}$$

where G is the constant power gains factor introduced by the amplifier and antenna, and $(d_{i,j}(t))^{-\alpha}$ is the pathloss. We define the set of UAVs and CUs that share subchannel k with UAV i as $W_i=\{m|\psi_{m,k}(t)=1,\forall m\in\mathcal{N}_e\setminus i\}\cup\{m|\phi_{m,k}(t)=1\}$. The received signal at UAV j over subchannel k is then given by

$$y_{i,j}^{k}(t)=\sqrt{P_{i,j}^{k}(t)}+\sum_{m\in W_i}\sqrt{P_{m,j}^{k}(t)}+n_j^{k}(t), \tag{4.36}$$

where $P_m^k(t)$ is the received power at UAV j from the UAVs and CUs in W_i, and $n_j^k(t)$ is the AWGN with zero mean and σ^2 variance. The interference from UAV m to UAV j over subchannel k is shown as

$$I_{m,UAV}^{k}(t)=(\phi_{m,k}(t)+\psi_{m,k}(t))P_U G(d_{m,j}(t))^{-\alpha}. \tag{4.37}$$

According to the channel reciprocity, the interference from CU m to UAV j over subchannel k can be expressed as

$$I_{m,C}^{k}(t)=\phi_{m,k}(t)\frac{P_U}{10^{PL_{avg,j}^{m}(t)/10}}, \tag{4.38}$$

where $PL_{avg,j}^{m}(t)$ is the average pathloss from UAV j to CU m, which can be derived from (4.22) through (4.25). The SINR at UAV j over subchannel k is shown as

$$\gamma_{i,j}^{k}(t)=\frac{P_U G(d_{i,j}(t))^{-\alpha}}{\sigma^2+\sum_{m=1,m\neq i}^{N_l+N_h}I_{m,UAV}^{k}(t)+\sum_{m=1}^{M}I_{m,C}^{k}(t)}. \tag{4.39}$$

When UAV i transmits its data to UAV j over subchannel k via U2U transmission, the data rate is given by

$$R_{i,j}^{k}(t)=\log_2(1+\gamma_{i,j}^{k}(t)). \tag{4.40}$$

4.3.2 Problem Formulation

In the following, we first formulate the joint subchannel allocation and UAV speed optimization problem and prove that the optimization problem is NP-hard, which cannot be solved directly within polynomial time. Therefore, in the next part, we decouple it into three subproblems and elaborate on them separately.

Joint Subchannel Allocation and UAV Speed Optimization Problem Formulation

Because all the data collected by the UAVs needs to be sent to the BS, the uplink sum rate of this network is one key metric to evaluate the performance of this network.[7] In time slot t, we denote the set of UAVs that have not completed the task along their trajectories by $\lambda(t)$. We aim to maximize the uplink sum rate of the UAVs in $\lambda(t)$ and the CUs by optimizing the subchannel allocation and UAV speed variables $\mathbf{\Phi}(t)$, $\mathbf{\Psi}(t)$, and $v_i(t)$. The joint subchannel allocation and UAV speed optimization problem can be formulated as follows:

$$\max_{\substack{\{v_i(t)\},\{\mathbf{\Phi}(t)\},\\ \{\mathbf{\Psi}(t)\}}} \sum_{k=1}^{K} \sum_{\substack{i=1\\ i\in\lambda(t)}}^{N_h+M} \phi_{i,k}(t) R_{i,BS}^{k}(t), \tag{4.41a}$$

$$s.t. \sum_{k=1}^{K} \psi_{i,k}(t) R_{i,j}^{k}(t) \geq R_0, \quad \forall i,j \in \mathcal{N}, \xi_{i,j}=1, \tag{4.41b}$$

$$v_i(t) \leq v_{\max}, \quad \forall i \in \mathcal{N}, \tag{4.41c}$$

$$\sum_{t=1}^{T} v_i(t) \geq L_i, \quad \forall i \in \mathcal{N}, \tag{4.41d}$$

$$\sum_{i=1}^{N_h+M} \phi_{i,k}(t) \leq 1, \quad \forall k \in \mathcal{K}, \tag{4.41e}$$

$$\sum_{k=1}^{K} \phi_{i,k}(t) \leq \chi_{\max}, \quad \forall i \in \mathcal{N}_h(t) \cup \mathcal{M}, \tag{4.41f}$$

$$\sum_{k=1}^{K} \psi_{i,k}(t) \leq \chi_{\max}, \quad \forall i \in \mathcal{N}_l(t), \tag{4.41g}$$

$$\phi_{i,k}(t), \psi_{i,k}(t) \in \{0,1\}, \forall i \in \mathcal{N} \cup \mathcal{M}, k \in \mathcal{K}. \tag{4.41h}$$

The minimum U2U transmission rate satisfies constraint (4.41b). Equation (4.41c) is the maximum speed constraint for the UAVs, and (4.41d) shows that the task completion time of each UAV is no more than T time slots. Constraint (4.41e) implies that each subchannel can be allocated to at most one UAV in the U2N mode or CU. Each UAV and CU can be paired with at most $\chi_{\max}$ subchannels, which is given in constraints (4.41f) and (4.41g). In the following theorem, we will prove that optimization problem (4.41) is NP-hard.

THEOREM 4.4 *Equation (4.41) is NP-hard.*

Proof We prove that (4.41) is NP-hard even when we do not perform UAV speed optimization. We construct an instance of problem (4.41) where each subchannel can serve no more than one U2U link and one U2N or CU link simultaneously. Let $\mathcal{N}_c$, $\mathcal{N}_e$, and $\mathcal{K}$ be three disjoint sets of UAVs in the U2N mode and

CUs, UAVs in the U2U mode, and subchannels, respectively, with $|\mathcal{N}_c| = N_h$, $|\mathcal{N}_e| = N_l$, and $|\mathcal{K}| = K$. Set $\mathcal{N}_c$, $\mathcal{N}_e$, and $\mathcal{K}$ to satisfy $\mathcal{N}_c \cap \mathcal{N}_e = \varnothing, \mathcal{N}_c \cap \mathcal{K} = \varnothing$, and $\mathcal{N}_e \cap \mathcal{K} = \varnothing$. Let $\mathcal{P}$ be a collection of ordered triples $\mathcal{P} \subseteq \mathcal{N}_c \times \mathcal{N}_e \times \mathcal{K}$, where each element in $\mathcal{P}$ consists a CU/UAV that performs U2N transmission, a UAV that performs U2U transmission, and a subchannel (i.e., $P_i = (N_{c,i}, N_{e,i}, K_i) \in \mathcal{P}$). For convenience, we set $L = \min\{N_h, N_l, K\}$. There exists $\mathcal{P}' \subseteq \mathcal{P}$ that holds $|\mathcal{P}'| = \text{L}$ and for any two distinct triples $(N_{c,i}, N_{e,i}, K_i) \in \mathcal{P}'$ and $(N_{c,j}, N_{e,j}, K_j) \in \mathcal{P}'$, we have $i \neq j$. Therefore, $\mathcal{P}'$ is a three-dimension matching (3-DM). Because 3-DM problem has been proved to be NP-complete in [95], the constructed instance of problem is also NP-complete. Thus, (4.41) is NP-hard [96]. □

Problem Decomposition

Because (4.41) is NP-hard, to tackle this problem efficiently, we decouple (4.41) into three subproblems: U2N and CU subchannel allocation, U2U subchannel allocation, and UAV speed optimization. In the U2N and CU subchannel allocation subproblem, the U2U subchannel matching matrix $\mathbf{\Psi}(t)$ and the UAV speed $\{v_i(t)\}$ are considered to be fixed. Therefore, the U2N and CU subchannel allocation subproblem is written as

$$\max_{\mathbf{\Phi}(t)} \sum_{k=1}^{K} \sum_{\substack{i=1 \\ i \in \lambda(t)}}^{N_h+M} \phi_{i,k}(t) R_{i,BS}^{k}(t), \tag{4.42a}$$

$$s.t. \sum_{i=1}^{N_h+M} \phi_{i,k}(t) \leq 1, \quad \forall k \in \mathcal{K}, \tag{4.42b}$$

$$\sum_{k=1}^{K} \phi_{i,k}(t) \leq \chi_{\max}, \quad \forall i \in \mathcal{N}_h(t) \cup \mathcal{M}, \tag{4.42c}$$

$$\phi_{i,k}(t) \in \{0,1\}, \quad \forall i \in \mathcal{N}_h(t) \cup \mathcal{M}, k \in \mathcal{K}. \tag{4.42d}$$

Given the U2N and CU subchannel pairing matrix $\mathbf{\Phi}(t)$ and the UAV speed $\{v_i(t)\}$, the U2U subchannel allocation subproblem can be written as

$$\max_{\mathbf{\Psi}(t)} \sum_{k=1}^{K} \sum_{\substack{i=1 \\ i \in \lambda(t)}}^{N_h+M} \phi_{i,k}(t) R_{i,BS}^{k}(t), \tag{4.43a}$$

$$s.t. \sum_{k=1}^{K} \psi_{i,k}(t) R_{i,j}^{k}(t) \geq R_0, \quad \forall i, j \in \mathcal{N}_l(t), \xi_{i,j} = 1, \tag{4.43b}$$

$$\sum_{k=1}^{K} \psi_{i,k}(t) \leq \chi_{\max}, \quad \forall i \in \mathcal{N}_l(t), \tag{4.43c}$$

$$\psi_{i,k}(t) \in \{0,1\}, \quad \forall i \in \mathcal{N}_l(t), k \in \mathcal{K}. \tag{4.43d}$$

Similarly, when the subchannel pairing matrices $\mathbf{\Phi}(t)$ and $\mathbf{\Psi}(t)$ are given, the UAV speed optimization subproblem can be expressed by

$$\max_{\{v_i(t)\}} \sum_{k=1}^{K} \sum_{\substack{i=1 \\ i\in\lambda(t)}}^{N_h+M} \phi_{i,k}(t) R_{i,BS}^{k}(t), \tag{4.44a}$$

$$\sum_{k=1}^{K} \psi_{i,k}(t) R_{i,j}^{k}(t) \geq R_0, \quad \forall i,j \in \mathcal{N}, \xi_{i,j} = 1, \tag{4.44b}$$

$$v_i(t) \leq v_{\max}, \quad \forall i \in \mathcal{N}, \tag{4.44c}$$

$$\sum_{t=1}^{T} v_i(t) \geq L_i, \quad \forall i \in \mathcal{N}. \tag{4.44d}$$

4.3.3 Joint Subchannel Allocation and UAV Speed Optimization

In the following, we propose an effective method (i.e., ISASOA) to obtain a suboptimal solution of (4.41) by solving its three subproblems (4.42), (4.43), and (4.44) iteratively. The U2N and CU subchannel allocation subproblem (4.42) can be relaxed to a standard linear programming problem, which can be solved by existing convex techniques, such as, the Matlab Software for Disciplined Convex Programming (CVX). We then utilize the branch-and-bound method to solve the nonconvex U2U subchannel allocation subproblem (4.43). For the UAV speed optimization subproblem (4.44), we discuss the feasible region and convert it into a convex problem, which can be solved by existing convex techniques. Iterative solutions of the three subproblems are found until the objective function converges to a constant. In the following, we first elaborate on the algorithms of solving the three subproblems. Afterward, we will provide the ISASOA and discuss its convergence and complexity.

U2N and CU Subchannel Allocation Algorithm

In this part, we give a detailed description of the U2N and CU subchannel allocation algorithm. As shown in Section 4.3.2, the decoupled subproblem (4.42) is an integer programming problem. To make the problem more tractable, we relax the variables $\mathbf{\Phi}(t)$ into continuous values, and the relaxed problem is expressed as

$$\max_{\mathbf{\Phi}(t)} \sum_{k=1}^{K} \sum_{\substack{i=1 \\ i\in\lambda(t)}}^{N_h+M} \phi_{i,k}(t) R_{i,BS}^{k}(t), \tag{4.45a}$$

$$s.t. \sum_{i=1}^{N_h+M} \phi_{i,k}(t) \leq 1, \quad \forall k \in \mathcal{K}, \tag{4.45b}$$

$$\sum_{k=1}^{K} \phi_{i,k}(t) \leq \chi_{\max}, \quad \forall i \in \mathcal{N}_h(t) \cup \mathcal{M}, \tag{4.45c}$$

$$0 \leq \phi_{i,k}(t) \leq 1, \quad \forall i \in \mathcal{N}_h(t) \cup \mathcal{M}, k \in \mathcal{K}. \tag{4.45d}$$

When we substitute (4.28), (4.29), (4.33), and (4.34) into (4.45a), it can be observed that the pairing matrix $\boldsymbol{\Phi}(t)$ is not relevant with $R^k_{i,BS}(t)$. Therefore, $R^k_{i,BS}(t)$ is fixed in this subproblem. Note that function (4.45a) is linear with respect to the optimization variables $\boldsymbol{\Phi}(t)$, and (4.45b), (4.45c), and (4.45d) are all linear. Thus, (4.45) is a standard linear programming problem, which can be solved efficiently by utilizing existing optimization techniques such as CVX [97]. In what follows, we will prove that the solution of the relaxed problem (4.45) is also the one of the original problem (4.42).

THEOREM 4.5 *All the variables in $\boldsymbol{\Phi}(t)$ are met with zero or one for the solution of* (4.45).

Proof We assume that the solution of (4.45) contains a variable $\phi_{i,k}(t)$ with $0 < \phi_{i,k}(t) < 1$. For simplicity, we denote the slope of $\phi_{i,k}(t)$ in the objective function (4.45a) by $X_{i,k} = \log_2(1 + \frac{P^k_{i,BS}(t)}{\sigma^2 + I_{k,U2U}(t)})$, where $X_{i,k} > 0, \forall i \in N, k \in K$. When the objective function is maximized, at least one of the constraints between (4.45b) and (4.45c) is met with equality. In the following, we separate the problem into two conditions and discuss them successively.

- *Only one constraint is met with equality:* Without loss of generality, we assume that only (4.45b) is met with equality. Because $\phi_{i,k}(t)$ is not an integer, there exists another variable $\phi_{j,k}(t)$ that is also a noninteger to meet the constraint equality of (4.45b). We assume that $X_{i,k} > X_{j,k}$. When we increase $\phi_{i,k}(t)$ and decrease $\phi_{j,k}(t)$ within the constraint, the objective function will be improved. Thus, the solution with $0 < \phi_{i,k}(t) < 1$ is not the optimal solution.
- *Both* (4.45b) *and* (4.45c) *are met with equality:* When both (4.45b) and (4.45c) are met with equality, there are at least three more variables that are nonintegers to meet the constraint equality. We denote the other three variables by $\phi_{j,k}(t)$, $\phi_{i,m}(t)$, and $\phi_{j,m}(t)$. If $X_{i,k} + X_{j,m} > X_{i,m} + X_{j,k}$, when we increase $\phi_{i,k}(t)$ and $\phi_{j,m}(t)$, and decrease $\phi_{j,k}(t)$ and $\phi_{i,m}(t)$, the objective function will be improved. If $X_{i,k} + X_{j,m} < X_{i,m} + X_{j,k}$, the opponent adjustment will improve the objective function. As a result, the current solution is not the optimal one.

In conclusion, the solution that contains $0 < \phi_{i,k}(t) < 1$ is not the optimal one. When the optimal solution of (4.45) is achieved, all the variables in $\boldsymbol{\Phi}(t)$ are either zero or one.

□

As shown in Theorem 4.5, the solution of (4.45) is either zero or one, which satisfies the constraint (4.42d) of the original problem. Therefore, the relaxation of variable $\phi_{i,k}(t)$ does not affect the solution of the subproblem (4.42). The solution of the relaxed problem (4.45) with the CVX method is equivalent to the solution of (4.42).

U2U Subchannel Allocation Algorithm

Now, we focus on solving the U2U subchannel allocation subproblem (4.43). We first substitute (4.27), (4.28), (4.29), (4.33), and (4.34) into (4.43a), and the objective function is given by

$$\max_{\Psi(t)} \sum_{k=1}^{K} \left(\sum_{\substack{i=1 \\ i \in \lambda(t)}}^{N_h} \phi_{i,k}(t) \log_2 \left(1 + \frac{P_{i,BS}^k(t)}{\sigma^2 + \sum_{\substack{j=1 \\ j \in \lambda(t)}}^{N_l} \psi_{j,k}(t) P_{j,BS}^k(t)} \right) \right. \tag{4.46}$$

$$\left. + \sum_{i=N_h+1}^{N_h+M} \phi_{i,k}(t) \log_2 \left(1 + \frac{P_{i,C}^k(t)}{\sigma^2 + \sum_{\substack{j=1 \\ j \in \lambda(t)}}^{N_l} \psi_{j,k}(t) P_{j,BS}^k(t)} \right) \right).$$

When substituting (4.39) and (4.40) into constraint (4.43b), it can be expanded as

$$R_{U2U,i} = \sum_{k=1}^{K} \psi_{i,k}(t) \log_2 \left(1 + \frac{P_U G(d_{i,j}(t))^{-\alpha}}{A+B} \right) \geq R_0, \quad \forall i, j \in N, \xi_{i,j} = 1, \tag{4.47}$$

where $A = \sigma^2 + \sum_{n=1, n \in \lambda(t)}^{N_h} \phi_{n,k}(t) I_{m,UAV}^k(t) + \sum_{m=N_h+1}^{N_h+M} \phi_{m,k}(t) I_{m,C}^k(t)$ is fixed in this subproblem, and $B = \sum_{m=1, m \neq i}^{N_l} \psi_{m,k}(t) P_U G(d_{m,j}(t))^{-\alpha}$. Equation (4.43) is a *0-1* programming problem, which has been proved to be NP-hard [98]. In addition, due to the interference from different U2U links, the continuity relaxed problem of (4.43) is still nonconvex with respect to $\Psi(t)$. Therefore, (4.43) cannot be solved by the existing convex techniques. To solve (4.43) efficiently, we utilize the branch-and-bound method [99].

To facilitate understanding of the branch-and-bound algorithm, we first introduce important concepts of fixed and unfixed variables.

DEFINITION 4.6 *When the value of a variable that corresponds to the optimal solution is ensured, we define it as a* ***fixed variable****. Otherwise, it is an* ***unfixed variable****.*

The solution space of the U2U subchannel pairing matrix $\Psi(t)$ can be considered a binary tree. Each node of the binary tree contains the information of all the variables in $\Psi(t)$. At the root node, all the variables in $\Psi(t)$ are unfixed. The value of an unfixed variable at a parent node can be either zero or one, which branches the node into two child nodes. Our objective is to search the binary tree for the optimal solution of (4.43). The key idea of the branch-and-bound method is to prune the infeasible branches and approach the optimal solution efficiently.

At the beginning of the algorithm, we obtain a feasible solution of (4.43) by a proposed low-complexity feasible solution searching (LFSS) method and set it as the lower bound of the solution. We then start to search the optimal solution of (4.43) in the binary tree from its root node. On each node, the branch-and-bound method consists two steps: bound calculation and variable fixation. In the bound calculation step, we evaluate the upper bound of the objective function and the bounds of the constraints separately to prune the branches that cannot achieve a feasible solution above the lower bound of the solution. In the variable fixation step, we fix the variables that have only one feasible value that satisfies the bound requirements in the bound calculation step. We then search the node that contains the newly fixed variables and continue the two steps of bound calculation and variable fixation. The algorithm terminates when we obtain a node with all the variables fixed. In what follows, we first

Algorithm 4.5 Initial feasible solution for U2U subchannel allocation

```
1  begin
2      Each UAV in the U2U mode calculates its data rate over every subchannel
         with U2N and CU interference;
3      Each UAV sorts the subchannels in descending order of achievable rate;
4      Assign the UAVs with their most preferred subchannel;
5      Calculate the data rate of each U2U link with U2N, CU, and U2U
         interference;
6      while The data rate of an UAV does not satisfy U2U rate constraint (4.43c)
         do
7          Assign the UAV to its most preferred subchannel that has not been paired;
8      end
9      Set the current U2U-subchannel pairing result as the initial feasible solution;
10 end
```

introduce the LFSS method to achieve the initial feasible solution and then describe the bound calculation and variable fixation process at each node in detail. Finally, we summarize the branch-and-bound method.

Initial Feasible Solution Search

In what follows, we propose the LFSS method to obtain a feasible solution of (4.43) efficiently. Each UAV in the U2U mode requests a subchannel until its minimum U2U rate threshold is satisfied and the BS assigns the requested subchannel to the corresponding UAV in the LFSS. The detailed description is shown in Algorithm 4.5.

Given the U2N and CU subchannel assignment, each UAV in the U2U mode can make a list of data rates that it may achieve from every subchannel without considering the potential U2U interference. The UAVs then sort the subchannels in descending order of achievable rate. We then calculate the data rate of each U2U link with U2N, CU, and U2U interference when the UAVs are assigned to their most preferred subchannels. If the data rate of an UAV is still below the minimum threshold, the UAV will be assigned to its most preferred subchannel that has not been paired. The subchannel assignment ends when the minimum U2U rate threshold (4.43c) is satisfied by every UAV in the U2U mode. Finally, we adopt the current U2U subchannel pairing result as the initial feasible solution.

Bound Calculation

In this part, we describe the process of bound calculation at each node. After the initialization step, we start bound calculation from the root node, in which all the variables in $\mathbf{\Psi}(t)$ are unfixed; that is, the value of each $\psi_{i,k}(t)$ in the optimal solution is unknown. We first define a branch pruning operation that is performed in the following bound calculation step.

DEFINITION 4.7 *When a node is **fathomed**, its child nodes cannot be the optimal solution of the problem.*

We calculate the bounds of the objective function and the constraints separately. For simplicity, we denote the objective function with U2U subchannel matrix by $f(\mathbf{\Psi}(t))$ and the lower bound of the solution by f^{lb}. In what follows, we will elaborate on the detailed steps of the bound calculation at each node.

In step 1 (objective bound calculation), the upper bound of the objective function (4.43a) is given as

$$\bar{f} = \sum_{k=1}^{K} \left(\sum_{\substack{i=1 \\ i \in \lambda(t)}}^{N_h} \phi_{i,k}(t) \log_2 \left(1 + \frac{P_{i,BS}^k(t)}{\sigma^2 + \sum_{j=1}^{N_l} \psi_{j,k}^F(t) P_{j,BS}^k(t)} \right) \right. \\ \left. + \sum_{i=N_h+1}^{N_h+M} \phi_{i,k}(t) \log_2 \left(1 + \frac{P_{i,C}^k(t)}{\sigma^2 + \sum_{j=1}^{N_l} \psi_{j,k}^F(t) P_{j,BS}^k(t)} \right) \right), \tag{4.48}$$

where $\psi_{j,k}^F(t)$ is the fixed variables in the current node. That is, we ignore the U2U interference of the unfixed variables. If the upper bound of the current node is below the lower bound of the solution ($\bar{f} < f^{lb}$), we fathom the current node and backtrack to an unfathomed node with an unfixed variable. If the current node is not fathomed by the objective function bound calculation, we move to step 2 to check the bounds of the constraints.

In step 2 (constraint bounds calculation), for each UAV in the U2U mode, the upper bound of its U2U rate needs to be larger than the minimum U2U rate threshold. The upper bound of U2U rate for UAV i is achieved when we set all the unfixed variables of UAV i as 1, and all the unfixed variables of other UAVs as 0, which can be expressed as

$$\bar{R}_{U2U,i} = \sum_{k=1}^{K} \psi_{i,k}(t)|_{\{\psi_{i,k}^U(t)=1\}} \log_2 \left(1 + \frac{P_U G(d_{i,j}(t))^{-\alpha}}{A + B^F} \right), \quad \forall i, j \in N, \xi_{i,j} = 1, \tag{4.49}$$

where $\psi_{m,k}^U(t)$ is the unfixed variables in the current node, and

$$B^F = \sum_{m=1, m \neq i}^{N_l} \psi_{m,k}^F(t) P_U G(d_{m,j}(t))^{-\alpha}. \tag{4.50}$$

If $\exists \xi_{i,j} = 1, \bar{R}_{U2U,i} < R_0$, the minimum U2U rate threshold cannot be satisfied, and the current node is fathomed. Moreover, if there exists a UAV i that does not satisfy constraint (4.43c) (i.e., $\sum_{k=1}^{K} \psi_{i,k}(t) > \chi_{\max}, \forall i \in N$), the current node is also fathomed. We then backtrack to an unfathomed node in the binary tree and perform bound calculation at the new node.

In the bound calculation procedure, if the objective function of a U2U subchannel pairing matrix $f(\tilde{\mathbf{\Psi}}(t))$ is found to be larger than the lower bound of the solution f^{lb}, and $\tilde{\mathbf{\Psi}}(t)$ satisfies all the constraints, we replace the lower bound of the solution with $f^{lb} = f(\tilde{\mathbf{\Psi}}(t))$ to improve the algorithm efficiency. A higher lower bound of the solution helps us to prune the infeasible branches more efficiently.

Variable Fixation

For a node that is not fathomed in the bound calculation steps, we try to prune the branches by fixing the unfixed variables as follows. The variable fixation is completed in two steps: objective fixation and U2U constraint fixation.

In step 1, the objective fixation process, we denote the reduction of the upper bound when fixing a free variable $\psi_{i,k}(t)$ at 0 or 1 by $p^0_{i,k}$ and $p^1_{i,k}$, respectively. For each unfixed variable $\psi_{i,k}(t)$, we compute $p^0_{i,k}$ and $p^1_{i,k}$ associated with the upper bound $\bar{f}$. If $\bar{f} - p^0_{i,k} \leq f_{opt}$, it means that when we set $\psi_{i,k}(t) = 0$, the upper bound of the child node will fall below the temporary feasible solution. Therefore, we prune the branch of $\psi_{i,k}(t) = 0$, and fix $\psi_{i,k}(t) = 1$. Similarly, if $\bar{f} - p^1_{i,k} \leq f_{opt}$, we prune the branch of $\psi_{i,k}(t) = 1$, and fix $\psi_{i,k}(t) = 0$.

In step 2, the U2U constraint fixation process, we denote the U2U rate upper bound reduction for UAV i when fixing a free variable $\psi_{i,k}(t)$ at 0 by $q^0_{i,k}$. If inequality $\bar{R}_{U2U,i} - q^0_{i,k} < R_0$ is satisfied, it means that only when subchannel k is assigned to UAV i, the minimum U2U rate threshold of UAV i can be satisfied. Therefore, we prune the branch of $\psi_{i,k}(t) = 0$ and fix $\psi_{i,k}(t) = 1$.

In the objective fixation and U2U constraint fixation steps, variable $\psi_{i,k}(t)$ may be fixed at different values, which implies that neither of the two child nodes satisfy the objective bound relation and the constraint bound relation simultaneously. Therefore, we fathom the current node and backtrack to an unfathomed node with an unfixed variable.

After performing the variable fixation step of the current node, if at least one unfixed variable is fixed at a certain value during this procedure, we move to the corresponding child node and continue the algorithm by performing bound calculation and variable fixation at the new node. Otherwise, we generate two new nodes by setting an unfixed variable at $\psi_{i,k}(t) = 0$ and $\psi_{i,k}(t) = 1$, respectively. We then move to one of the two nodes and continue the algorithm. The branch-and-bound algorithm is accomplished when all variables have been fixed, and the fixed variables are the final solution.

The branch-and-bound method that solves the U2U subchannel allocation subproblem (4.43) is summarized as Algorithm 4.6.

UAV Speed Optimization Algorithm

In the following, we introduce how to solve the UAV speed optimization subproblem (4.44). The problem is difficult to optimize directly due to the complicated expression of the air-to-ground transmission model and the change in interference caused by the movement of the UAVs. In the following, we first raise two rational assumptions that simplify this problem and then propose an efficient solution that gives an approximate solution for (4.44).

Two Basic Assumptions

In this part, we give two assumptions to simplify the UAV speed optimization problem.

Algorithm 4.6 Branch-and-bound method for U2U subchannel allocation

Input: The U2N subchannel allocation matrix $\mathbf{\Phi}(t)$, and the UAV trajectories $\omega(t)$.

Output: The U2U subchannel allocation matrix $\mathbf{\Psi}(t)$.

```
1  begin
2      Initialization: Compute an initial feasible solution Ψ(t) to (4.43) and set it as
         the lower bound of the solution;
3      Perform bound calculation and variable fixation at the root node;
4      while Not all variables have been fixed do
5          Bound calculation;
6          if The bound constraints cannot be satisfied then
7              Fathom the current node and backtrack to an unfathomed node with
                 unfixed variable;
8          end
9          Variable fixation;
10         if At least one variable can be fixed then
11             Go to the node with newly fixed variable;
12         else
13             Generate two new nodes by setting an unfixed variable ψ_{i,k}(t) = 0
                 and ψ_{i,k}(t) = 1;
14             Go to one of the two nodes firstly;
15         end
16     end
17     The fixed variables are the final output of Ψ(t);
18 end
```

We first assume that the pathloss variables $PL_{LoS,i}(t)$ and $PL_{NLoS,i}(t)$ change much faster than the LoS probability variables $P_{LoS,i}(t)$ and $P_{NLoS,i}(t)$ with the movement of a UAV.

Proof When the elevation angle of a UAV changes by $\Delta\theta$, such as from θ to $\theta+\Delta\theta$, with $\theta \gg \Delta\theta$, the change of the transmission distance can be approximated as $d_i(t)\tan\theta\Delta\theta$ (Fig. 4.17). According to (4.24), the rate of change of the LoS probability to the elevation angle is given as

$$\frac{\Delta P_{LoS,i}(t)}{\Delta\theta} = \frac{ab\exp(-b(\theta-a))}{(1+a\exp(-b(\theta-a)))^2}. \tag{4.51}$$

The relation between the pathloss and the transmission distance is shown in (4.22), and the rate of change of the pathloss to the elevation angle is

$$\begin{aligned}\frac{\Delta PL_{LoS,i}(t)}{\Delta\theta} &= \frac{20\log(d_i(t)\tan\theta\Delta\theta)-20\log(d_i(t))}{\Delta\theta}\\ &= \frac{20\log(\tan\theta\Delta\theta)}{\Delta\theta}.\end{aligned} \tag{4.52}$$

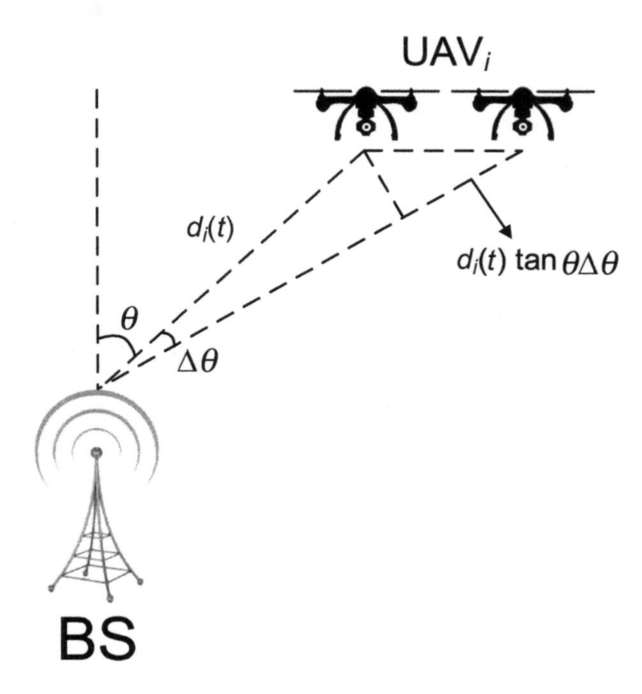

Figure 4.17 Transmission model variation with UAV movement

When substituting the typical value of a, b, and θ into (4.51) and (4.52), we have $\frac{\Delta P_{LoS,i}(t)}{\Delta\theta} \ll \frac{\Delta PL_{LoS,i}(t)}{\Delta\theta}$. Therefore, the channel pathloss varies much faster than the LoS probability with the movement of a UAV. □

Second, we assume that the U2U transmission distance is much larger than the moving distance of a UAV in one time slot (i.e., $d_{i,j}(t) \gg v_{\max}$).

With these two assumptions, we then introduce the solution that solves (4.44) efficiently. Note that in problem (4.44), the speed optimization of a pair of U2U transmitting and receiving UAVs are related to constraint (4.44b), but the speed optimization of the UAVs in the U2N mode that do not receive U2U transmissions are irrelevant to constraint (4.44b). Therefore, the speed optimization of the UAVs can be separated into two types: non-U2U participated UAVs and U2U participated UAVs that contain the transmitting UAVs and the corresponding receiving UAVs, respectively.

Non-U2U Participated UAV Speed Optimization

For non-U2U participated UAVs, constraint (4.44b) is not considered. We denote the length of trajectory that UAV i has flown before time slot t by $\mathscr{L}_i(t)$. To satisfy constraints (4.44c) and (4.44d), the length of the trajectory that UAV i needs to move along in the following time slots should be no more than the number of following time slots $T-t-1$ times the maximum UAV speed $v_{\max}$; that is, $L_i - \mathscr{L}_i(t+1) < v_{\max} \times (T-t-1)$. Therefore, the feasible range of UAV i's speed in time slot t

is $\min\{0, L_i - \mathscr{L}_i(t) - v_{\max} \times (T - t - 1)\} \le v_i(t) \le v_{\max}$. Equation (4.44) can be simplified as

$$\max_{v_i(t)} \sum_{k=1}^{K} \sum_{\substack{i=1 \\ i \in \lambda(t)}}^{N_h+M} \phi_{i,k}(t) R_{i,BS}^{k}(t), \tag{4.53a}$$

$$\min\{0, L_i - \mathscr{L}_i(t) - v_{\max} \times (T - t - 1)\} \le v_i(t) \le v_{\max}. \tag{4.53b}$$

With these two basic assumptions, we can assume that the probability of the LoS and NLoS connections do not change prominently in a single time slot, and the uplink rate is determined by the LoS and NLoS pathloss the given in (4.22) and (4.23). Therefore, (4.53) is approximated as a convex problem and can be solved with existing convex optimization methods.

U2U Participated UAV Speed Optimization

In this part, we introduce the speed optimization of a pair of UAVs: UAV i and UAV j, with $\xi_{i,j} = 1$. In time slot t, UAV i performs U2U transmission and sends the collected data to UAV j. Then, UAV j receives the data from UAV i and performs U2N transmission simultaneously. Similarly, constraints (4.44c) and (4.44d) can be simplified as $\min\{0, L_i - \mathscr{L}_i(t) - v_{\max} \times (T - t - 1)\} \le v_i(t) \le v_{\max}$ and $\min\{0, L_j - \mathscr{L}_j(t) - v_{\max} \times (T - t - 1)\} \le v_j(t) \le v_{\max}$ for UAV i and UAV j, respectively. Given the subchannel pairing matrices $\mathbf{\Phi}(t)$ and $\mathbf{\Psi}(t)$, the U2U rate constraint (4.44b) can be transformed to a distance constraint. When substituting (4.39) and (4.40) into (4.44b), the U2U rate constraint can be shown as

$$d_{i,j}(t) \le \frac{P_U G}{\left(\sigma^2 + \sum_{m=1, m \ne i}^{N_l+N_h} I_{m,UAV}^{k}(t) + \sum_{m=1}^{M} I_{m,C}^{k}(t)\right)} \frac{1}{\left(2^{\frac{R_0}{\sum_{k=1}^{K} \psi_{i,k}(t)}} - 1\right)}. \tag{4.54}$$

Given the second basic assumption, the U2U interference can be approximated to a constant in each single time slot. Therefore, the right side of (4.54) can be regarded as a constant, denoted by $d_{i,j}^{\max}$ for simplicity. Given the feasible speed range of UAV i and the maximum distance between UAV i and UAV j (i.e., $d_{i,j}^{\max}$), a feasible speed range of UAV j in time slot t can be obtained, which is written as $v_j(t)^{\min} \le v_j(t) \le v_j(t)^{\max}$. The UAV speed optimization subproblem is reformulated as

$$\max_{v_i(t)} \sum_{k=1}^{K} \sum_{\substack{i=1 \\ i \in \lambda(t)}}^{N_h+M} \phi_{i,k}(t) R_{i,BS}^{k}(t), \tag{4.55a}$$

$$\min\{0, L_i - \mathscr{L}_i(t) - v_{\max} \times (T - t - 1)\} \le v_i(t) \le v_{\max}, \tag{4.55b}$$

$$\min\{0, L_j - \mathscr{L}_j(t) - v_{\max} \times (T - t - 1)\} \le v_j(t) \le v_{\max}, \tag{4.55c}$$

$$v_j(t)^{\min} \le v_j(t) \le v_j(t)^{\max}. \tag{4.55d}$$

Similar to (4.53), (4.55) can also be considered a convex problem, which can be solved with the existing convex optimization methods.

Algorithm 4.7 Iterative subchannel allocation and UAV speed optimization algorithm

1 **begin**
2 **Initialization:** Set $r = 0$, $\mathbf{\Phi}^0(t) = \{0\}$, $\mathbf{\Psi}^0(t) = \{0\}$, $\omega_i^0(t) = \{0\}, \forall i \in I(t)$;
3 **while** $\mathscr{R}\Big(\mathbf{\Phi}^r(t), \mathbf{\Psi}^r(t), \omega^r(t)\Big) - \mathscr{R}\Big(\mathbf{\Phi}^{r-1}(t), \mathbf{\Psi}^{r-1}(t), \omega^{r-1}(t)\Big) > \epsilon$ **do**
4 $r = r + 1$;
5 Solve U2N and CU subchannel allocation subproblem (4.42), given $\mathbf{\Psi}^{r-1}(t)$ and $\mathbf{v}^{r-1}(t)$;
6 Solve U2U subchannel allocation subproblem (4.43), given $\mathbf{\Phi}^r(t)$ and $\mathbf{v}^{r-1}(t)$;
7 Solve UAV speed optimization subproblem (4.44), given $\mathbf{\Phi}^r(t)$ and $\mathbf{\Psi}^r(t)$;
8 **end**
9 **Output:** $\mathbf{\Phi}^r(t), \mathbf{\Psi}^r(t), \mathbf{v}^r(t)$;
10 **end**

Iterative Subchannel Allocation and UAV Speed Optimization Algorithm
In this subsubsection, we introduce the ISASOA to solve (4.41), where U2N and CU subchannel allocation, U2U subchannel allocation, and UAV speed optimization subproblems are solved iteratively. In time slot t, we denote the optimization objective function after the rth iteration by $\mathscr{R}(\mathbf{\Phi}^r(t), \mathbf{\Psi}^r(t), \mathbf{v}^r(t))$. In iteration r, the U2N and CU subchannel allocation matrix $\mathbf{\Phi}(t)$, the U2U subchannel allocation matrix $\mathbf{\Psi}(t)$, and the UAV speed variable of UAV i are denoted by $\mathbf{\Phi}^r(t)$, $\mathbf{\Psi}^r(t)$, and $v_i^r(t)$, respectively. The process of the iterative algorithm for each single time slot is summarized in Algorithm 4.7.

In time slot t, we first set the initial condition, where all the subchannels are vacant, and the speed of all the UAVs are given as a fixed value v_0: $\mathbf{\Phi}^0(t) = \{0\}$, $\mathbf{\Psi}^0(t) = \{0\}$, and $v_i^0(t) = \{v_0\}, \forall i \in \mathscr{N}$. We then perform iterations of subchannel allocation and UAV speed optimization until the objective function converges. In each iteration, the U2N and CU subchannel allocation is performed first, with the U2U subchannel pairing and UAV speed results given in the last iteration, and the U2N and CU subchannel pairing variables are updated. Next, the U2U subchannel allocation is performed as shown in Section 4.3.3, with the UAV speed obtained in the last iteration and the U2N and CU subchannel pairing results. Afterward, we perform UAV speed optimization as described in Section 4.3.3, given the subchannel pairing results. When an iteration is completed, we compare the values of the objective function obtained in the last two iterations. If the difference between the values is less than a preset error tolerant threshold ϵ, the algorithm terminates and the results of subchannel pairing and UAV speed optimization are obtained. Otherwise, the ISASOA will continue.

In the following, we discuss the convergence and complexity of the proposed ISASOA.

THEOREM 4.8 *The proposed ISASOA is convergent.*

Proof In the $(r+1)$th iteration, we first perform U2N and CU subchannel allocation, and the optimal U2N and CU subchannel allocation solution is obtained with the given $\mathbf{\Psi}^r(t)$ and $v_i^r(t)$. Therefore, we have

$$\mathscr{R}\Big(\mathbf{\Phi}^{r+1}(t), \mathbf{\Psi}^r(t), \mathbf{v}^r(t)\Big) \geq \mathscr{R}\Big(\mathbf{\Phi}^r(t), \mathbf{\Psi}^r(t), \mathbf{v}^r(t)\Big). \tag{4.56}$$

That is, the total rate of U2N and CU transmissions does not decrease with the U2N and CU subchannel allocation in the $(r+1)$th iteration. When solving U2U subchannel allocation, we give the optimal solution of $\mathbf{\Psi}^{r+1}(t)$ with $\mathbf{\Phi}^{r+1}(t)$ and $\mathbf{v}^r(t)$. The relation between $\mathscr{R}(\mathbf{\Phi}^{r+1}(t), \mathbf{\Psi}^{r+1}(t), \mathbf{v}^r(t))$ and $\mathscr{R}(\mathbf{\Phi}^{r+1}(t), \mathbf{\Psi}^r(t), \mathbf{v}^r(t))$ can then be expressed as

$$\mathscr{R}\Big(\mathbf{\Phi}^{r+1}(t), \mathbf{\Psi}^{r+1}(t), \mathbf{v}^r(t)\Big) \geq \mathscr{R}\Big(\mathbf{\Phi}^{r+1}(t), \mathbf{\Psi}^r(t), \mathbf{v}^r(t)\Big). \tag{4.57}$$

The optimal speed for the UAVs with $\mathbf{\Phi}^r(t)$ and $\mathbf{\Psi}^r(t)$ are obtained in the UAV speed optimization algorithm, which can be expressed as

$$\mathscr{R}\Big(\mathbf{\Phi}^{r+1}(t), \mathbf{\Psi}^{r+1}(t), \mathbf{v}^{r+1}(t)\Big) \geq \mathscr{R}\Big(\mathbf{\Phi}^{r+1}(t), \mathbf{\Psi}^{r+1}(t), \mathbf{v}^r(t)\Big). \tag{4.58}$$

In the $(r+1)$th iteration, we have the following inequation:

$$\begin{aligned} \mathscr{R}\Big(\mathbf{\Phi}^{r+1}(t), \mathbf{\Psi}^{r+1}(t), \mathbf{v}^{r+1}(t)\Big) &\geq \mathscr{R}\Big(\mathbf{\Phi}^{r+1}(t), \mathbf{\Psi}^{r+1}(t), \mathbf{v}^r(t)\Big) \\ &\geq \mathscr{R}\Big(\mathbf{\Phi}^{r+1}(t), \mathbf{\Psi}^r(t), \mathbf{v}^r(t)\Big) \geq \mathscr{R}\Big(\mathbf{\Phi}^r(t), \mathbf{\Psi}^r(t), \mathbf{v}^r(t)\Big). \end{aligned} \tag{4.59}$$

As shown in (4.59), the objective function does not decrease in each iteration. It is known that such a network has a capacity bound, and the uplink sum rate cannot increase unlimitedly. Therefore, the objective function has an upper bound and will converge to a constant after limited iterations (i.e., the proposed ISASOA is convergent). □

THEOREM 4.9 *The complexity of the proposed ISASOA is* $O((N_h(t)+M)\times 2^{N_l(t)})$.

Proof The complexity of the proposed ISASOA is the number of iterations times the complexity of iteration. As shown in Algorithm 4.7, the objective function increases for at least ϵ in each iteration. We denote the average uplink sum rate of the initial solution by $\bar{R}_0(N_h(t), M)$, and the average uplink sum rate of the ISASOA by $\bar{R}(N_h(t), M)$. The number of iteration is no more than $(\bar{R}(N_h(t), M) - \bar{R}_0(N_h(t), M))/\epsilon$. In addition, the increment of the uplink sum rate can be expressed as $(\bar{R}(N_h(t), M) - \bar{R}_0(N_h(t), M)) = (N_h(t)+M)\log_2(\frac{1+\bar{\gamma}_I}{1+\bar{\gamma}_0})$, where $\bar{\gamma}_I$ is the average SNR of the UAVs in the U2N mode and CUs with ISASOA, and $\bar{\gamma}_0$ is the average SNR of the UAVs in the U2N mode and CUs with the initial solution. Therefore, the number of iterations is given as $C \times (N_h(t)+M)$, where C is a constant.

In each iteration, the U2N subchannel allocation is solved directly with convex problem solutions. The U2U subchannel allocation is solved using a branch-and-bound method, with the complexity being $O(2^{N_l(t)})$. The speeds of different UAVs are

Table 4.2 Simulation parameters

Parameter	Value
Number of subchannels K	10
Number of UAVs in the U2U mode N_l	5
Number of UAVs N	20
Number of CUs M	5
Transmission power P_U	23 dBm
Noise variance σ^2	−96 dBm
Center frequency	1 GHz
Power gains factor G	−31.5 dB
Maximum number of subchannels used by one user $\chi_{\max}$	2
Algorithm convergence threshold ϵ	0.1
U2N channel parameter η_{LoS}	1
U2N channel parameter η_{NLoS}	20
U2N channel parameter a	12
U2N channel parameter b	0.135
U2U pathloss coefficient α	2
Maximum UAV speed $v_{\max}$	10 m/time slot
Length of trajectory L_i	300 m
Minimum U2U rate R_0	10 bit/(s×Hz)
SNR threshold γ_{th}	10 dB

optimized using convex optimization methods, with a complexity of $O(N_h(t)+N_l(t))$. Therefore, the complexity of each iteration is $O(2^{N_l(t)})$, and the complexity of the proposed ISASOA is $O((N_h(t)+M)\times 2^{N_l(t)})$. □

4.3.4 Performance Evaluation

In this section, we evaluate the performance of the proposed ISASOA. The selection of the simulation parameters is based on the existing works and 3GPP specifications [74]. In this simulation, the locations of the UAVs are randomly and uniformly distributed in a 3D area of 2 km × 2 km × $h_{\max}$, where $h_{\max}$ is the maximum possible height for the UAVs. To study the impact of UAV height on the performance of this network, we simulate two scenarios with $h_{\max}$ being 100 m and 200 m, respectively. The direction of the predetermined trajectory for each UAV is given randomly. All curves are generated with over 1,000 instances of the proposed algorithm. The simulation parameters are listed in Table 4.2. We compare the proposed algorithm with the greedy subchannel allocation algorithm proposed in [69]. In the greedy algorithm scheme, the subchannel allocation is performed based on matching theory, and the UAV speed is the same as the proposed ISASOA scheme. The maximum possible height for the UAVs in the greedy algorithm is set at 200 m.

Figure 4.18 depicts the uplink sum rate with a different number of UAVs in the U2N mode. In the proposed ISASOA, the difference between $T = 50$ and $T = 30$ in terms of the uplink sum rate is about 7%. It is shown that a larger task completion time

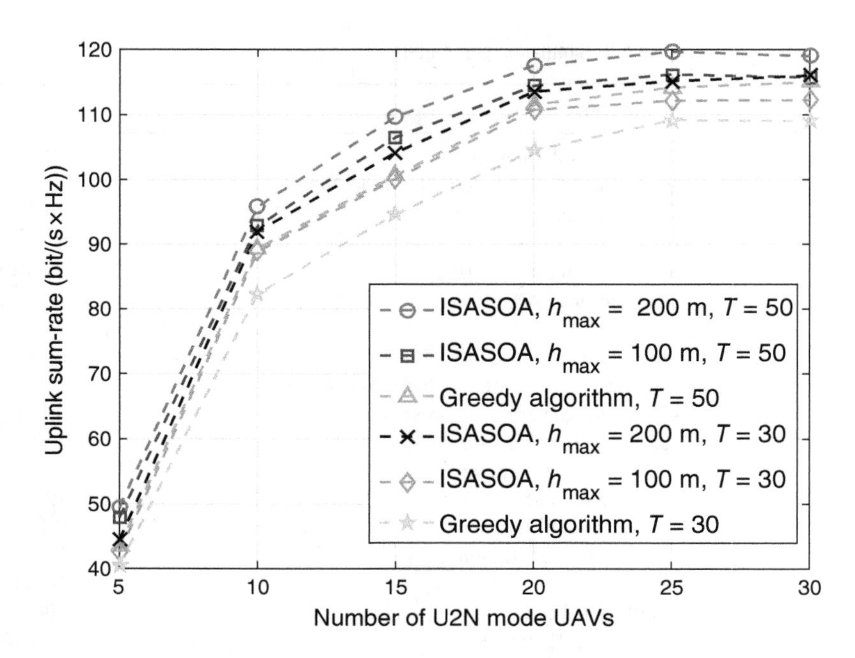

Figure 4.18 Number of UAVs in the U2N mode versus the uplink sum rate

T corresponds to a higher uplink sum rate, because the UAVs have a larger degree of freedom on the optimization of their speeds with a looser time constraint. The scenario with $h_{\max} = 200$ m has about 3% higher uplink sum rate than the scenario with $h_{\max} = 100$ m. The performance gap between the two scenarios is mainly affected by the U2N pathloss caused by different LoS and NLoS probabilities. The uplink sum rate with the ISASOA is 10% larger than that of the greedy algorithm on average due to the efficient U2N and U2U subchannel allocation. All six curves show that the uplink sum rate of U2N and CU transmissions increases with the number of UAVs, and the growth slows as N increases due to the saturation of network capacity.

Figure 4.19 shows the uplink sum rate with a different U2U-UAV/UAV ratio when the number of UAVs is set at 20. It is shown that the uplink sum rate decreases with more UAVs in the U2U mode in the network, and the descent rate is larger with more UAVs in the U2U mode. A larger U2U-UAV/UAV ratio not only reduces the number of UAVs in the U2N mode but also leads to a larger number of U2U receiving UAVs. Therefore, more UAVs in the U2N mode are restricted by the U2U transmission rate constraint and cannot move with the speed that corresponds to the maximum rate for the U2N links.

Figure 4.20 illustrates the relation between the U2U-UAV/UAV ratio and the sum rate for U2U transmissions, with the number of UAVs set at 20. The total U2U transmission rate increases with a larger U2U-UAV/UAV ratio, but the rate of the increment decreases with a larger U2U-UAV/UAV ratio. That is, the average U2U transmission rate decreases with more UAVs in the U2N mode in the network. The reason is that, with the increment of UAVs in the U2N mode, the U2U-to-U2U interference

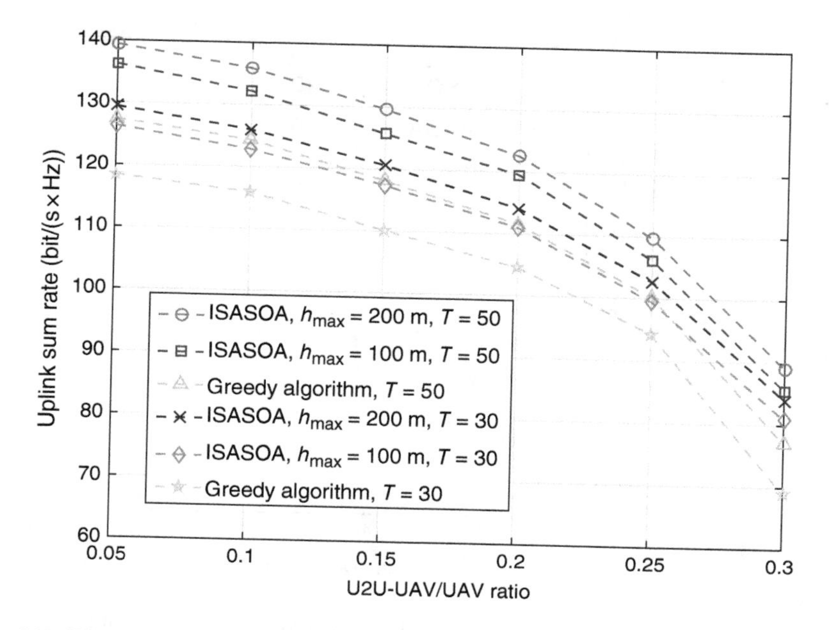

Figure 4.19 U2U-UAV/UAV ratio versus the uplink sum rate

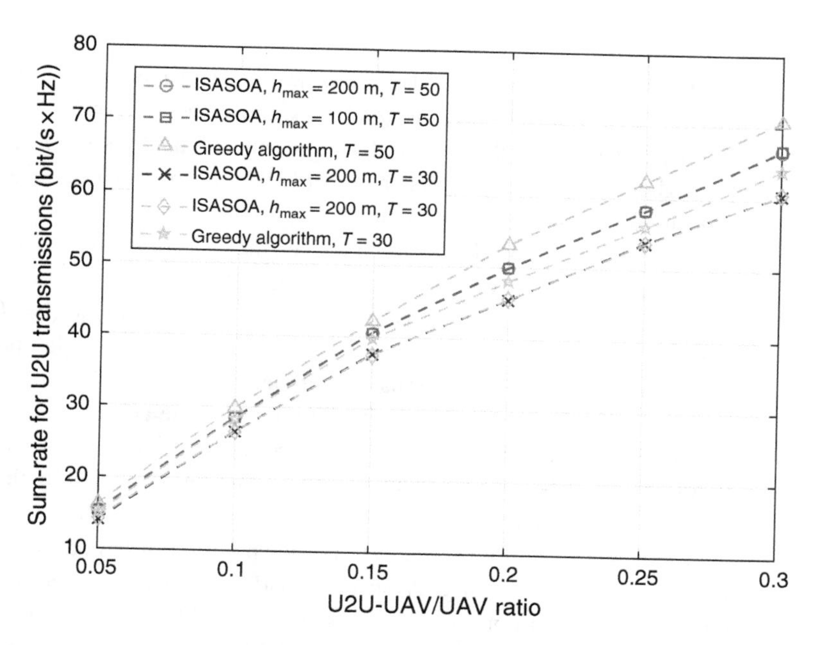

Figure 4.20 U2U-UAV/UAV ratio versus the sum rate for U2U transmissions

rises rapidly, which reduces the data rate for a U2U link. There is no significant difference between the ISASOA with a different $h_{\max}$ in terms of the total U2U transmission rate because the U2U transmission rate for each link is only determined

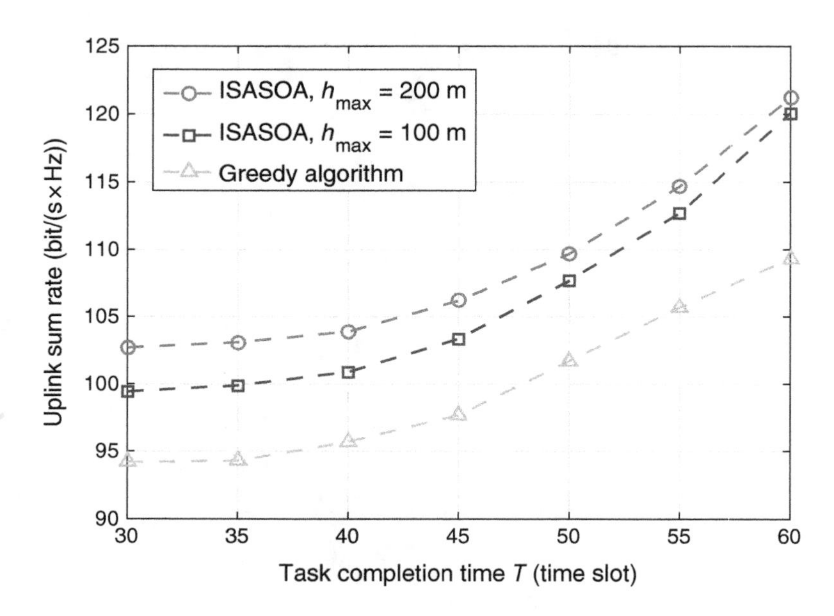

Figure 4.21 Minimum task completion time T versus the uplink sum rate

by the distance between the U2U transmitting and receiving UAVs. Note that the average U2U transmission rate is always above the U2U rate threshold within the simulation range. For the greedy algorithm scheme, the total U2U transmission rate is 5% higher than the ISASOA, but a higher U2U transmission rate squeezes the network capacity for the U2N transmissions.

In Fig. 4.21, we give the relation between the task completion time T and the uplink sum rate. The uplink sum rate increases with a larger minimum task completion time T, and the rate of change increases with T. The scheme with $h_{max} = 200$ m has a larger uplink sum rate than the scheme with $h_{max} = 100$ m due to a higher probability of LoS U2N transmission. The performance gap decreases when T becomes larger because the UAVs with $h_{max} = 100$ can stay for a longer time at the locations with relatively high LoS transmission probability. The greedy algorithm is about 10% lower than the ISASOA. It can be inferred that the uplink sum rate is affected by the delay tolerance of the data collection.

In Fig. 4.22, the uplink sum rate is shown with a different maximum UAV speed v_{max}. A larger maximum UAV speed provides the UAVs a larger degree of freedom on the UAV speed optimization. It is shown that the uplink sum rate increases significantly with the maximum UAV speed when $v_{max} \leq 20$ m/time slot. The uplink sum rate becomes stable in the $v_{max} > 30$ m/time slot because speed is not the main restriction on the uplink sum rate when the maximum UAV speed is sufficiently large. The difference between the $h_{max} = 200$ and $h_{max} = 100$ schemes decreases with the increment of v_{max}, and the greedy algorithm is about 10% lower than the ISASOA within the simulation range.

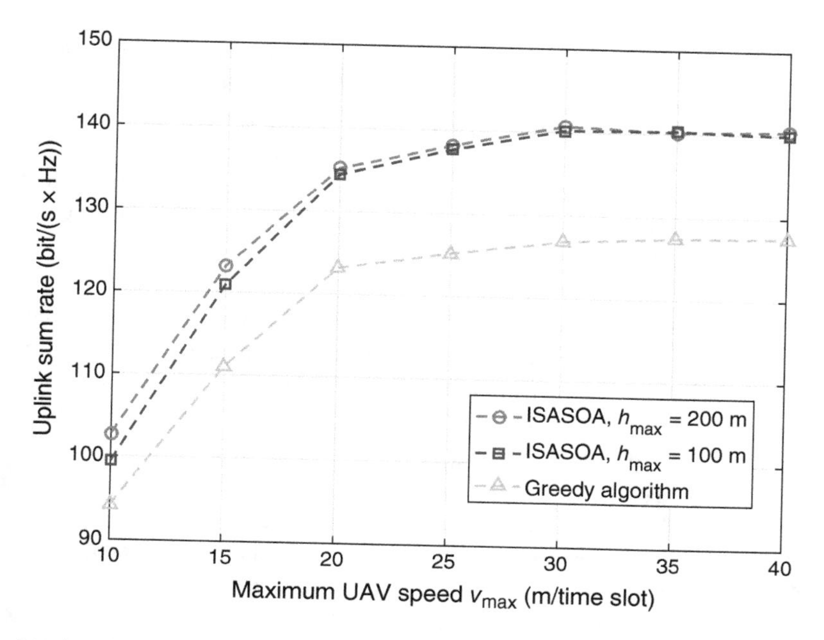

Figure 4.22 Maximum UAV speed v_{max} versus the uplink sum rate

4.3.5 Conclusion

In this section, we introduced U2X communications into the cellular internet of UAVs. As an example, we formulated a joint subchannel allocation and UAV speed optimization problem to improve the uplink sum rate in a single cell multi-UAV network, where multiple UAVs upload their collected data to the BS via U2N and U2U transmissions. To solve the NP-hard problem, we decoupled it into three subproblems: U2N and CU subchannel allocation, U2U subchannel allocation, and UAV speed optimization. These three subproblems were then solved with optimization methods, and the novel ISASOA was proposed to obtain a convergent solution of this problem. Simulation results have shown that the uplink sum rate decreases with a tighter task completion time constraint, and the proposed ISASOA can achieve about 10% more uplink sum rate than the greedy algorithm.

4.4 Application: Aerial-Ground Air Quality Sensing System

A recent report from the World Health Organization [100] explained that air pollution has become the world's largest environmental health risk, as one in eight of global deaths are caused by air pollution exposure each year. Air pollution is caused by gaseous pollutants that are harmful to humans and the ecosystem, and it is especially concentrated in the urban areas of developing countries. Thus, reducing air pollution would save millions of lives, and many countries have invested significantly on efforts to monitor and reduce the emission of air pollutants. Government agencies use the air quality index (AQI) to quantify the degree of air pollution. The index is calculated based on the concentration of a number of air pollutants (e.g., the concentration of $PM_{2.5}$, PM_{10} particles, and so on in developing countries). A higher value of AQI indicates that air quality is "heavily" or "seriously" polluted, which means a greater proportion of the population may experience harmful health effects [14]. To intuitively reflect AQI value of locations in either 2D or 3D area, an AQI map is defined to offer such convenience [101].

Monitoring can be completed by sensors at governmental static observation stations, generating an AQI map in a local area (e.g., a city [102]). However, these static sensors only obtain a limited number of measurement samples in the observation area and may often induce high costs. For example, there are only 28 monitoring stations in Beijing. The distance between two nearby stations is typically tens of thousands of meters, and the AQI is monitored every 2 hours [103]. To provide more flexible monitoring and reduce the cost, mobile devices such as cell phones, cars, and balloons are used to carry sensors and process real-time measurements. Crowdsourced photos contributed by massive numbers of cell phones can help depict the 2D AQI map in a large geographical region in Beijing [104], with a range of 4 km × 4 km. Mobile nodes equipped with sensors can provide 100 m × 100 m 2D on-ground concentration maps with relatively high resolution [105–107]. Sensors carried by tethered balloons can build the height profile of an AQI at a fixed observation height within 1,000 m [108]. A mobile system with sensors equipped in cars and drones can help monitor $PM_{2.5}$ in open 3D space [109], with 200 m per measurement.

Even though current mobile sensing approaches can provide relatively accurate and real-time AQI monitoring data, they are spatially coarse-grained because two measurements are separated by few hundreds of meters in horizontal or vertical directions in the 3D space. However, an AQI intrinsically changes from meter to meter, and it is preferred to perform AQI monitoring in the 3D space surrounding an office building or throughout a university campus, rather than city-wide [110, 111]. The AQI distribution in meter-sliced areas, called fine-grained areas, would be desirable for people, particularly those living in urban areas. A fine-grained AQI map can help with the design of building ventilation systems, which for example, can guide teachers and students to stay away from the pollution sources on campus [112].

Due to the high power consumption of mobile devices, one can only measure a limited number of locations in a large space. To avoid an exhaustive measurement, using an estimation model to approximate the value an unmeasured area has been widely adopted. In [113], the prediction model is based on a few public air quality stations and meteorological data, taxi trajectories, road networks, and point of interests (POIs). However, because they estimate AQI using a feature set based on historical data, their model cannot respond in real time to the change in pollution concentration at an hourly granularity, leading to large errors at times. In [109], the random walk model is used for prediction by dividing the whole space into different cube shapes. However, the model may not reflect the physical dispersion of particles [114, 115], and all locations are measured without considering battery life constraints when mobile devices are

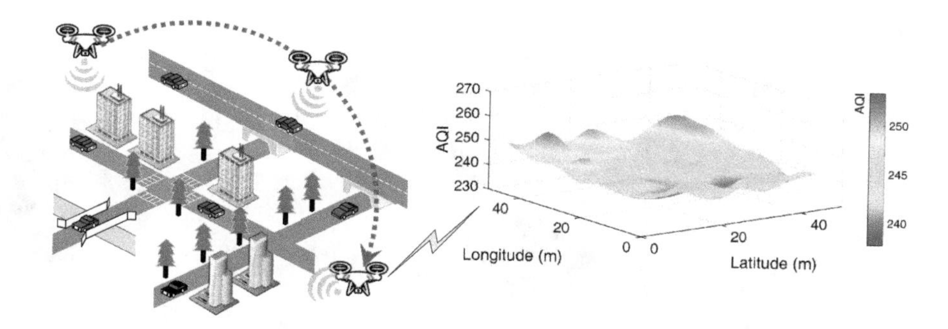

Figure 4.23 AQI measurement using mobile sensing over UAV

used. Mobile sensor nodes used in [105] employ a regression model as well as graph theory to estimate the AQI value at unmeasured locations. However, they mainly focus on a 2D area and can hardly produce a 3D fine-grained map. Neural networks (NN) are also used for forecasting the AQI distribution [116–119]. However, their performance in a fine-gained area is not satisfied without considering the physical characteristic of the real AQI distribution.

In this section, we design a mobile sensing system based on UAVs, called *ARMS*, that can effectively catch AQI variance at the meter level and profile the corresponding fine-grained distribution. ARMS is a real-time monitoring system as illustrated in Fig. 4.23. It can generate a current AQI map within a few minutes, whereas previous methods took a few hours. With ARMS, the fine-grained AQI map construction can be decomposed into two parts. First, we propose a novel AQI distribution model, called the Gaussian Plume model embedding neural network (GPM-NN), that combines physical dispersion and nonlinear NN structure to predict unmeasured areas. Second, we detail the adaptive monitoring algorithm as well as address its applications in a few typical scenarios. By measuring only selected locations in different scenarios, GPM-NN is used to estimate AQI value at unmeasured locations and generate real time AQI maps, which can save the battery life of mobile devices while maintaining high accuracy in AQI estimation.

The rest of this section is organized as follows. In Section 4.4.1, we briefly introduce our UAV sensing system. In Section 4.4.2, we present our fine-grained AQI distribution model. The adaptive monitoring algorithm is addressed in Section 4.4.3. In Sections 4.4.4 and 4.4.5, we present two typical application scenarios and performance analysis of ARMS, respectively. Finally, this section is summarized in Section 4.4.6.

4.4.1 Preliminaries of UAV Sensing System

In this part, we first provide a brief introduction of ARMS [21], and then we show how to construct a dataset using ARMS. To confirm the reliability of the collected dataset, we compare the collected data and the official AQI measured by the nearest Beijing government's monitoring station: the Haidian station [120]. To determine the

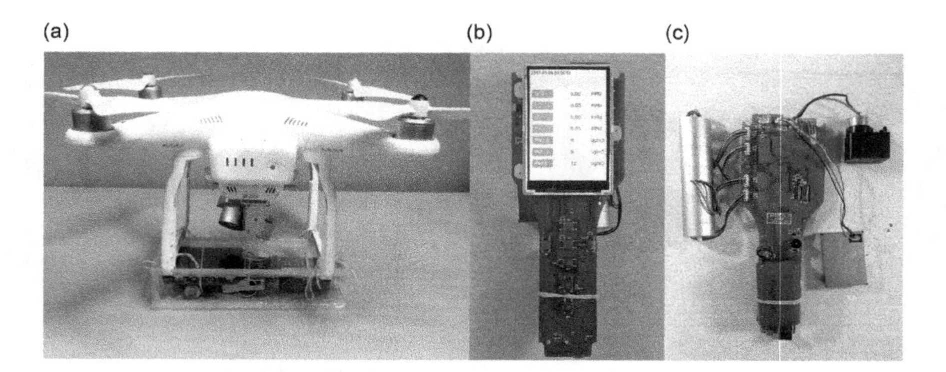

Figure 4.24 ARMS system (a) UAV, (b) front and (c) back of the sensor board

parameters of our model, we test possible factors that may influence the AQI, such as wind and locations, and remove factors that have small correlations with the AQI in the fine-grained scenarios from our model.

System Overview

The architecture of ARMS includes an UAV and an air quality sensor boarded on the UAV, as shown in Fig. 4.24. The sensor is fixed in a plastic box with vent holes and bundlled on the bottom of UAV. The sensor uses a laser-based AQI detector [121], which can provide the concentration within ±3% monitor error for common pollutants in AQI calculation, such as $PM_{2.5}$, PM_{10}, CO, NO, SO_2, and O_3. The values of these pollutants are recorded in real time, and we use them to calculate the corresponding AQI value at measuring locations.

For the UAV, we selected the DJI Phantom 3 Quadcopter [122] as the mobile sensing device. The UAV can keep hosting for at most 15 minutes due to the battery constraint, which restricts the longest continuous duration within one measurement. The GPS sensor on the UAV can provide the real-time 3D position. During one measurement, the UAV is programmed with a trajectory, including all locations that need to be measured. Following this trajectory, the UAV hovers for 10 seconds to collect sufficient data to derive the AQI value at each stop before moving to the next one.

During one monitoring process, ARMS measures all target locations and records the corresponding AQI values. After the measuring process is completed, the data is then sent to the offline PC and put into the GPM-NN model to construct the real-time AQI map. Thus, the map construction process is offline.

Dataset Description

Data collected by ARMS are then arranged as a dataset.[8] As shown in Fig. 4.23, we conducted a measurement study in both typical 2D and 3D scenarios (i.e., a roadside

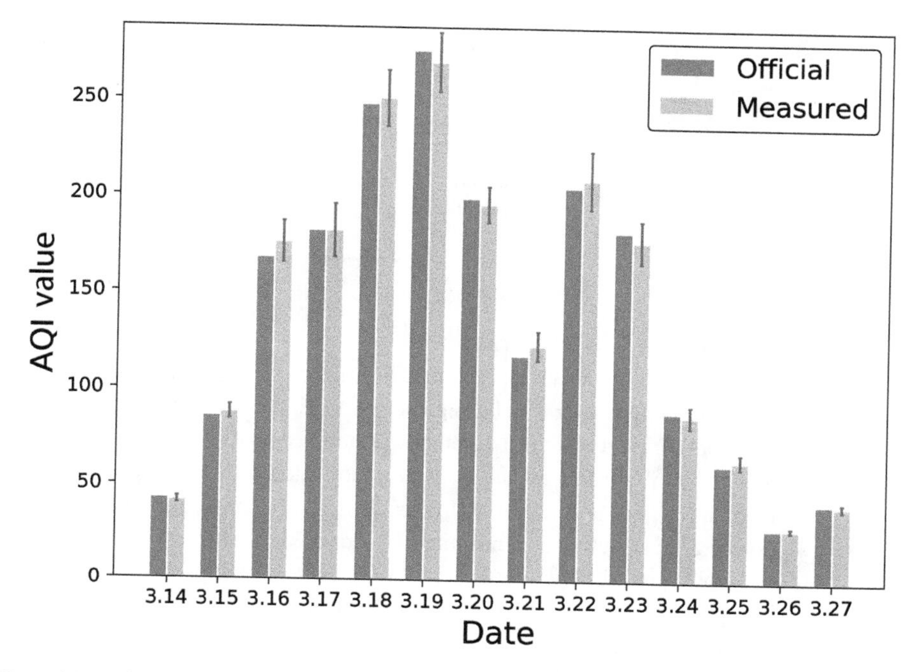

Figure 4.25 AQI value comparison between official data and data we collected for 14 days in March 2017

park and the courtyard of an office building at Peking University, respectively) from February 11 to July 1, 2017, for more than 100 days to collect sufficient data [123].

In the dataset, each text file includes one complete measurement over a day in one typical scenario. In each text file, each sample has four parameters: 3D coordinates (x, y, z) and an AQI value. Each value represents the measured AQI, and its coordinates in the matrix reflect the position in different scenarios. In the 2D scenario, we assume $z = 0$, while measuring at an interval of 5 m in x and y directions. In the 3D scenario, every row represents a fixed position in the xy plane, and every column represents the height at an interval of 5 m in the z direction.

Data Reliability

To verify that there is no measurement error, we show the results of the relationship between our collected data and the official data (i.e., Haidian station [120]), in Fig. 4.25. Note that the official data is limited and only for the 2D space, whereas our system is mobile and suitable for the 3D space profiling. We selected 14 consecutive days for about 60 instances of monitoring from March 14 to March 27, 2017 to verify the reliability of our measurement. We use the two-tailed hypothesis test [124]: $\mathscr{H}_0$: $\mu_1 = \mu_2$ versus $\mathscr{H}_1 : \mu_1 \neq \mu_2$, where μ_1 denotes our average measured data for all days and μ_2 is the average for the official ones. The test result, $P = 0.9999 \gg 0.05$, indicates that there is no significant difference between the two values, which confirms the reliability of our measurements.

Table 4.3 Result of the hypothesis test

Tested parameter	P value
Wind	7.5693×10^{-5} ($\ll 0.05$)
Location	2.0981×10^{-5} ($\ll 0.05$)
Temperature	0.9070 ($\gg 0.05$)
Humidity	0.6996 ($\gg 0.05$)

Selection of Model Parameters

According to the previous AQI monitoring results for coarse-grained scenarios [115], AQI is related to wind (including speed and direction), temperature, humidity, altitude, and spatial locations. But for fine-grained scenarios, correlations between AQI and these spatial parameters need to be reconsidered due to the heterogenous diffusion in both vertical and horizontal directions in a small-scale area. In this test, all these potential parameters are measured by our ARMS with different sensors. To evaluate the real correlation between these parameters, we adopt the spatial regression according to [125] and test the coefficient for each parameter. Mathematically, the spatio-temporal model is given by

$$C(s_i) = \boldsymbol{z}(s_i)\boldsymbol{\beta}^{\mathrm{T}} + \varepsilon(s_i), \tag{4.60}$$

where $C(s_i)$ is the particle concentration at position s_i, $\boldsymbol{z}(s_i) = (z_1(s_i), \ldots, z_n(s_i))$ denotes the vector of n parameters at s_i, and $\boldsymbol{\beta} = (\beta_1, \ldots, \beta_n)$ is the coefficient vector. The Gaussian white-noise process is $\varepsilon(s_i) \sim N(0, \sigma^2)$.

Based on our data, we use the least square regression and implement a hypothesis test for each coefficient β_j, as $\mathscr{H}_0 : \beta_j = 0$. The results in Table 4.3 indicate that wind and location are highly related to AQI distribution, whereas temperature and humidity are not.

4.4.2 Fine-Grained AQI Distribution Model

In this part, we provide a prediction model considering both physical particle dispersion and NN structure. We first introduce the physical dispersion model for the fine-grained scenario. Then, we provide a brief introduction to the NN we adopt in modeling, which can adapt to complicated cases, such as the nonlinearity introduced by extreme weather. Finally, we embed the dispersion model in the NN to design our model.

Physical Particle Dispersion Model

We first address the physical particle dispersion model for fine-grained scenarios. Specifically, we ignore the influence of temperature and humidity according to discussions in Section 4.4.1, and select the Gaussian Plume model (GPM) in the particle movement theory [126] to describe the particle's dispersion. The GPM is widely used to describe particles' physical motion [114, 127], and its robustness has been proved in a small-scale system [128]. The model is expressed as

$$C(x,y,z) = \frac{Q}{2\pi\sigma_y\sigma_z u}\exp\left(-\frac{(z-H)^2}{2\sigma_z^2}\right)\exp\left(-\frac{y^2}{2\sigma_y^2}\right), \quad (4.61)$$

where Q is the point source strength, u is the average wind speed, and H denotes the height of source.

To adopt GPM into the fine-grained scenario, the GPM is revised as

$$\begin{aligned} C(\boldsymbol{x},u) &= \int_{-\frac{L}{2}}^{\frac{L}{2}} \frac{\lambda}{2\pi\sigma_y\sigma_z u}\exp\left(-\frac{(z-H)^2}{2\sigma_z^2}-\frac{y^2}{2\sigma_y^2}\right)dy \\ &= \frac{\lambda\exp\left(-\frac{(z-H)^2}{2\sigma_z^2}\right)}{2\pi\sigma_z u}\int_{-\frac{L}{2\sigma_y}}^{\frac{L}{2\sigma_y}}\exp\left(-\frac{\gamma^2}{2}\right)d\gamma \\ &= \frac{\lambda}{\sqrt{2\pi}\sigma_z u}\exp\left(-\frac{(z-H)^2}{2\sigma_z^2}\right)\left[1-2Q\left(\frac{L}{2\sigma_y}\right)\right], \end{aligned}$$

where $C(\boldsymbol{x},u)$ is the AQI value at location $\boldsymbol{x}$, u is the real wind speed at different locations in the entire space, and H denotes a variable that reflects the influence of wind direction, which presents severely polluted areas along the z axis. Pollution mainly derives as a line source aligned the y axis, and L denotes the length of polluted source, λ denotes the particle density at the source. The diffusion parameters are σ_y and σ_z in y and z directions, which are both empirically given. The dispersion model in (4.62) can reflect physical characteristics but can hardly deal with unpredictable complicated changes, such as the nonlinearity introduced by extreme weather.

Neural Network Model

The neural network model, especially multilayer perceptron (MLP), has been widely adopted to do estimation for air quality [116–119]. Models are usually trained using a huge amount of data to achieve decent performance. All possible influential factors are involved as the neural network input variables for network training. Other types of NNs [129, 130] are proposed for better classification with more complex structures. As it has been proved that a three-layer neural network can compute any arbitrary function [131–133], a NN is able to present the complicated changes in fine-grained scenario. However, without considering the physical characteristics of AQI, the NN model may overfit and perform worse on the test data than on the training data [116].

GPM-NN Model

In order to utilize the advantages of both the GPM and NN, we embed the revised GPM in the NN and put forward the GPM-NN model.

Model Description

As shown in Fig. 4.26, the model structure contains a linear part (the physical dispersion model) and a nonlinear part (the NN structure) for fine-grained AQI distribution,

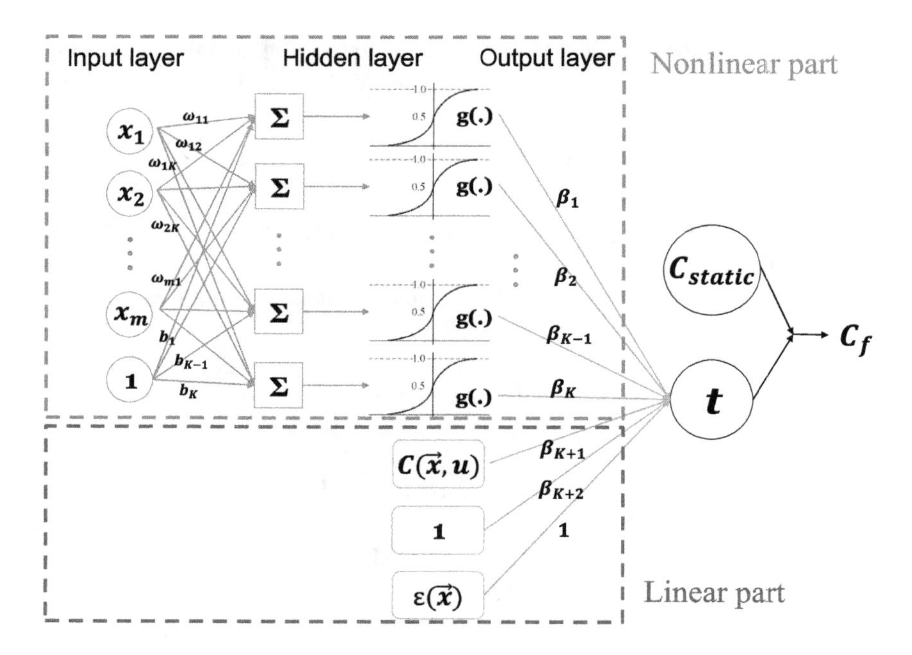

Figure 4.26 Model structure of GPM-NN

respectively. Let N be the total data collected by ARMS, which is represented by a pair $(\boldsymbol{X}_j, t_j)$, where $\boldsymbol{X}_j = [x_1\ x_2\ \ldots\ x_m]^{\mathrm{T}}$ is the jth sample with a dimensionality of m variables and t_j is the measured AQI value.

- In the nonlinear NN part, let K denote the total number of neurons in the hidden layer. The weights for these neurons are denoted by $\boldsymbol{W} = [\boldsymbol{W}_1\ \boldsymbol{W}_2\ \ldots\ \boldsymbol{W}_K]$, where $\boldsymbol{W}_i = [\omega_{i1}\ \omega_{i2}\ \ldots\ \omega_{im}]$ is the m-dimensional weight vector containing the weights between the components of input vectors and the ith neuron in the hidden layer. The bias term of the ith neuron is $\boldsymbol{b} = [b_1\ b_2\ \ldots\ b_K]$. The non-linear part with K neurons in the hidden layer will have $\boldsymbol{\beta} = [\beta_1\ \beta_2\ \ldots\ \beta_K]$ as weights for output layer, and $g(\cdot)$ is the activation function.
- In the linear part, we use $C(\boldsymbol{x}, u)$, a constant value and a Gaussian process as inputs, to reflect the influence of the physical model. The regression weights are correspondingly determined as β_{K+1}, β_{K+2}, and 1.

Thus, the mathematical expression of the proposed model can be written as

$$t(\boldsymbol{x}, u) = \sum_{i=1}^{K} \beta_i g(\boldsymbol{W}_i \boldsymbol{X}_j + b_i) + \beta_{K+1} C(\boldsymbol{x}, u) \\ + \beta_{K+2} + \varepsilon(\boldsymbol{x}), \qquad j = 1, 2, \ldots, N, \tag{4.62}$$

where $t(\boldsymbol{x}, u)$ is the estimated value of t_j and it represents the model's output. The output of the dispersion model in (4.62) is $C(\boldsymbol{x}, u)$ and β_i are regression coefficients. The measurement error defined by a Gaussian white-noise process is $\varepsilon(\boldsymbol{x}) \sim N(0, \sigma^2)$.

Because there is a risk that the NN part will overfit and perform worse on the test data than training data, the estimated AQI value is expressed as

$$C_f(\boldsymbol{x},u) = C_{static} + t(\boldsymbol{x},u), \tag{4.63}$$

where C_{static} is the average value of our measured AQI in a day, which is an invariant to quantify basic distribution characteristics.

Parameter Estimation

As shown in (4.62), GPM-NN has $(K+3)$ parameters, H, β_1, β_2, ..., β_{K+2}, which need to be estimated based on data collected by ARMS. We use 50 days of data to train the nonlinear part of GPM-NN. We use the least-square regression to estimate the parameters. Let $\mathscr{S}$ denote the residual error as

$$\mathscr{S} = \sum_{i=1}^{N} \left\| \hat{C}_f(\boldsymbol{x}_i,u_i) - \beta_{K+2} - \beta_{K+1}C(\boldsymbol{x},u) - \sum_{j=1}^{K} \beta_j g_j \right\|^2, \tag{4.64}$$

where i denotes the measuring sample of the ith observation point, and $g_j = g(\boldsymbol{W}_j \boldsymbol{X}_i + b_j)$.

PROPOSITION 4.2 *Equation (4.64) has a unique minimum point for estimated parameters β_1, β_2, ..., β_{K+2} and H, when $\sigma_z^2 > \max\{2z_i^2, 2H_0^2\}$.*

Proof For β_j where $j \in [1, K+2]$, we have

$$\frac{\partial^2 \mathscr{S}}{\partial \beta_j^2} = \begin{cases} 2\sum_{i=1}^{N} g_j^2 > 0, & 1 \le j \le K, \\ 2\sum_{i=1}^{N} C^2(\boldsymbol{x}_i,u_i) > 0, & j = K+1, \\ 2\sum_{i=1}^{N} 1 = 2N > 0, & j = K+2. \end{cases} \tag{4.65}$$

Hence, $\partial\mathscr{S}/\partial\beta_j$ are all convex functions, with $j \in [1, K+2]$.

As for variable H, the second-order partial derivative can be calculated as

$$\begin{aligned} \frac{\partial^2 \mathscr{S}}{\partial H^2} = -2\sum_{i=1}^{N} \beta' \Bigg[&- \frac{\beta'}{\sigma_z^6 u_i^2}(z_i - H)^2 \exp\left(-\frac{(z_i-H)^2}{\sigma_z^2}\right) \\ &- \left(C - \frac{\beta' \exp\left(-\frac{(z_i-H)^2}{2\sigma_z^2}\right)}{u_i \sigma_z} \right) \frac{1}{u_i \sigma_z^3} \exp\left(-\frac{(z_i-H)^2}{2\sigma_z^2}\right) \\ &+ \left(C - \frac{\beta' \exp\left(-\frac{(z_i-H)^2}{2\sigma_z^2}\right)}{u_i \sigma_z} \right) \frac{(z_i-H)^2 \exp\left(-\frac{(z_i-H)^2}{2\sigma_z^2}\right)}{u_i \sigma_z^5} \Bigg], \end{aligned} \tag{4.66}$$

where $C = (\hat{C}_f(\boldsymbol{x}_i,u_i) - C_{static} - \beta_{K+2} - \sum_{j=1}^{K} \beta_j g_j)$, and $\beta' = \frac{\lambda}{\sqrt{2\pi}}\beta_{K+1}(1 - 2Q(\frac{L}{2\sigma_y}))$. Then we have

$$\frac{\partial^2 \mathscr{S}}{\partial H^2} = 2\sum_{i=1}^{N}\Bigg[\left(\frac{2\beta'^2(z_i - H)^2}{\sigma_z^6 u_i^2} - \frac{\beta'^2}{\sigma_z^4 u_i^2}\right)\exp\left(-\frac{(z_i - H)^2}{\sigma_z^2}\right)$$
$$+ C\left(\frac{\beta'}{\sigma_z^3 u_i} - \frac{\beta'(z_i - H)^2}{u_i \sigma_z^5}\right)\exp\left(-\frac{(z_i - H)^2}{2\sigma_z^2}\right)\Bigg]. \tag{4.67}$$

Let $t_i = \exp(-\frac{(z_i-H)^2}{2\sigma_z^2})$, where each item of the summation is equivalent to a quadratic function $Q_i(t_i) = a_i t_i^2 + b_i t_i$. Note that $t_i \in (0,1]$, and $t_i = 0$ is one zero point of $Q_i(t_i)$. To satisfy the proposition that $\partial^2\mathscr{S}/\partial H^2$ always has positive value, the problem becomes

$$\begin{cases} a_i = \dfrac{2\beta'^2(z_i - H)^2}{u_i^2\sigma_z^6} - \dfrac{\beta'^2}{u_i^2\sigma_z^4} < 0, \\ b_i = C\left(\dfrac{\beta'}{u_i\sigma_z^3} - \dfrac{\beta'(z_i - H)^2}{u_i\sigma_z^5}\right) > 0, \end{cases} \quad \forall i \in [1,N], \tag{4.68}$$

which can be simplified as

$$\begin{cases} \sigma_z^2 > \max\limits_i \ 2(z_i - H)^2, \\ \sigma_z^2 > \max\limits_i \ (z_i - H)^2. \end{cases} \tag{4.69}$$

We define $H \in [0, H_0]$, where H_0 is the upper bound for a fine-grained measurement. Hence, by choosing appropriate diffusion parameter σ_z as $\sigma_z^2 > \max\{2z_i^2, 2H_0^2\}$, we have

$$\frac{\partial^2 \mathscr{S}}{\partial H^2} = 2\sum_{i=1}^{N} Q_i(t_i) = 2\sum_{i=1}^{N}(a_i t_i^2 + b_i t_i)$$
$$= 2\sum_{i=1}^{N}\left(\frac{b_i^2}{4|a_i|} - |a_i|\left(t_i + \frac{b_i}{2a_i}\right)^2\right) > 0, \quad \forall t_i \in (0,1]. \tag{4.70}$$

Therefore, $\partial\mathscr{S}/\partial H$ is also a convex function, which indicates that (4.64) has a minimum as well as a unique value. □

To find the minimum point of the residual error function $\mathscr{S}(H, \beta_1, \ldots, \beta_{K+2})$, we use the Newton method [134] to solve the following equations, which have analytical solutions that do not exist, as

$$\begin{cases} \dfrac{\partial \mathscr{S}}{\partial H} = 0, \\ \dfrac{\partial \mathscr{S}}{\partial \beta_j} = 0, \quad j = 1, 2, \ldots, K+2. \end{cases} \tag{4.71}$$

When the estimation value of H (denoted as H^*) is determined, $C(\boldsymbol{x}, u)$ is correspondingly determined. Denote

$$
\boldsymbol{J} = \begin{bmatrix} g(\boldsymbol{W}_1\boldsymbol{X}_1+b_1) & \cdots & g(\boldsymbol{W}_K\boldsymbol{X}_1+b_K) & C(\boldsymbol{x}_1,u_1) & 1 \\ g(\boldsymbol{W}_1\boldsymbol{X}_2+b_1) & \cdots & g(\boldsymbol{W}_K\boldsymbol{X}_2+b_K) & C(\boldsymbol{x}_2,u_2) & 1 \\ \vdots & \ddots & \vdots & \vdots & \vdots \\ g(\boldsymbol{W}_1\boldsymbol{X}_N+b_1) & \cdots & g(\boldsymbol{W}_K\boldsymbol{X}_N+b_K) & C(\boldsymbol{x}_N,u_N) & 1 \end{bmatrix}_{N\times(K+2)}
$$

as the model output matrix, and similarly

$$
\boldsymbol{\beta} = \begin{bmatrix} \beta_1 \\ \beta_2 \\ \vdots \\ \beta_K \\ \beta_{K+1} \\ \beta_{K+2} \end{bmatrix}_{(K+2)\times 1}
$$

is the vector that needs to be estimated. Hence, the estimated value of N samples can be written as

$$
\boldsymbol{T} = \boldsymbol{J}\boldsymbol{\beta}. \tag{4.72}
$$

Note that $\boldsymbol{J}$ is both a row-column full rank matrix, which has a corresponding generalized inverse matrix [135]. Because we have proved (4.64) has a unique minimum point, we then have

$$
\begin{aligned}
\boldsymbol{\beta} &= (\boldsymbol{J}^{\mathrm{T}}\boldsymbol{J})^{-1}\boldsymbol{J}^{\mathrm{T}}\boldsymbol{J}\boldsymbol{\beta} \\
&= (\boldsymbol{J}^{\mathrm{T}}\boldsymbol{J})^{-1}\boldsymbol{J}^{\mathrm{T}}\boldsymbol{T} \\
&= \boldsymbol{J}^{\dagger}\boldsymbol{T},
\end{aligned} \tag{4.73}
$$

where $\boldsymbol{J}^{\dagger} = (\boldsymbol{J}^{\mathrm{T}}\boldsymbol{J})^{-1}\boldsymbol{J}^{\mathrm{T}}$ is known as the Moore-Penrose pseudo inverse of $\boldsymbol{J}$. This equation is the least-squares solution for an over-determined linear system and is proved to have the unique minimum solution [136]. Thus, this equation is equal to the multivariate equation in (4.71), by which we can find the minimum value point of $\mathscr{S}$.

Performance Evaluation

To determine the initial value of the weights $\boldsymbol{W}$ and biases $\boldsymbol{b}$ for the hidden layer, we use the training data to do preprocessing and acquire the optimal values. Hence, the model can be completely determined for describing the AQI distribution in fine-grained scenarios.

For evaluating the performance of GPM-NN, we use average estimation accuracy (AEA) as the merit, expressed as

$$
\overline{AEA} = \frac{1}{n}\sum_{i=1}^{n}\left(1 - \frac{|\hat{C}_f(i) - C_f(i)|}{C_f(i)}\right), \tag{4.74}
$$

where n denotes the total locations in the scenario, $\hat{C}_f(i)$ denotes the estimation AQI value in the ith location, and $C_f(i)$ denotes the real measured value. In Sections 4.4.4 and 4.4.5, we compare the accuracy of AQI map constructed by our GPM-NN and other existing models.

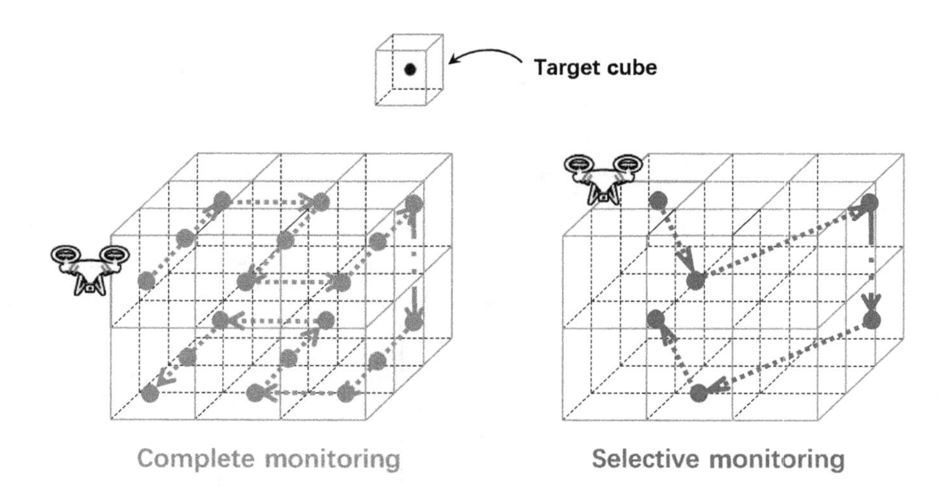

Figure 4.27 Example of the adaptive monitoring algorithm, with complete and selective monitoring

4.4.3 Adaptive AQI Monitoring Algorithm

In the following, we provide the adaptive monitoring algorithm of ARMS. Intuitively, a larger number of measurement locations introduces a higher accuracy of the AQI map. However, based on the physical characteristic of particle dispersion in GPM-NN, we can build a sufficiently accurate AQI map by regularly measuring only a few locations. This process can effectively save energy and thus improve the efficiency of the system. Specifically, an AQI monitoring is decomposed into two steps – *complete monitoring* and *selective monitoring* – for efficiency and accuracy (Fig. 4.27). We first trigger *complete monitoring* everyday for one time to establish a baseline distribution. Then ARMS periodically (e.g., every one hour) measures only a small set of observation points, which are acquired by analyzing the characteristic of the established AQI map. This process, called *selective monitoring*, is based on GPM-NN to update the real-time AQI map. By accumulating current measurements with the previous map, a new AQI map is generated in a timely manner. Every time selective monitoring is done, ARMS compares the newly measured results and the most recent measurement. If there is a large discrepancy between them, which indicates that the AQI experiences severe environmental changes, we would again trigger the complete monitoring to rebuild the baseline distribution. Thus, ARMS can effectively reduce the measurement effort as well as cope with the unpredictable spatio-temporal variations in the AQI values.

Complete Monitoring

The *complete monitoring* is designed to obtain a baseline characteristic of the AQI distribution in a fine-grained area and is triggered at daily intervals.

The entire space can be divided into a set of 5 m × 5 m × 5 m cubes. In the complete monitoring process, ARMS measures all cubes continuously and builds a baseline AQI

Algorithm 4.8 Operation of monitoring algorithm

```
1  /* Complete Monitoring: triggered between days */
2  for i = 1 to sum(Cube) do
3  |   measure the AQI value of Cube_i and record;
4  |   move to the next cube;
5  end
6  generate baseline 3D AQI map B;
7  /* Selective Monitoring: triggered between hours */
8  for i = 1 to sum(Cube) do
9  |   calculate PDT_cubei;
10 |   if PDT_cubei ≥ PDT || PDT_cubei ≤ δ then
11 |   |   add Cube_i to M;
12 |   end
13 end
14 generate min trajectory D of M;
15 forall p_i ∈ D do
16 |   measure the AQI value of Cube_i and record;
17 end
18 update the realtime AQI map M based on previous B and D;
19 if M deviates B by a large σ then
20 |   enter the complete monitoring period;
21 end
```

map using GPM-NN. The process is of high dissipation and thus is triggered over a long observation period.

Selective Monitoring

To reflect changes in the AQI distribution in a small-scale space over time (e.g., between each hour in a day) [109], ARMS uses *selective monitoring* to capture such dynamics. The selective monitoring makes use of the previous AQI map, analyzing its physical characteristics, to reduce the monitoring overhead in the next survey and maintain the real-time AQI map accordingly.

In the selective monitoring process, ARMS measures the AQI value of only a small set of selected cubes and generates an AQI map over the entire fine-grained area. To deal with the inherent tradeoff between measurement consumption and accuracy, we put forward an important index called the partial derivative threshold (PDT) to guide system selecting specific cubes. PDT is defined as

$$PDT_i = \frac{\left|\frac{\partial C_f}{\partial x_i}\right| - \left|\frac{\partial C_f}{\partial x_i}\right|_{\min}}{\left|\frac{\partial C_f}{\partial x_i}\right|_{\max} - \left|\frac{\partial C_f}{\partial x_i}\right|_{\min}}, \tag{4.75}$$

where x_i denotes the ith variable in GPM-NN ($i = 1,2,\dots,m$), and $C_f = C_f(\boldsymbol{x},u)$ denotes the entire distribution in a small-scale area. The minimum and maximum values of the partial derivative for parameter x_i are denoted by $|\partial C_f/\partial x_i|_{\min}$ and $|\partial C_f/\partial x_i|_{\max}$, respectively. Note that $\partial C_f/\partial x_i$ describes the upper bound of dynamic change degrees we can tolerate, expressed as

$$\frac{\partial C_f}{\partial x_i} = PDT_i \cdot \left(\left| \frac{\partial C_f}{\partial x_i} \right|_{\max} - \left| \frac{\partial C_f}{\partial x_i} \right|_{\min} \right) + \left| \frac{\partial C_f}{\partial x_i} \right|_{\min}.$$

$$0 \le PDT_i \le 1. \tag{4.76}$$

For each parameter, there is one corresponding PDT. In general, PDT reflects the threshold for dynamic change degrees in a fine-grained area. An area where the model's parameters have a large change rate would have a larger PDT value, indicating more drastic changes. When given a specific PDT, any cube with a $\partial C_f/\partial x_i$ that is above the threshold of (4.76) will be moved into a set $\mathscr{M}$. Moreover, when PDT_i is too small (less than a small constranit δ), the corresponding ith cube will also be added into $\mathscr{M}$. Mathematically, set $\mathscr{M}$ is given as

$$\mathscr{M} = \{i \mid PDT_i \ge PDT\} \cup \{i \mid PDT_i \le \delta\}. \tag{4.77}$$

Remark 4.2 Elements in $\mathscr{M}$ can be the severe changing areas in a small-scale space (e.g., a tuyere or abnormal building architecture) or typically the lowest or the highest value that can reflect basic features of the distribution. These elements are sufficient to depict the entire AQI map and hence are needed to be measured between two measurements. Thus, by only measuring cubes in $\mathscr{M}$, ARMS can generate a real-time AQI map implemented by GPM-NN, while greatly reducing the measurement overhead.

In general, PDT is adjusted manually for different scenarios. When PDT is low, the threshold for abnormal cubes declines, indicating the measuring cubes will increase and the estimation accuracy is relatively high. However, it requires high battery consumption. On the other hand, as PDT is high, the measuring cubes will decrease. This can cause a decline in accuracy but can highly reduce consumption. In summary, the tradeoff between accuracy and consumption should be studied to obtain better system performance.

Trajectory Optimization

When target cubes in set $\mathscr{M}$ are determined, the total network can be modelled as a 3D graph $G = (V,E)$ with a number of $|V|$ target cubes. Hence, finding the minimum trajectory over these cubes is equal to finding the shortest Hamiltonian cycle in a 3D graph. This problem is known as the traveling salesman problem (TSP), which is NP-hard [137].

To solve TSP in this case, we propose a greedy algorithm to find the suboptimal trajectory. In the fine-grained scenario, ARMS has a limited power budget and can monitor no more than n cubes over one measurement. To find the corresponding trajectory, we focus on how to determine the next measuring cube based on the current location of ARMS. Let $\mathbb{Z} = \{O_0, O_1, \ldots, O_{|V|-1}\}$ be the set of coverage cubes, where O_i denotes every observation cube. The aim is to acquire as many target cubes as possible over the trajectory for higher AQI estimation accuracy. Considering the significant physical characteristic of PDT, our greedy solution can be formulated by maximizing the next cube's PDT and minimizing the traveling cost from the current location to next cube. Hence, finding the optimal trajectory in this case is equal to an iteration of solving the following optimization problem:

$$\begin{aligned} i^* = \quad & \arg\max_i \left| \frac{PDT_i}{cost(i)} \right| \\ s.t. \quad & O_i \in \mathscr{M}, \\ & O_i \cap \bigcup \{O_0, O_1, \ldots, O_{i-1}\} = \varnothing, \end{aligned} \tag{4.78}$$

where $cost(i)$ is the consumption for the UAV to traverse from the $(i-1)$th cube to the ith cube, and PDT_i is acquired by analyzing the characteristic of latest AQI map.

For every current location i, the selection of the next target cube follows (4.78). Note that there are limited target cubes in $\mathscr{M}$, which are also determined by (4.76), so the objective function aims to generate the trajectory point by point. Thus, using the solution of (4.78), the greedy algorithm can effectively select key cubes and generate the suboptimal trajectory for ARMS in different scenarios.

To analyze the complexity of our algorithm, there are V target cubes in total that need to be added from $\mathscr{M}$. When the current location of ARMS is at the ith cube, it needs to compare another $|V - i|$ edges in G to determine the next measuring cube. Note that every target cube contains m parameters ($m = 4$ in our model) and $O(V) = O(n)$. Thus, the total operation time is $O(m \sum_{i=1}^{V-1} |V - i|) = O(n^2)$.

Algorithm 4.8 describes the whole process of the monitoring algorithm. Complete monitoring is triggered between days, and selective monitoring is triggered between hours. When the monitoring area experiences severe environmental changes such as a gale, ARMS compares the result of the map built by selective monitoring and the map built previously. If there is a large deviation σ between them, ARMS would again trigger the complete monitoring to rebuild the baseline distribution.

4.4.4 Example I: Performance Analysis in Horizontal Open Space

In this section, we implement the adaptive monitoring algorithm in a typical 2D scenario: the horizontal open space. We analyze the performance of GPM-NN and adaptive monitoring algorithm in this typical scenario.

Scenario Description

When the 3D space has a limited range in height, ARMS needs to cover target cubes in nearly the same horizontal plane. Two distant cubes at the same height may have a low correlation because the wind may create different concentrations of pollutants in a horizontal plane. This is considered a typical 2D scenario and often with a horizontal open space (e.g., a roadside park), as shown in Fig. 4.28.

Figure 4.28 A typical application scenario of ARMS in 2D space (a roadside park)

Performance Analysis

In this section, we first compare the accuracy of GPM-NN with other existing models using the experimental result in Fig. 4.29. Then, Fig. 4.30 illustrates the influence of different numbers of neurons in the hidden layer. To study GPM-NN's performance when the AQI varies, in Fig. 4.31, we show the relationship between different AQI values and the corresponding estimation accuracy. In Fig. 4.32, we present the performance of our monitoring algorithm versus other selection algorithms. Finally, Fig. 4.33 shows the tradeoff between system battery consumption and estimation accuracy via different PDTs.

Model Accuracy

In Fig. 4.29, we compare three prediction models – our regression model GPM-NN, linear interpolation (LI) [125], and classical multivariable linear regression (MLR) [125] – using different values of PDT. Linear interpolation uses interpolation to estimate the AQI value of undetected cubes by other measured cubes, whereas MLR uses multiple parameters (wind, humidity, temperature, etc.) of measured cubes to do regression and estimation.

In the horizontal open space scenario, we can find that GPM-NN achieves the highest accuracy. In each curve, we can see that the average estimation accuracy decreases as the PDT value increases. As discussed before, when PDT has a higher threshold, target cubes in set $\mathcal{M}$ decline (i.e., the total cubes measured by ARMS decrease). Thus, the estimation accuracy correspondingly drops. When $PDT = 0.1$, GPM-NN performs the best among the three models, which proves our model is robust and precise. Moreover, as PDT increases (e.g., $PDT = 0.75$), GPM-NN still maintains a high accuracy (almost 80%), whereas others experience a rapid decrease. This implies that our model is suitable for adaptive energy saving monitoring in a fine-grained area.

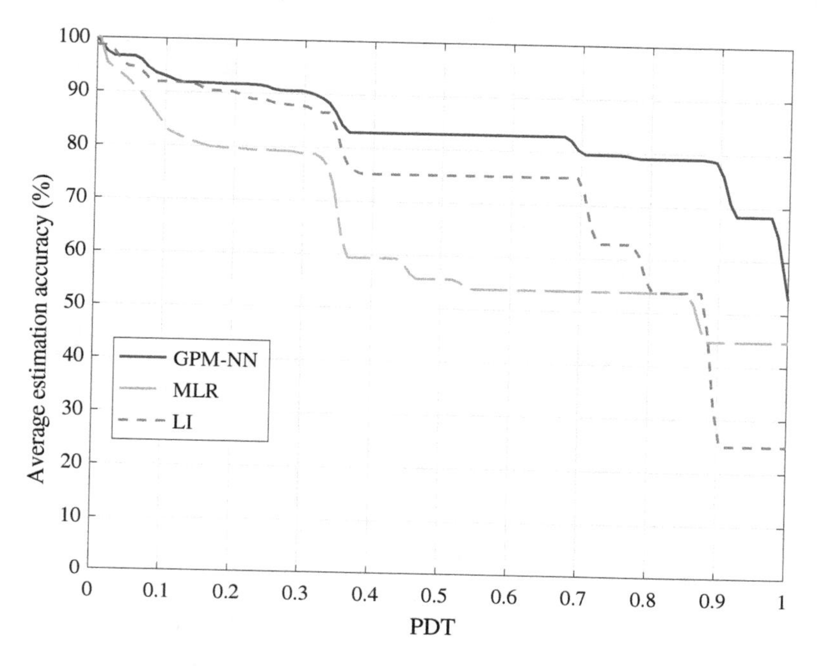

Figure 4.29 Comparison of estimation accuracy between GPM-NN, MLR, and LI in 2D scenario

Effects of Neuron Numbers

As we adopt the NN structure to introduce the non-linear part for our GPM-NN model, the number of neurons in the hidden layer can greatly impact the estimation results. In Fig. 4.30, we plot the estimation accuracy of different number of neurons in GPM-NN via PDT, to study their influence.

From Fig. 4.30, when $PDT < 0.1$, the monitoring contains all cubes. When the number of neurons is 0, our model is equal to the physical model in (4.62) with regression, which only contains the linear part. By comparing this curve with others, we find out that cases with zero neurons are worse than those with nonzero neurons. By adding the nonlinear part (NN structure), GPM-NN performs better with higher accuracy. Moreover, the curve with fewer numbers of neurons (e.g., 10 neurons) performs worse than with more neurons (e.g., 500 neurons). In this scenario, we can find that 1,000 neurons can achieve the highest estimation accuracy. We ignore the situation where the number of neurons is greater than 1,000 because too many neurons in the hidden layer can cause overfitting.

Effects of Various AQI

In Fig. 4.31, we plot the estimation accuracy of GPM-NN with different AQI values (i.e., AQI $\leq$ 50, 50 $\leq$ AQI $\leq$ 200, and AQI $\geq$ 200 [120]), via different PDTs. From the curves, we can find that in a 2D scenario, GPM-NN performs the best when AQI $\geq$ 200. As 50 $\leq$ AQI $\leq$ 200, GPM-NN also maintains high accuracy,

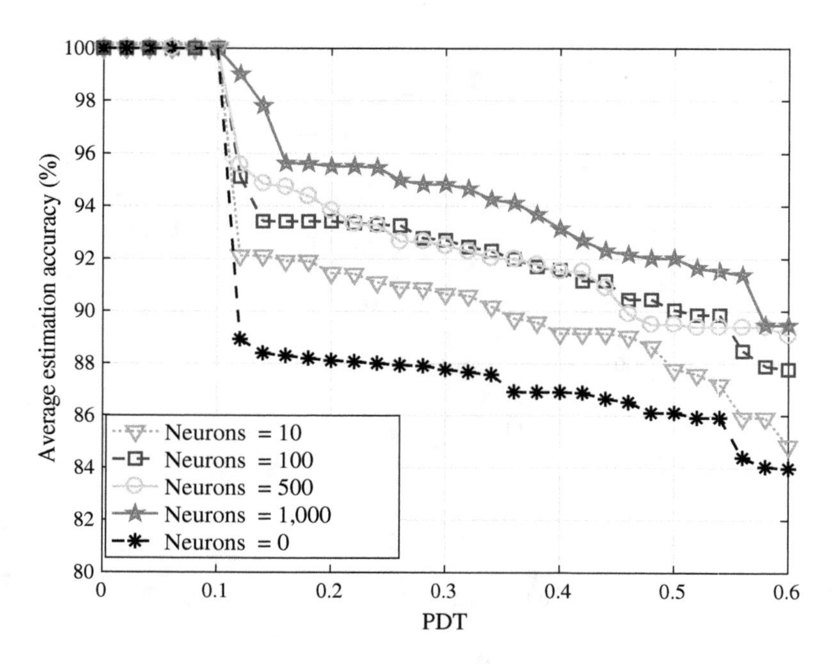

Figure 4.30 Impact of the number of neurons in the nonlinear part in a 2D scenario

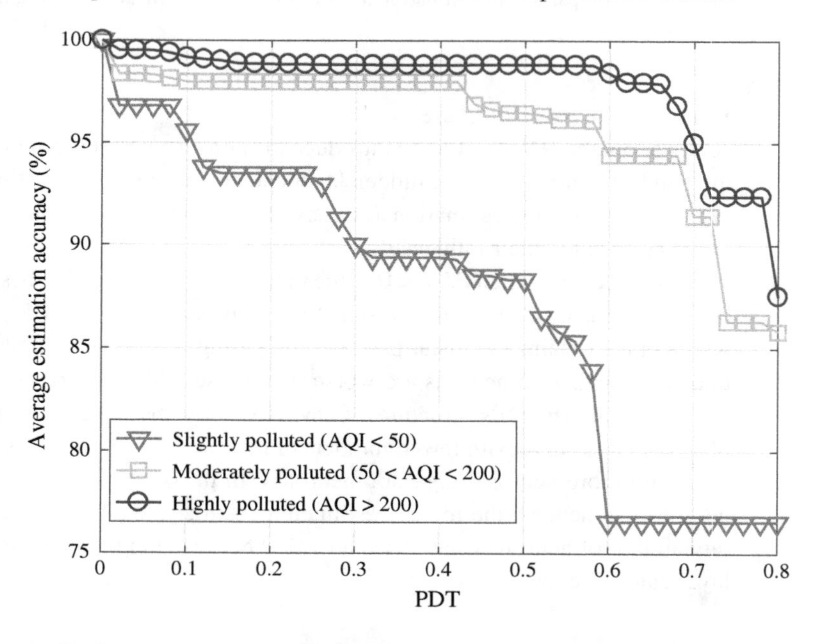

Figure 4.31 Performance of GPM-NN with different AQI values in a 2D scenario

although accuracy is relatively worse when the AQI is low. This indicates that our model is better at predicting on moderately and highly polluted days, which has great significance in forecasting severe pollution as well as prevention. This characteristic is

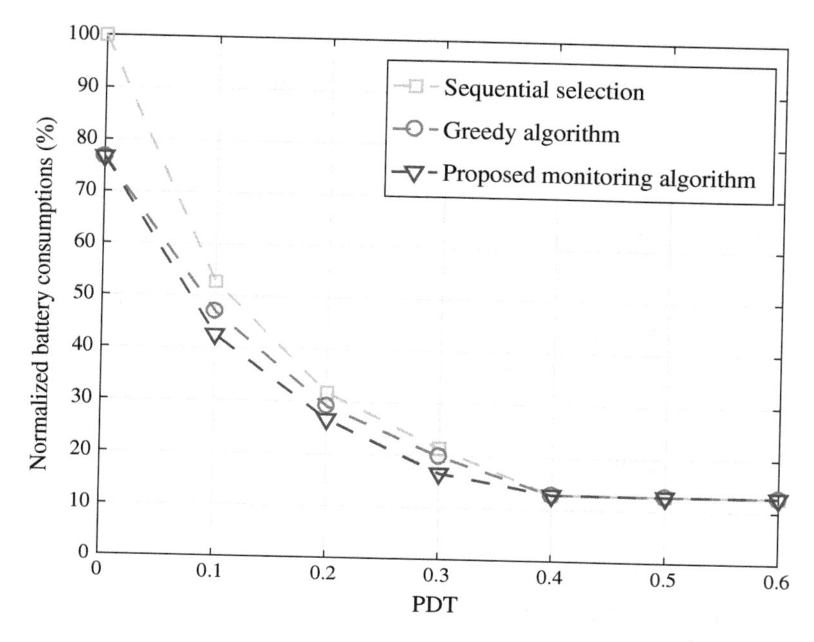

Figure 4.32 Comparison of the adaptive monitoring algorithm, greedy algorithm and sequential selection

also suitable for the adaptive monitoring algorithm when AQI is high. Note that, even though GPM-NN does not perform as well when the AQI is low, it still outperforms other models.

Performance of Adaptive Monitoring Algorithm

In this part, we compare the results of the proposed monitoring algorithm for trajectory planning against those of other algorithms such as the greedy algorithm and sequential selection by plotting their battery consumptions over one measurement in Fig. 4.32. The greedy algorithm aims to select the nearest target cube in $\mathcal{M}$ to generate the trajectory [107], whereas sequential selection is done by selecting cubes from the bottom (or left) to the top (or right) in order [109].

In the typical horizontal open space, we plot the normalized battery consumption achieved by the three algorithms in Fig. 4.32 using different PDTs. The normalized consumption is the cost percentage achieved by each monitoring method of one total battery charge (i.e., 15 minutes). As PDT increases, the consumption would correspondingly decrease because there are fewer target cubes in $\mathcal{M}$. By comparing the three curves, we can see that sequential selection is the most consuming method. Our monitoring algorithm performs the best and is better than the normal greedy algorithm, while $0.1 \leq PDT \leq 0.4$. After PDT reaches 0.4, the consumption of three methods becomes equal, because there is no difference in using these algorithms when the number of target cubes in $\mathcal{M}$ is so low. Hence, the adaptive monitoring

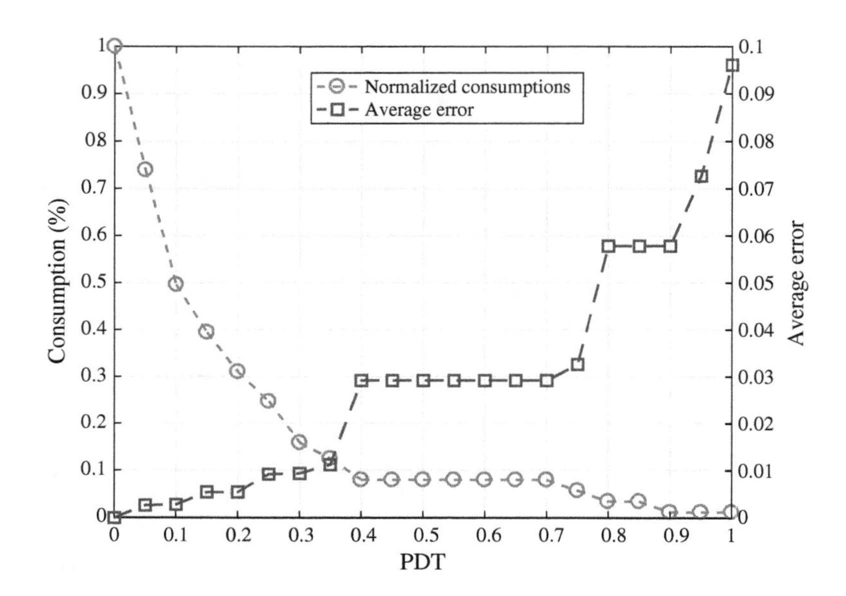

Figure 4.33 Tradeoff between system battery consumption and estimation accuracy in a 2D scenario

algorithm can relatively reduce the power consumption for monitoring the AQI in the 2D scenario.

Tradeoff between Consumption and Accuracy

In Fig. 4.33, we illustrate the tradeoff between the battery consumption and estimation accuracy. To better illustrate the tradeoff, we use *average error* as a merit, expressed as

$$\overline{ERR} = \frac{1}{n} \sum_{i=1}^{n} \left(\frac{\hat{C}_f(i) - C_f(i)}{C_f(i)} \right)^2, \tag{4.79}$$

where n, $\hat{C}_f(i)$ and $C_f(i)$ are the same in (4.74). We plot the curves of the system's power consumption and average estimation error versus PDT.

Figure 4.33 illustrates the relationship between the accuracy and the battery consumption. Intuitively, a larger PDT introduces less power consumption, which proves that with a higher PDT, consumption declines as the number of measured cubes decreases. Moreover, when $PDT \geq 0.4$, the total consumption of the whole system can be reduced by 90%. The rapid decline of consumption is also related to the high redundancy of data in the typical 2D space such as the roadside park. On the other hand, the average error of ARMS increases as PDT becomes larger, which confirms the existence of the tradeoff between power consumption and estimation accuracy. Under this circumstance, choosing $PDT = 0.41$ can achieve a relatively high predicting accuracy (over 80%) while greatly reducing the battery consumption of the system.

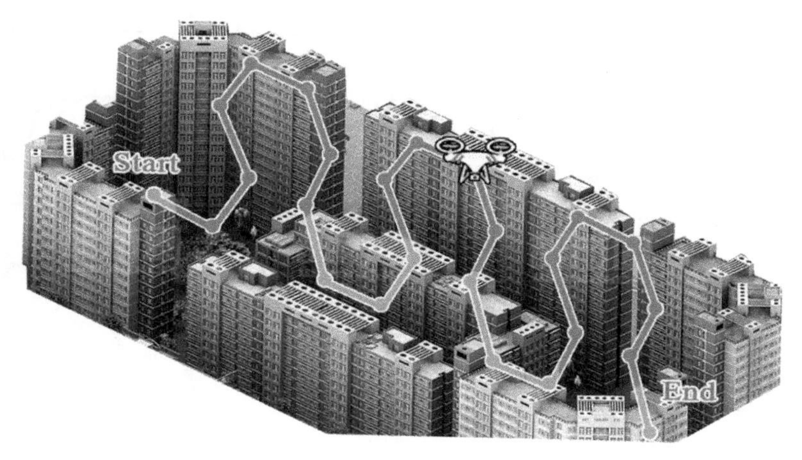

Figure 4.34 Typical application scenario of ARMS in 3D space (courtyard inside a high-rise building)

4.4.5 Example II: Performance Analysis in Vertical Enclosed Space

In this part, we implement the adaptive monitoring algorithm in a typical 3D scenario: a vertical enclosed space. We then analyze the performance of the GPM-NN and the adaptive monitoring algorithm in this typical scenario.

Scenario Description

In the typical 3D scenario, the 3D space has target cubes of various heights. In this type of scenario, the planar area is relatively limited (e.g., the courtyard inside a high-rise building). As shown in Fig. 4.34, in such a vertical enclosed space, there is no significant difference in AQI values between two horizontally neighboring cubes, but the wind may create a discrepancy in the pollutant concentration for two cubes at different heights. Hence, the benefit of selecting more cubes vertically outweighs the cost of traversing between distant cubes at the same heights.

Performance Analysis

In this part, we analyze the performance of ARMS for typical 3D scenario.

In Fig. 4.35, we compare three prediction models. In the vertical enclosed space scenario, GPM-NN still maintains the highest accuracy among the three models using different PDTs. Compared to the 2D scenario, LI decreases rapidly as PDT increases, which indicates that the 3D AQI distribution is heterogenous. Moreover, when $PDT = 0.8$, GPM-NN would experience a violent decline. This phenomenon is caused by the inherent characteristic of PDT. When PDT is high, the corresponding number of target cubes in $\mathscr{M}$ is so low that the predicting accuracy can significantly drop, even if only one point is unmeasured (e.g., 10 cubes with $PDT = 0.75$ and 9 cubes with $PDT = 0.8$). This result can provide the basis for choosing the suitable PDT value.

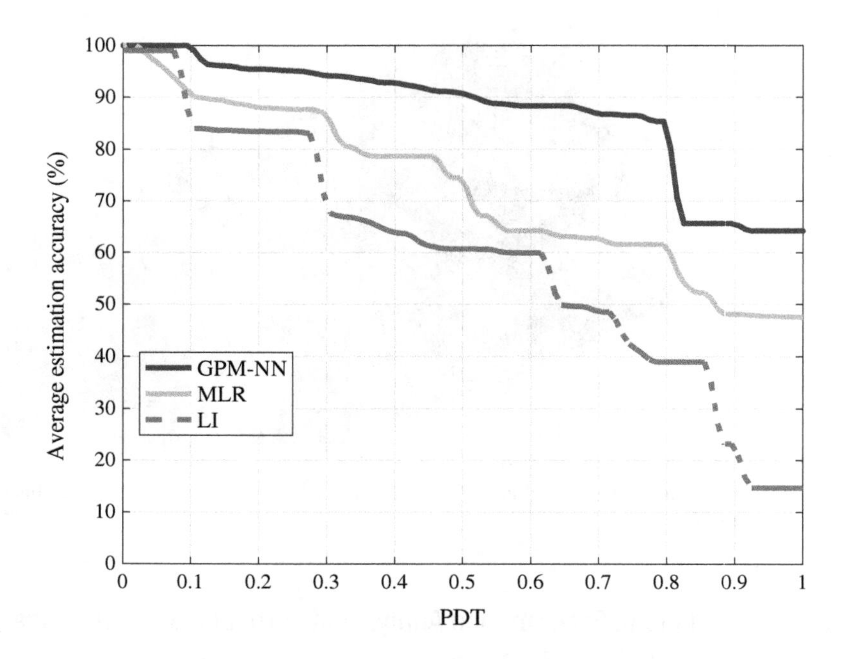

Figure 4.35 Comparison of estimation accuracy between GPM-NN, MLR, and LI in a 3D scenario

In conclusion, GPM-NN performs better in both 2D and 3D fine-grained scenarios, with high estimation accuracy even if there are few measuring cubes.

In Fig. 4.36, we study the effects of the number of neurons in a typical 3D scenario. When $PDT < 0.1$, the result is the same as in the 2D scenario, in that each curve performs the best. As PDT increases, the curve with zero neurons declines most rapidly like that in Fig. 4.30. Also, curves with fewer numbers of neurons (e.g., 100 neurons) perform worse than those with more neurons (e.g., 100 or 1,000 neurons) as well. In this scenario, we can find that 500 neurons achieve the highest estimation accuracy, which is different from the result in the 2D scenario.

Therefore, our GPM-NN model (with a combination of linear and nonlinear parts) is robust and better than only linear models. Moreover, the number of neurons in the hidden layer can effectively influence the model's performance, and the optimal value is different in various scenarios.

In Fig. 4.37, we again plot the estimation accuracy of GPM-NN with different AQI values in the 3D sceanrio. From the curves, we can find that GPM-NN also performs the best when the air is moderately and highly polluted, although it performs relatively worse when the AQI is low.

This verifies that GPM-NN can maintain better estimation accuracy when the AQI value is moderate and high, which is suitable for the operation of our ARMS.

In the 3D scenario, Fig. 4.38 shows the consumption of the three algorithms using different PDTs. From the figure, we can see when PDT is low, sequential selection consumes much more than our method and the greedy algorithm. This indicates that

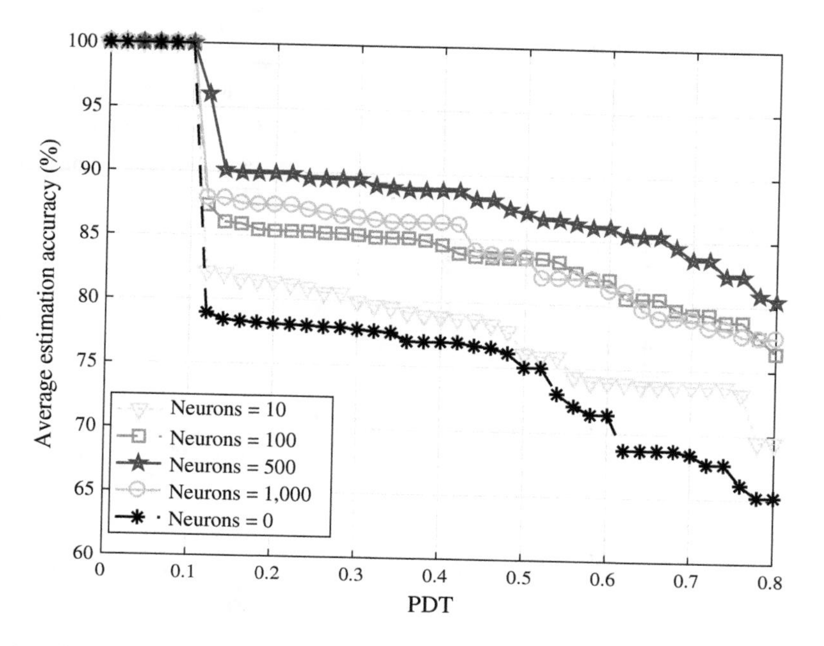

Figure 4.36 Impact of the number of neurons in the nonlinear part in a 3D scenario

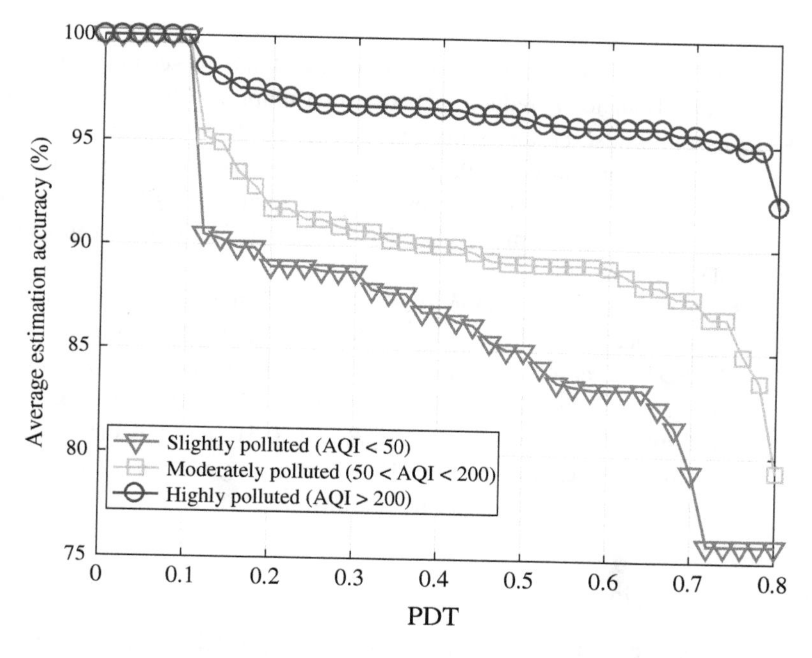

Figure 4.37 Performance of GPM-NN with different AQI values in a 3D scenario

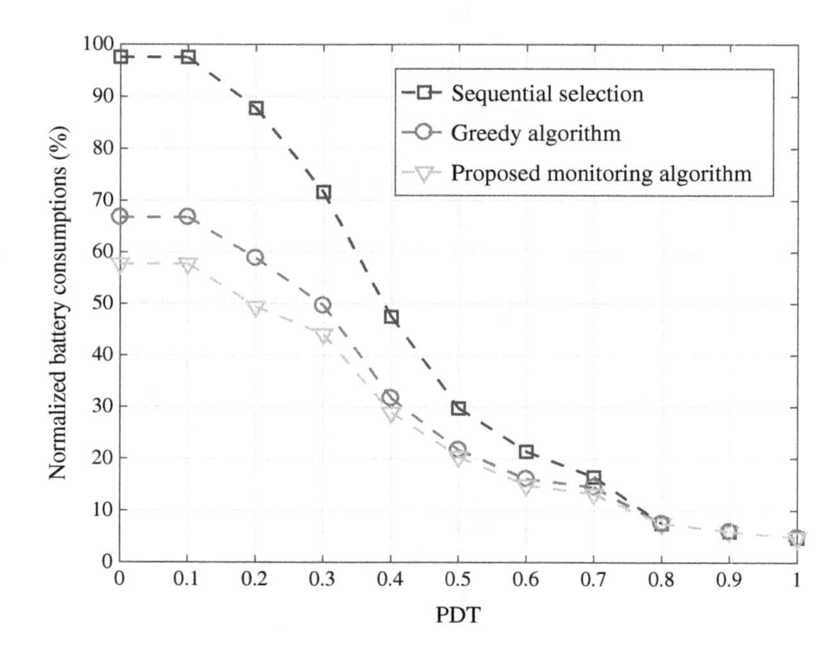

Figure 4.38 Comparison of the adaptive monitoring algorithm, greedy algorithm, and sequential selection

the cube selection can be more complicated in a 3D scenario, and a suitable selection method can highly reduce the battery consumption. Moreover, the adaptive monitoring algorithm also performs the best among the three methods, and it is better than the greedy algorithm when $PDT \leq 0.8$. As PDT increases, the normalized consumption of the three algorithms is closer and is equal when $PDT \geq 0.8$. Thus, the adaptive monitoring algorithm can effectively save the battery life for monitoring the AQI in a 3D scenario.

In Fig. 4.39, we plot the tradeoff in the 3D scenario of a horizontal enclosed space. This typical 3D scenario is more common, and hence the result is more instructive. As PDT increases, the average error grows rapidly as consumption can drop fairly. Given the average error, for example, when $\overline{ERR} = 0.04$ (average estimation accuracy is about 80%), the corresponding $PDT = 0.51$, and thus the power consumption can be reduced to as little as 37%. Hence, choosing a suitable PDT value for monitoring can greatly reduce the measuring effort.

4.4.6 Conclusion

In this section, we designed a UAV sensing system, ARMS, to construct fine-grained AQI maps. A novel fine-grained AQI distribution model GPM-NN has been proposed based on a NN and physical model to help generate a real-time AQI map with data collected by ARMS. To reduce the battery consumption of ARMS, we proposed the adaptive monitoring algorithm to efficiently update real-time AQI maps. For the 2D

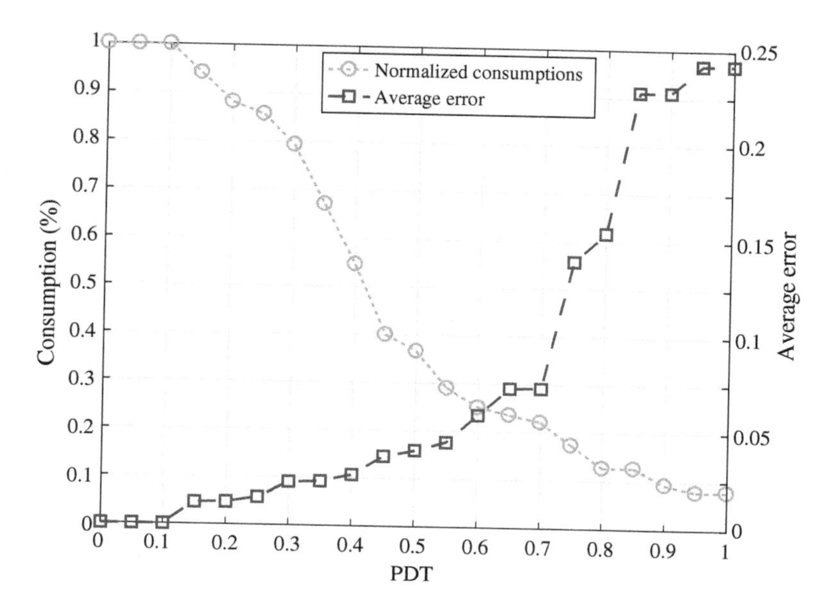

Figure 4.39 Tradeoff between system battery consumption and estimation accuracy in a 3D scenario

and 3D scenarios, we have applied the adaptive monitoring algorithm. By using the proposed index PDT, the system can balance the intrinsic tradeoff between the estimation accuracy and power consumption. Experimental results showed that GPM-NN can achieve a higher accuracy in AQI map construction than other existing models, and the number of neurons in the hidden layer of GPM-NN should be adjusted in various scenarios to acquire better performance. Moreover, the adaptive monitoring algorithm can generate the trajectory while greatly saving the battery life of the UAV, and ARMS can balance the tradeoff between the accuracy of the AQI map and battery consumption.

4.5 Summary

In this chapter, we first presented the system for the cellular internet of UAVs and proposed centralized and decentralized sense-and-send protocols to coordinate multiple UAVs. Then, we introduced the major challenge of a trajectory design in the cellular internet of UAVs and proposed a reinforcement-learning-based method to solve the problem. Next, we proposed a more flexible networking called U2X communications to satisfy the divergent requirements for different sensing applications and addressed key challenges such as resource management and trajectory design. Finally, we reviewed an application of the cellular internet of UAVs on the air-quality sensing and proposed a novel fine-grained AQI distribution model to generate a real-time AQI map.

[illegible] system battery consumption [illegible]

[illegible] scenarios, we have applied the adaptive monitoring [illegible] the system can balance the tradeoff between [illegible] accuracy and power consumption [illegible] higher accuracy [illegible] number of neurons in the hidden layer of [illegible] NN [illegible] better performance [illegible] trajectory while [illegible] balance the tradeoff between [illegible]

4.6 Summary

[illegible] we first presented the system [illegible] of [illegible] centralized and [illegible] sense [illegible] Then, we introduced the main challenges [illegible] of UAVs and proposed [illegible] based [illegible] Next, we proposed a more flexible [illegible] the different requirements [illegible] [illegible]

Part III

HAP Communication Networks

5 Deployment for High Altitude Platform Systems with Perturbation

A high-altitude platform (HAP) is a communication platform deployed in the stratosphere at an altitude of approximately 20 km [138]. It has the characteristics of wide coverage, flexible deployment, low-cost operation, long flight time, residence, and good channel conditions. Therefore, HAP communication systems have attracted considerable attention from academia and industry. Example HAP projects and innovations include Google Loon, HAPSMobile, and Airbus Zephyr [5]. Deployment is an important issue in HAP systems. Authors in [139] jointly optimized resource allocation and HAP deployment to minimize the power consumption in the Internet of remote things (IoRT) networks. In [140], the authors studied HAP deployment design to maximize of the coverage ratio. Authors in [141] proposed a multiobjective deployment method for HAP communication networks to match user demands. However, most of prior works assume that the HAP is stable, which needs more practical consideration.

In practice, because the environment of the stratosphere layer is complicated, the HAP is inevitably affected by short-term airflow, leading to HAP random perturbation with respect to location and angular movements and an undesirable decline in QoS. Several works have studied HAP perturbation, which mainly focuses on the influence on handover performance. In [142], the authors established a coverage geometry model for a HAP in the swing state and analyzed the impact of the swinging state on handover. The authors in [143] discussed both the swing and vertical movements of HAP systems and calculated the handover probability of the two movement modes. Different from the literature, we consider the performance of a single HAP system under HAP perturbation instead of the handover among HAP systems, where the HAP is equipped with an antenna array to serve users on the ground. We aim to maximize the system sum rate by jointly optimizing the location of the HAP and the beamformer.

In this chapter, we first propose a moment-based uncertainty model without a priori knowledge of the perturbation distribution except for the first and second-order statistics for the HAP system in Section 5.1 and investigate the distributionally robust optimization problem for sum-rate maximization under the information transmission chance constraint in Section 5.2. In Section 5.3, we develop an algorithm to tackle the formulated problem efficiently. Finally, we provide a performance evaluation in Section 5.4 and a chapter conclusion in Section 5.5.

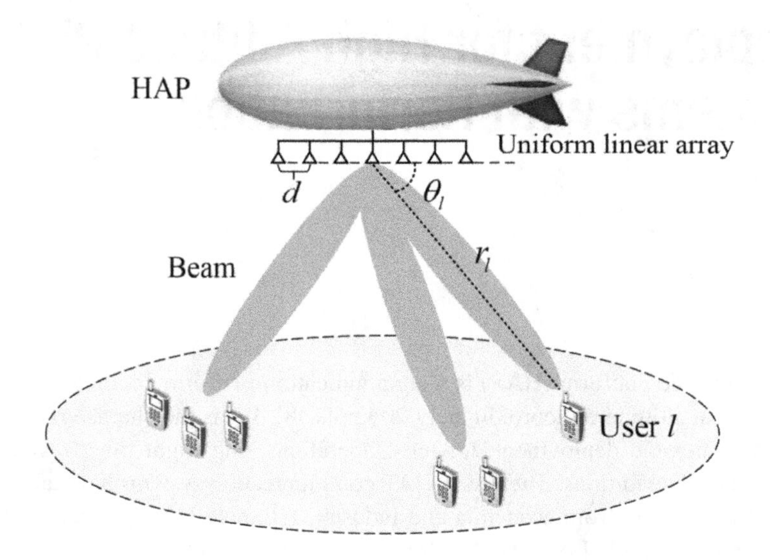

Figure 5.1 HAP-based communication system

5.1 System Model

As illustrated in Fig. 5.1, we consider a downlink HAP-based communication syste where one HAP equipped with a uniform linear array (ULA) is deployed to ser L single-antenna users on the ground, denoted by $\mathscr{L} = \{1,2,...,L\}$. We partition L users into G clusters, where the set of users in cluster g is denoted by $\mathscr{L}_g$, w $\bigcup_{g=1}^{G} \mathscr{L}_g = \mathscr{L}$ and $\sum_{g=1}^{G} |\mathscr{L}_g| = L$. The Cartesian coordinate system is consider where the horizontal coordinates of the HAP and user l are expressed as $\boldsymbol{x}_0 = (x_0, y_0$ and $\boldsymbol{x}_l = (x_l, y_l)^T$, respectively. Moreover, we assume that the HAP hovers at a fix altitude H_0 above the ground.

We assume that the HAP employs a ULA with N_T antenna elements. The downli propagation is composed of both LoS and NLoS components. Therefore, the chan vector between the HAP and user l can be modelled as [144]

$$\mathbf{h}_l = \sqrt{\alpha_l}\left(\sqrt{\frac{K_l}{1+K_l}}\mathbf{a}(\theta_l) + \sqrt{\frac{1}{1+K_l}}\mathbf{b}_l\right), \quad (5$$

where K_l is the Rician factor of user l. The large-scale fading factor is α_l $(4\pi r_l/\lambda)^{-2}$, where $r_l = \sqrt{\|\boldsymbol{x}_0 - \boldsymbol{x}_l\|_2^2 + H_0^2}$ is the distance between the HAP a user l, and λ is the carrier wavelength. Each element of NLoS components $\mathbf{b}_l$ obe $\mathscr{CN}(0,1)$, and the LoS component is given by

$$\mathbf{a}(\theta_l) = \left[1, \exp\left(-j2\pi\frac{d}{\lambda}\cos\theta_l\right), \ldots, \exp\left(-j2\pi(N_T-1)\frac{d}{\lambda}\cos\theta_l\right)\right]^T, \quad (5$$

where θ_l is the angle of departure (AoD) of the path between the ULA and user l, a d denotes the interelement spacing of the antenna array. According to the chan model, the perturbation of the HAP has an impact on the QoS as r_l and θ_l vary with time.

The HAP serves users in each cluster through multicast communication, and thus there is no mutual interference among users in a cluster. For user l in cluster g, the received signal can be modeled as

$$y_l = \mathbf{h}_l^H \mathbf{w}_{gl} s_l + \sum_{l' \in \mathcal{L}_{g'},\, g' \neq g} \mathbf{h}_l^H \mathbf{w}_{g'l'} s_{l'} + n_l, \tag{5.3}$$

where s_l is the information symbol for user l, $\mathbf{w}_{gl}$ is the lth column of the precoder for cluster g $\mathbf{W}_g$, and $n_l \sim \mathcal{CN}(0, \sigma_l^2)$ denotes additive white Gaussian noise. According to (5.3), the received signal to interference plus noise ratio (SINR) at user l can be expressed by

$$\Gamma_l = \frac{\left|\mathbf{h}_l^H \mathbf{w}_{gl}\right|^2}{\sum_{l' \in \mathcal{L}_{g'},\, g' \neq g} \left|\mathbf{h}_l^H \mathbf{w}_{g'l'}\right|^2 + \sigma_l^2}. \tag{5.4}$$

The rate expression for user l in cluster g is then given by

$$R_l = \log_2 (1 + \Gamma_l). \tag{5.5}$$

5.2 Problem Formulation

Our goal is to maximize the system sum rate by jointly optimizing the location of the HAP and the beamformer. In practice, the stability of the HAP is inevitably affected by the random nature of stratospheric wind. In this section, we consider the HAP's pitch and horizontal displacement [145]. The horizontal location of the HAP and the AoD between the HAP and user l with the perturbation can be denoted by $\boldsymbol{x}_0' = \boldsymbol{x}_0 + \Delta \boldsymbol{x} = (x_0 + \Delta x, y_0 + \Delta y)^T$ and $\theta_l' = \theta_l + \Delta\theta$, respectively, where $\Delta \boldsymbol{x} = (\Delta x, \Delta y)^T$ are the $x - y$ location movements, and $\Delta\theta$ denotes the varying pitch angle. However, it is challenging to get the explicit distribution of the HAP perturbation, and only the statistical characteristics can be known. To be specific, we have no prior knowledge of the exact distribution of the HAP random perturbation, and only the first and second-order moments are known [146]. We define an ambiguity set $\mathscr{P}$, which contains all probability distributions on the support of Δx, Δy, and $\Delta\theta$, supported on the same means and variances, as follows:

$$\mathscr{P} = \left\{ \mathbb{P} : \begin{array}{l} \mathbb{E}_{\mathbb{P}}(\Delta x) = \mu_x,\ \mathbb{E}_{\mathbb{P}}(\Delta y) = \mu_y,\ \mathbb{E}_{\mathbb{P}}(\Delta\theta) = \mu_\theta, \\ \mathbb{E}_{\mathbb{P}}[(\Delta x - \mu_x)(\Delta x - \mu_x)^T] = \sigma_x^2, \\ \mathbb{E}_{\mathbb{P}}[(\Delta y - \mu_y)(\Delta y - \mu_y)^T] = \sigma_y^2, \\ \mathbb{E}_{\mathbb{P}}[(\Delta\theta - \mu_\theta)(\Delta\theta - \mu_\theta)^T] = \sigma_\theta^2, \end{array} \right\}, \tag{5.6}$$

where μ_x, μ_y, μ_θ, σ_x^2, σ_y^2, and σ_θ^2 denote the means and variances of Δx, Δy, and $\Delta\theta$ under distribution $\mathbb{P}$, respectively, and $\mathbb{P}$ is an arbitrary distribution as long as its distribution satisfies the mean and variance constraints in (5.6).

We develop a distributionally robust optimization problem with the QoS requirements and HAP perturbation, which can be formulated as

$$\max_{\boldsymbol{x}_0, \mathbf{w}_{gl}} \quad \sum_{l \in \mathscr{L}} R_l \tag{5.7a}$$

$$\text{s.t.} \quad \Pr_{\mathbb{P}}[R_l \geq R_{l,\min}] \geq 1 - \gamma_l, \quad \forall l \in \mathscr{L}, \tag{5.7b}$$

$$\sum_{l \in \mathscr{L}} \mathrm{Tr}(\mathbf{w}_{gl}\mathbf{w}_{gl}^H) \leq P, \tag{5.7c}$$

$$\forall (\Delta x, \Delta y, \Delta\theta) \sim \mathbb{P} \in \mathscr{P}, \tag{5.7d}$$

where $R_{l,\min}$ denotes the minimum data rate required by user l, $\gamma_l \in (0,1)$ is the outage probability of transmission, and P is the total power budget at the HAP. Equation (5.7b) is the information transmission distributionally robust chance constraint, and (5.7c) reflects the power constraint for the HAP.

5.3 Algorithm Design

The optimization problem in (5.7) in uncertain set $\mathscr{P}$ is not computationally tractable. Then, we decompose it into two subproblems: location optimization and beamformer design. In the following, we solve these two subproblems to obtain a suboptimal solution of (5.7).

5.3.1 Location Optimization

In the location optimization subproblem, we do not consider the random perturbation of the HAP. Given precoders $\mathbf{w}_{gl}$, (5.7) can be converted into

$$\max_{\boldsymbol{x}_0} \quad \sum_{l \in \mathscr{L}} R_l \tag{5.8a}$$

$$\text{s.t.} \quad R_l \geq R_{l,\min}, \quad \forall l \in \mathscr{L}. \tag{5.8b}$$

Note that (5.8) is not a convex optimization problem due to the nonconvex objective function and constraint. Therefore, a successive convex optimization approach is applied to obtain the optimal solution approximately. For constraint (5.8b), it can be converted into the difference of two convex functions with respect to $\|\boldsymbol{x}_0 - \boldsymbol{x}_l\|_2^2$ as

$$\begin{aligned} R_l &= \log_2(1 + \Gamma_l) \\ &= \log_2\left(1 + \frac{\frac{(\lambda/4\pi)^2}{\|\boldsymbol{x}_0 - \boldsymbol{x}_l\|_2^2 + H_0^2} E_l}{\frac{(\lambda/4\pi)^2}{\|\boldsymbol{x}_0 - \boldsymbol{x}_l\|_2^2 + H_0^2} E_{l'} + \sigma_l^2}\right) = \tilde{R}_l - \hat{R}_l, \end{aligned} \tag{5.9}$$

where

$$\tilde{R}_l = \log_2\left(\frac{(\lambda/4\pi)^2}{\|\boldsymbol{x}_0 - \boldsymbol{x}_l\|_2^2 + H_0^2}(E_l + E_{l'}) + \sigma_l^2\right), \tag{5.10}$$

$$\hat{R}_l = \log_2\left(\frac{(\lambda/4\pi)^2}{\|\boldsymbol{x}_0 - \boldsymbol{x}_l\|_2^2 + H_0^2} E_{l'} + \sigma_l^2\right), \tag{5.11}$$

$$E_l = \left|\left(\rho_l \mathbf{a}(\theta_l) + \sqrt{1-\rho_l^2}\mathbf{b}_l\right)^H \mathbf{w}_{gl}\right|^2, \tag{5.12}$$

$$E_{l'} = \sum\nolimits_{l' \in \mathcal{L}_{g'}, g' \neq g} \left|\left(\rho_l \mathbf{a}(\theta_l) + \sqrt{1-\rho_l^2}\mathbf{b}_l\right)^H \mathbf{w}_{g'l'}\right|^2, \tag{5.13}$$

$$\rho_l = \sqrt{K_l/(1+K_l)}. \tag{5.14}$$

We define the HAP location in the rth iteration as $\boldsymbol{x}_0^r$, and E_l^r and $E_{l'}^r$ are constants. We note that $\tilde{R}_l$ is convex with respect to $\|\boldsymbol{x}_0 - \boldsymbol{x}_l\|_2^2$, although it is not concave with respect to $\boldsymbol{x}_0$. Thus, with the local point $\boldsymbol{x}_0^r$ in the rth iteration, $\tilde{R}_l$ is lower-bounded by its first-order Taylor expansion as

$$\tilde{R}_l \geq -C_l^r\left(\|\boldsymbol{x}_0 - \boldsymbol{x}_l\|_2^2 - \|\boldsymbol{x}_0^r - \boldsymbol{x}_l\|_2^2\right) + D_l^r \triangleq \tilde{R}_l^{[lb]}, \tag{5.15}$$

where

$$C_l^r = \frac{\frac{(\lambda/4\pi)^2}{(\|\boldsymbol{x}_0^r - \boldsymbol{x}_l\|_2^2 + H_0^2)^2}\left(E_l^r + E_{l'}^r\right)\log_2(e)}{\frac{(\lambda/4\pi)^2}{\|\boldsymbol{x}_0^r - \boldsymbol{x}_l\|_2^2 + H_0^2}\left(E_l^r + E_{l'}^r\right) + \sigma_l^2}, \tag{5.16}$$

$$D_l^r = \log_2\left(\frac{(\lambda/4\pi)^2}{\|\boldsymbol{x}_0^r - \boldsymbol{x}_l\|_2^2 + H_0^2}\left(E_l^r + E_{l'}^r\right) + \sigma_l^2\right), \tag{5.17}$$

are constants. With (5.9) and (5.15), (5.8b) can be transformed into

$$\tilde{R}_l^{[lb]} - \hat{R}_l \geq R_{l,\min}. \tag{5.18}$$

The left side of Equation (5.18) is still nonconvex due to $\hat{R}_l$. Introducing slack variables $\mathbf{S} = \{S_l \leq \|\boldsymbol{x}_0 - \boldsymbol{x}_l\|_2^2, \forall l\}$, (5.18) can be reformulated as

$$\tilde{R}_l^{[lb]} - \log_2\left(\frac{(\lambda/4\pi)^2}{S_l + H_0^2} E_{l'}^r + \sigma_l^2\right) \geq R_{l,\min}. \tag{5.19}$$

Equation (5.19) is a convex constraint because it is jointly concave with respect to $\boldsymbol{x}_0$ and S_l. Because $\|\boldsymbol{x}_0 - \boldsymbol{x}_l\|_2^2$ is convex with respect to $\boldsymbol{x}_0$, its lower bound can be obtained via the first-order Taylor expansion. That is, with the given point $\boldsymbol{x}_0^r$, we have

$$\|\boldsymbol{x}_0 - \boldsymbol{x}_l\|_2^2 \geq \|\boldsymbol{x}_0^r - \boldsymbol{x}_l\|_2^2 + 2(\mathbf{x}_0^r - \boldsymbol{x}_l)^T(\boldsymbol{x}_0 - \boldsymbol{x}_0^r). \tag{5.20}$$

Then, S_l should satisfy

$$S_l \leq \|\boldsymbol{x}_0^r - \boldsymbol{x}_l\|_2^2 + 2(\mathbf{x}_0^r - \boldsymbol{x}_l)^T(\boldsymbol{x}_0 - \boldsymbol{x}_0^r). \tag{5.21}$$

Therefore, with given the HAP location $\mathbf{x}_0^r$ obtained in the rth iteration, (5.8) can be approximated as (5.22) in the $(r+1)$th iteration, which is convex and can be solved by classical optimization techniques.

$$\max_{\boldsymbol{x}_0, \mathbf{S}} \quad \sum_{l\in\mathcal{L}} \left(\tilde{R}_l^{[lb]} - \log_2 \left(\frac{(\lambda/4\pi)^2}{S_l + H_0^2} E_{l'}^r + \sigma_l^2 \right) \right) \tag{5.22}$$
$$s.t. \quad (5.19), (5.21).$$

5.3.2 Beamformer Design Algorithm

Given the HAP stationary location $\boldsymbol{x}_0$, the beamformer design subproblem can be written by

$$\max_{\mathbf{w}_{gl}} \quad \sum_{l\in\mathcal{L}} R_l \tag{5.23}$$
$$s.t. \quad (5.7b), (5.7c), (5.7d).$$

The LoS component with perturbation is given by

$$\mathbf{a}(\theta_l') = \left[1, e^{-j2\pi\frac{d}{\lambda}\cos(\theta_l+\Delta\theta)}, \ldots, e^{-j2\pi(N_T-1)\frac{d}{\lambda}\cos(\theta_l+\Delta\theta)}\right]^T. \tag{5.24}$$

We note that $\mathbf{a}(\theta_l')$ is a nonlinear function with respect to $\Delta\theta$, which complicates a robust algorithm design. To tackle this problem, because $\Delta\theta$ is generally small, we approximate $\mathbf{a}(\theta_l')$ by applying the first-order Taylor expansion:

$$\mathbf{a}(\theta_l') \approx \mathbf{a}^{(0)}(\theta_l) + \mathbf{a}^{(1)}(\theta_l)(\theta_l' - \theta_l), \tag{5.25}$$

where

$$\mathbf{a}^{(0)}(\theta_l) = \left[1, e^{-j2\pi\frac{d}{\lambda}\cos\theta_l}, \ldots, e^{-j2\pi(N_T-1)\frac{d}{\lambda}\cos\theta_l}\right]^T, \tag{5.26}$$

$$\mathbf{a}^{(1)}(\theta_l) = \left[0, j2\pi\frac{d}{\lambda}\sin\theta_l, \ldots, j2\pi(N_T-1)\frac{d}{\lambda}\sin\theta_l\right]^T \circ \mathbf{a}^{(0)}(\theta_l). \tag{5.27}$$

Then, we have

$$\mathbf{a}_l' = \mathbf{a}_l + \Delta\mathbf{a}_l, \tag{5.28}$$

where

$$\mathbf{a}_l \triangleq \mathbf{a}^{(0)}(\theta_l) \text{ and } \Delta\mathbf{a}_l \triangleq \mathbf{a}^{(1)}(\theta_l)\Delta\theta_l. \tag{5.29}$$

The worst-case distributionally robust chance constraint can be represented as

$$\inf_{\mathbb{P}\in\mathcal{P}} \Pr_{\mathbb{P}}\left[\Gamma_l \geq \Gamma_{l,\min}\right] \geq 1 - \gamma_l, \tag{5.30}$$

where $\Gamma_{l,\min} = 2^{R_{l,\min}} - 1$, and

$$\Gamma_l = \frac{\frac{(\lambda/4\pi)^2}{\|\boldsymbol{x}_0' - \boldsymbol{x}_l\|_2^2 + H_0^2} \left| \left(\rho_l \mathbf{a}_l' + \sqrt{1-\rho_l^2}\mathbf{b}_l\right)^H \mathbf{w}_{gl} \right|^2}{\frac{(\lambda/4\pi)^2}{\|\boldsymbol{x}_0' - \boldsymbol{x}_l\|_2^2 + H_0^2} \sum_{\substack{l'\in\mathcal{L}_{g'},\\ g'\neq g}} \left| \left(\rho_l \mathbf{a}_l' + \sqrt{1-\rho_l^2}\mathbf{b}_l\right)^H \mathbf{w}_{g'l'} \right|^2 + \sigma_l^2}. \tag{5.31}$$

In order to transform constraint (5.30) into a tractable form, we decouple the uncertainty variables and rewrite $\Gamma_l \geq \Gamma_{l,\min}$ equivalently as

$$0 \geq \frac{\Gamma_{l,\min}\sigma_l^2}{(\lambda/4\pi)^2}\left(\|(\boldsymbol{x}_0+\Delta\boldsymbol{x})-\boldsymbol{x}_l\|_2^2+H_0^2\right) + \Gamma_{l,\min}\sum_{l'\in\mathcal{L}_{g'},g'\neq g}\left|\left(\rho_l\mathbf{a}_l'+\sqrt{1-\rho_l^2}\mathbf{b}_l\right)^H\mathbf{w}_{g'l'}\right|^2 - \left|\left(\rho_l\mathbf{a}_l'+\sqrt{1-\rho_l^2}\mathbf{b}_l\right)^H\mathbf{w}_{gl}\right|^2. \tag{5.32}$$

By applying (5.28) and (5.29), we can rewrite (5.32) as follows:

$$\begin{aligned}0 \geq{}& \frac{\Gamma_{l,\min}\sigma_l^2}{(\lambda/4\pi)^2}\left((\Delta\boldsymbol{x})^T\Delta\boldsymbol{x}+2(\boldsymbol{x}_0-\boldsymbol{x}_l)^T\Delta\boldsymbol{x}+(\boldsymbol{x}_0-\boldsymbol{x}_l)^T(\boldsymbol{x}_0-\boldsymbol{x}_l)+H_0^2\right)\\&+(\Delta\theta)^2\rho_l^2\left[\mathbf{a}^{(1)}(\theta_l)\right]^H\bar{\mathbf{W}}_l\mathbf{a}^{(1)}(\theta_l)\\&+2(\Delta\theta)\mathrm{Re}\left\{\left(\rho_l\mathbf{a}_l+\sqrt{1-\rho_l^2}\mathbf{b}_l\right)^H\bar{\mathbf{W}}_l\rho_l\mathbf{a}^{(1)}(\theta_l)\right\}\\&+\left(\rho_l\mathbf{a}_l+\sqrt{1-\rho_l^2}\mathbf{b}_l\right)^H\bar{\mathbf{W}}_l\left(\rho_l\mathbf{a}_l+\sqrt{1-\rho_l^2}\mathbf{b}_l\right),\end{aligned} \tag{5.33}$$

where $\bar{\mathbf{W}}_l = \Gamma_{l,\min}\sum_{l'\in\mathcal{L}_{g'},g'\neq g}\mathbf{W}_{g'l'}-\mathbf{W}_{gl}$ and $\mathbf{W}_{gl}=\mathbf{w}_{gl}\mathbf{w}_{gl}^H$. We replace the distributionally robust optimization problem with a tractable semidefinite programming (SDP) problem using the conditional value-at-risk (CVaR) based method [147].

PROPOSITION 5.1 *The worst-case CVaR constraint can be transformed into SDP reformulations.*

Proof We define CVaR of $f(\boldsymbol{\xi})$ at tolerance level $\epsilon \in (0,1)$ with respect to $\mathbb{P}$ as

$$\mathbb{P}-\mathrm{CVaR}_\epsilon\{f(\boldsymbol{\xi})\} = \inf_{\beta\in\mathbb{R}}\left\{\beta+\frac{1}{\epsilon}\mathbb{E}_{\mathbb{P}}[(f(\boldsymbol{\xi})-\beta)^+]\right\}, \tag{5.34}$$

where β is an auxiliary variable related to CVaR. The worst-case distributionally robust chance constraint can be approximated by the tractable worst-case CVaR constraint, which is expressed by

$$\sup_{\mathbb{P}\in\mathcal{P}}\mathbb{P}-\mathrm{CVaR}_\epsilon\{f(\boldsymbol{\xi})\}\leq 0 \Rightarrow \inf_{\mathbb{P}\in\mathcal{P}}\mathrm{Pr}_{\mathbb{P}}\{f(\boldsymbol{\xi})\leq 0\}\geq 1-\epsilon. \tag{5.35}$$

Let $f(\boldsymbol{\xi})=\boldsymbol{\xi}^T\boldsymbol{Q}\boldsymbol{\xi}+\boldsymbol{q}^T\boldsymbol{\xi}+q^0$ for some $\boldsymbol{Q}\in\mathbb{S}^n$, $\boldsymbol{q}\in\mathbb{R}^n$, and $q^0\in\mathbb{R}$. We can recast CVaR as a tractable SDP:

$$\begin{aligned}&\sup_{\mathbb{P}\in\mathcal{P}}\mathbb{P}-\mathrm{CVaR}_\epsilon\{f(\boldsymbol{\xi})\} = \min_{\beta,\boldsymbol{M}}\beta+\frac{1}{\epsilon}\mathrm{Tr}(\boldsymbol{\Omega}\boldsymbol{M})\\&\text{s.t. } \boldsymbol{M}\in\mathbb{S}^{n+1},\beta\in\mathbb{R},\\&\qquad \boldsymbol{M}\succeq\mathbf{0}, \boldsymbol{M}-\begin{bmatrix}\boldsymbol{Q} & \frac{1}{2}\boldsymbol{q}\\ \frac{1}{2}\boldsymbol{q}^T & q^0-\beta\end{bmatrix}\succeq\mathbf{0},\end{aligned} \tag{5.36}$$

where $\boldsymbol{\Omega}$ is a matrix defined as $\boldsymbol{\Omega}=\begin{bmatrix}\boldsymbol{\Sigma}+\boldsymbol{\omega}\boldsymbol{\omega}^T & \boldsymbol{\omega}\\ \boldsymbol{\omega}^T & 1\end{bmatrix}$, and $\boldsymbol{\omega}\in\mathbb{R}^n$ and $\boldsymbol{\Sigma}\in\mathbb{S}^n$ are the mean vector and covariance matrix of random vector $\boldsymbol{\xi}$, respectively. □

With the distributionally robust chance constraint in (5.7b), we define a random vector $\boldsymbol{\xi} = [(\Delta \boldsymbol{x})^T, \Delta\theta]^T$, and the loss function f can be obtained by (5.30) and (5.33). Based on the results in Proposition 5.1, the worst-case distributionally robust chance constraint can be transformed as the following tractable SDP:

$$\begin{cases} \beta + \dfrac{1}{\gamma_l}\mathrm{Tr}(\boldsymbol{\Omega M}) \le 0, \\ \boldsymbol{M} \in \mathbb{S}^4, \beta \in \mathbb{R}, \\ \boldsymbol{M} \succeq \boldsymbol{0}, \boldsymbol{M} - \begin{bmatrix} \boldsymbol{Q} & \frac{1}{2}\boldsymbol{q} \\ \frac{1}{2}\boldsymbol{q}^T & q^0 - \beta \end{bmatrix} \succeq \boldsymbol{0}, \end{cases} \tag{5.37}$$

where $\boldsymbol{M}$ and β are two auxiliary variables, $\boldsymbol{\Omega} = \begin{bmatrix} \boldsymbol{\Sigma} + \boldsymbol{\omega}\boldsymbol{\omega}^T & \boldsymbol{\omega} \\ \boldsymbol{\omega}^T & 1 \end{bmatrix}$, $\boldsymbol{\Sigma} = \begin{bmatrix} \sigma_x^2 & 0 & 0 \\ 0 & \sigma_y^2 & 0 \\ 0 & 0 & \sigma_\theta^2 \end{bmatrix}$, $\boldsymbol{\omega} = [\mu_x, \mu_y, \mu_\theta]^T$, and

$$\boldsymbol{Q} = \begin{bmatrix} \frac{\Gamma_{l,\min}\sigma_l^2}{(\lambda/4\pi)^2}\boldsymbol{I}_2 & \boldsymbol{0} \\ \boldsymbol{0} & \rho_l^2[\mathbf{a}^{(1)}(\theta_l)]^H \bar{\mathbf{W}}_l \mathbf{a}^{(1)}(\theta_l) \end{bmatrix}, \tag{5.38}$$

$$\boldsymbol{q} = \left[\frac{2\Gamma_{l,\min}\sigma_l^2}{(\lambda/4\pi)^2}(\boldsymbol{x}_0 - \boldsymbol{x}_l)^T, 2\mathrm{Re}\left\{\rho_l\left(\rho_l\mathbf{a}_l + \sqrt{1-\rho_l^2}\mathbf{b}_l\right)^H \bar{\mathbf{W}}_l \mathbf{a}^{(1)}(\theta_l)\right\}\right]^T, \tag{5.39}$$

$$\begin{aligned} q^0 = {} & \left(\rho_l\mathbf{a}_l + \sqrt{1-\rho_l^2}\mathbf{b}_l\right)^H \bar{\mathbf{W}}_l \left(\rho_l\mathbf{a}_l + \sqrt{1-\rho_l^2}\mathbf{b}_l\right) \\ & + \frac{\Gamma_{l,\min}\sigma_l^2}{(\lambda/4\pi)^2}\left(\|\boldsymbol{x}_0 - \boldsymbol{x}_l\|_2^2 + H_0^2\right). \end{aligned} \tag{5.40}$$

Furthermore, to immunize the objective function against the distributional ambiguity, a lower-bound $\sum_{l\in\mathscr{L}} R_{l,\min}$ should be considered a worst-case performance over the uncertainty set, which is admissible with respect to the distributionally robust chance constraints while maximizing the objective function $\sum_{l\in\mathscr{L}} R_l$ [148]. Therefore, the beamformer design subproblem can be reformulated as a data rate max-min problem, which is given by

$$\max_{\mathbf{w}_{gl}, R_{l,\min}, \mathbf{M}, \beta} \quad \sum_{l\in\mathscr{L}} R_{l,\min} \tag{5.41a}$$

$$\text{s.t.} \quad \sum_{l\in\mathscr{L}} \mathrm{Tr}(\mathbf{W}_{gl}) \le P, \tag{5.41b}$$

$$\mathbf{W}_{gl} \succeq 0, \ \forall l \in \mathscr{L}, \tag{5.41c}$$

$$(5.37).$$

The convex problem in (5.41) can be efficiently solved by standard convex solvers such as CVX.

5.3.3 Overall Algorithm

Because the stationary position of the HAP will remain unchanged for a period of time, while the beamforming updates are relatively frequent, we can adopt the two-step optimization strategy. To be specific, we solve the location optimization subproblem and the beamformer design subproblem sequentially. The algorithm is illustrated in Algorithm 5.1.

The time complexity of Algorithm 5.1 is polynomial because only two convex subproblems need to be solved. For the HAP location optimization subproblem, the complexity of the convex problem is $O(N)$. As for the beamformer design subproblem, the scale of computational complexity of the convex problem is $O(N_T^2 + \sqrt{Z}\log(\frac{1}{\tau}))$, where Z is the number of constraints, and τ is the inner tolerance threshold for convergence.

Algorithm 5.1 Overall algorithm

1 **Initialization:** Feasible $\boldsymbol{x}_0^0$ and $\mathbf{w}_{gl}^0$, initial iteration index $r = 0$;
2 **HAP location optimization:**
3 **repeat**
4   For given $\boldsymbol{x}_0^r$ and $\mathbf{w}_{gl}^0$, solve convex problem (5.22) to obtain the updated HAP location $\boldsymbol{x}_0^{r+1}$;
5   Update $r = r + 1$;
6 **until** *The increase of the objective function is below a predefined threshold* $\varepsilon > 0$;
7 **Beamformer design:**
8 For given $\boldsymbol{x}_0 = \boldsymbol{x}_0^r$, solve (5.41) to obtain precoders $\mathbf{w}_{gl}$.

5.4 Performance Evaluation

In this section, we evaluate the performance of Algorithm 5.1. The selection of the simulation parameters is based on the existing works [149, 150]. The height of the HAP is considered $H_0 = 20$ km, and the users are randomly distributed within a circle of radius 20 km. The maximum total transmit power P is given as 46 dBm, and the noise variance σ_l^2 is -96 dBm. We set the carrier center frequency to 2.4 GHz and set $d = 4\lambda$, $K_l = 10$ dB, $\mu_x = \mu_y = \mu_\theta = 0$, and $\gamma_l = 0.1$. The K-means clustering algorithm is adopted for user grouping, and the number of clusters is specified in each simulation.

Figure 5.2 illustrates the sum data rate versus σ_θ^2, for $G = 2$. Each cluster has two users, $\sigma_x^2 = \sigma_y^2 = 0.05$, and different numbers of transmit antennas N_T at the HAP. Here, we compare our proposed scheme with two schemes: the nonrobust scheme and the maximum ratio transmission (MRT) scheme. In the nonrobust scheme, the HAP is treated as stable, and the distributionally robust chance constraint will be transformed into determinate constraint. In the MRT scheme, the beamforming vector is set as

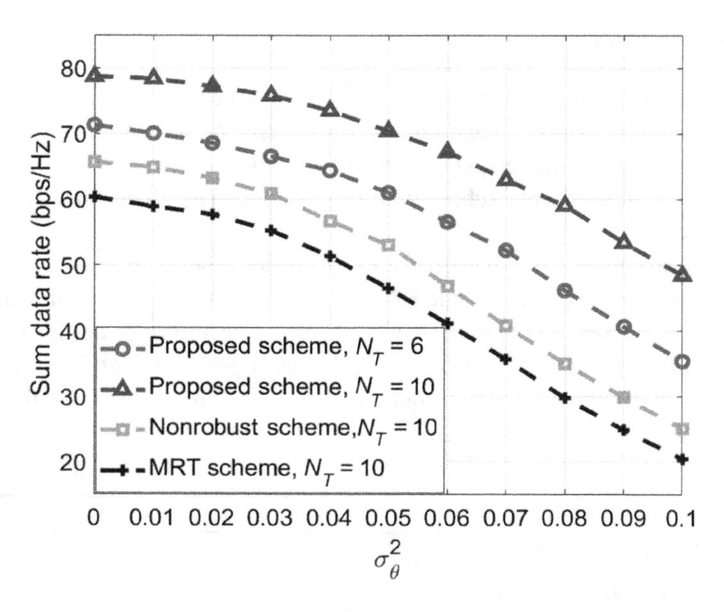

Figure 5.2 Sum data rate versus σ_θ^2

$\mathbf{w}_{gl} = \sqrt{p_{gl}}\mathbf{h}_l \|\mathbf{h}_l\|_2^{-1}$, where p_{gl} and $\mathbf{h}_l$ are the allocated power and the channel vector of user l, respectively. Then, the allocated power and the HAP 2D positioning vector are jointly optimized. We observe that the sum data rate decreases monotonically with increasing σ_θ^2. This is due to the fact that as σ_θ^2 increases, it is more difficult for the HAP to perform accurate downlink beamforming. Moreover, the sum data rate can be improved by increasing the number of HAP antennas because of the extra degrees of freedom provided by the additional antennas. Our proposed scheme outperforms the MRT scheme because the beamforming vector is partially fixed for the MRT scheme. The performance of our proposed scheme is better than that of the nonrobust method, which reveals the distributionally robust method can guarantee the system utility with HAP random perturbation.

Figure 5.3 compares the sum data rate versus G for $N_T = 6$, $L = 8$ and variances σ_x^2, σ_y^2, and σ_θ^2. We observe that for our proposed scheme and the nonrobust scheme the sum rate first rises and then decreases as G grows. It can be seen that there exists an optimal G, which is due to the fact that a maximum sum rate exists. We note that the sum rate is more sensitive to AoD variations because the sum rate in the case of $\sigma_x^2 = \sigma_y^2 = 0.05$, $\sigma_\theta^2 = 0.01$ is higher than that in the case of $\sigma_x^2 = \sigma_y^2 = 0.01$, $\sigma_\theta^2 = 0.05$. Both the nonrobust scheme and the MRT scheme result in a lower sumrate compared to our proposed robust scheme for the entire considered range of G.

Figure 5.4 depicts the sum data rate versus the number of users L for $G = 2$, $N_T = 6$, and $\sigma_\theta^2 = \sigma_x^2 = \sigma_y^2 = 0.05$. As we can see, the sum data rate first increases as more users get access to the network and then decreases due to the growth of the mutual interference brought on by the increased users. Moreover, the performance of our proposed scheme is better than that of the non-robust scheme and the MRT scheme.

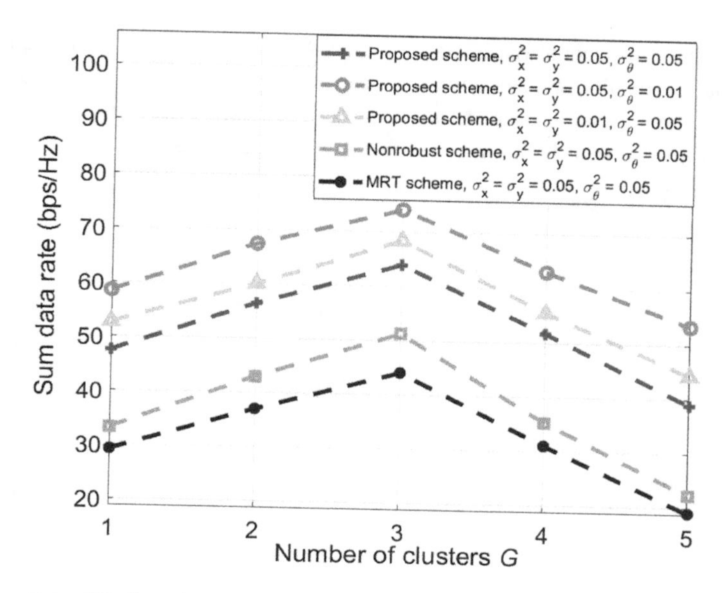

Figure 5.3 Sum data rate versus G

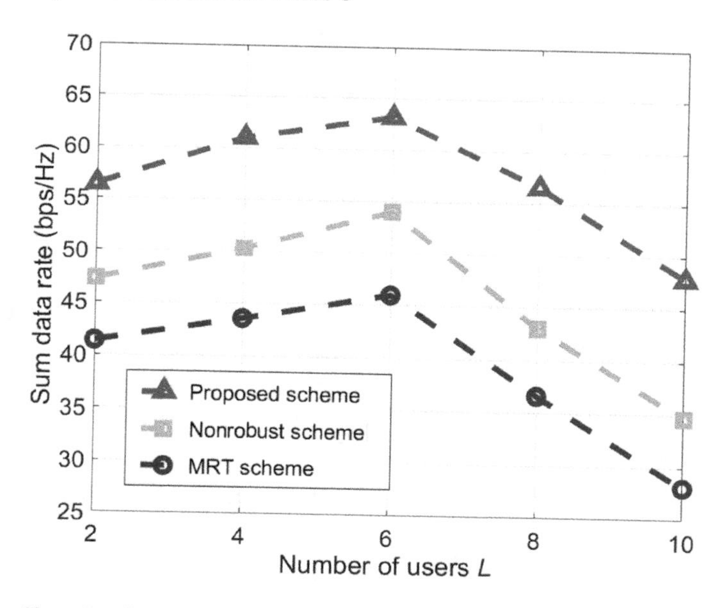

Figure 5.4 Sum data rate versus L

5.5 Conclusion

In this section, we studied a HAP system with HAP perturbation and addressed a distributionally robust optimization problem that contains HAP location optimization and beamformer design to maximize the system sum rate. Simulation results compared the robustness performance of our proposed scheme with that of the benchmark schemes, revealing that our scheme can guarantee QoS with HAP random perturbation. From the simulation results, we can conclude that there exists an optimal number of cluster G given the number of users. Moreover, the system sum rate is more sensitive to the AoD variation than the location uncertainty, which suggests that the AoD estimation is important.

6 Cooperative HAP and LEO Satellite Schemes for Data Collection and Transmission

In the past few years, 5G communication technology has entered commercial applications, and the 6G communication system has gained significant interest from both academia and industry [151–153]. For 6G technologies, the AAN provides global coverage and ubiquitous service for the remote user equipment in depopulated areas, such as oceans and deserts, that cannot be served by general ground base stations due to geographical factors [154, 155]. The AAN is composed of satellites in space, aircraft in the air, ground stations and a data-processing center (DPC) on the ground, and terrestrial users, especially in the remote areas. Data from remote area users can be collected and transmitted back to Earth by cooperating aircraft and satellites in an AAN.

In this chapter, we focus on the remote area users' data collection and transmission via the integration of HAPs and LEO satellites in an AAN, and Fig. 6.1 presents a concrete AAN scenario. Specifically, the data of terrestrial users collected by HAPs can be transmitted by the LEO satellite networks back to the ground DPC. Although LEO satellites can directly provide services for terrestrial users, the high-speed movement of LEO satellites results in intermittent and unstable connections between terrestrial users and LEO satellites. Consequently, we consider utilizing HAPs to fill the gap between remote area users and satellites. Note that the connections between HAPs and LEO satellites are dynamic due to the satellites' movements, and thus the storage capability of HAPs is leveraged to store the collected data if there is no visible LEO satellite for the HAP in the current time slot. The data is stored and carried by the HAP to the next time slot and then forwarded to the visible LEO satellite. Similarly, the intermittent connections among different LEO satellites should also be tackled: The data will be stored and carried by the LEO satellite if it lacks other visible satellites or ground stations for transmission in the current time slot, and the data will be forwarded to other visible satellites or ground stations in the next time slot. In short, the store-carry-forward mechanism [156] of HAPs and LEO satellites in an AAN is leveraged.

To maximize the total data received in the ground DPC while considering the contact plan restrictions, multiple resource capacity limitations, and corresponding flow constraints, we propose an Benders decomposition based optimal algorithm (BDOA) to cope with the original problem to acquire the optimal solution with limited complexity. To further alleviate the solution complexity in large-scale networks, we propose an acceleration algorithm, including an approximation algorithm for the master problem (AAMP) and a unit-flow-based shortest-path algorithm (UFSP) for

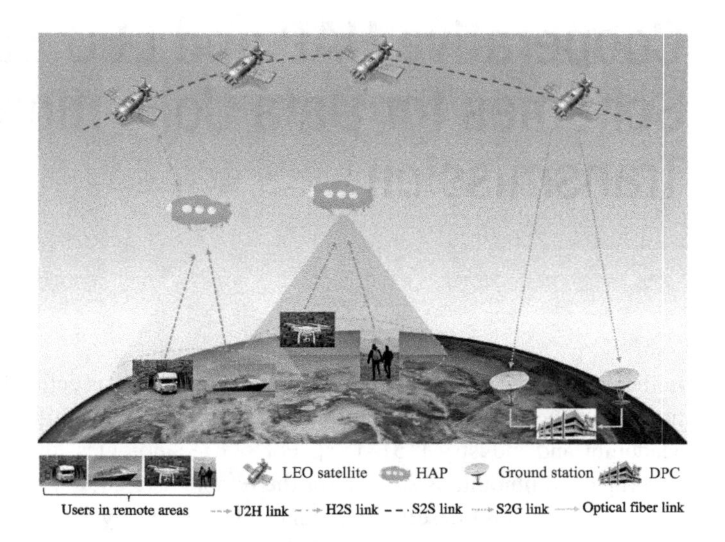

Figure 6.1 AAN scenario

the subproblem. Also, two tips including the ϵ-optimal mechanism and a limited iteration step are introduced in the acceleration algorithm.

The rest of this chapter is organized as follows. In Section 6.1, we present the system model and corresponding problem formulation. In Section 6.2, we detail the algorithm design and corresponding analysis. Simulations and numerical results are provided in Section 6.3. Finally, conclusions are drawn in Section 6.4.

6.1 System Model and Problem Formulation

In this section, we propose the AAN system model and corresponding problem formulation. In Section 6.1.1, the AAN scenario is described. Then, we present the time expanding graph (TEG) model in Section 6.1.2. Moreover, the specific channel model and energy cost model are discussed in Sections 6.1.3 and 6.1.4, respectively. Finally, the problem formulation is detailed in Section 6.1.5.

6.1.1 AAN Scenario

As shown in Fig. 6.1, we propose a comprehensive AAN scenario composed of HAPs, LEO satellites, ground stations, and a DPC to provide services for terrestrial users. The users include individuals, vehicles in depopulated areas, ships in the sea, low-altitude UAVs, and so forth. Data from these users are collected and transmitted by HAPs and LEO satellite networks to the ground station and are forwarded to the DPC by high-speed optical fiber cables.

Specifically, the AAN includes the set of LEO satellites $\mathscr{S}$, the set of HAPs $\mathscr{H}$, the set of remote area users $\mathscr{U}$, the set of ground stations $\mathscr{G}$, and DPC D. Data from user $u \in \mathscr{U}$ with a data size of σ_u will be collected and transmitted back to the DPC. In particular, the user data is first collected by HAPs through the user-to-HAP link (U2H), and then the HAP relays the received data to LEO satellites by the HAP-to-satellite link (H2S). The LEO satellite will transmit the received data to another LEO satellite leveraging the satellite-to-satellite link (S2S) or to the ground station through the satellite-to-ground link (S2G). Finally, the data that arrives at the ground station will be transmitted to the DPC by a high-speed wired mode, and the transmission capacity is deemed unlimited. Note that if it lacks an active H2S, S2S, or S2G link, the data is stored and carried by the HAP or LEO satellite until there is a chance for transmission, and it is the store-carry-and-forward mechanism in time-varying networks. In addition, key notations used in this work are listed in Table 6.1.

6.1.2 TEG Model

General Illustration of TEG

Due to the high-speed motion of LEO satellites, the connections between HAPs and LEO satellites, LEO satellites and LEO satellites, as well as LEO satellites and ground stations change with time. Fortunately, thanks to the periodic movement of LEO satellites, these connections are predictable, so we can leverage the TEG to represent the dynamic but predictable resources as well as demands in the AAN. In particular, we consider a time horizon $\mathscr{T}$ that is larger than the period of LEO satellites. A simple TEG case is shown in Fig. 6.2, in which $\mathscr{T}$ is divided into T time slots and the length of each time slot is τ, $t \in T$. The TEG in Fig. 6.2 is composed of two users u^1 and u^2, two HAPs h^1 and h^2, two LEO satellites s^1 and s^2, two ground stations g^1 and g^2, and a DPC D, within three continuous time slots. Besides, to precisely representing the dynamic AAN, the corresponding TEG is denoted as follows.

Here, $\mathbb{G}_{\mathscr{T}} = \{\mathscr{V}, \mathscr{A}\}$, including vertices $\mathscr{V} = \mathscr{V}_u \cup \mathscr{V}_h \cup \mathscr{V}_s \cup \mathscr{V}_g \cup \mathscr{V}_d$ and links $\mathscr{A} = \mathscr{A}_{uh} \cup \mathscr{A}_{hs} \cup \mathscr{A}_{ss} \cup \mathscr{A}_{sg} \cup \mathscr{A}_{gd} \cup \mathscr{A}_s$. The vertices $\mathscr{V}_s = \{i_t | i \in \mathscr{S}, 1 \le t \le T\}$, $\mathscr{V}_h = \{i_t | i \in \mathscr{H}, 1 \le t \le T\}$, $\mathscr{V}_u = \{i_t | i \in \mathscr{U}, 1 \le t \le T\}$, $\mathscr{V}_g = \{i_t | i \in \mathscr{G}, 1 \le t \le T\}$, and $\mathscr{V}_d = \{i_t | i \in D, 1 \le t \le T\}$ are the replicas of LEO satellites, HAPs, users, ground stations, and the DPC in the TEG, respectively. Links $\mathscr{A}_{uh} = \{(i_t, j_t) | (i_t, j_t) \in \text{U2H}, 1 \le t \le T\}$, $\mathscr{A}_{hs} = \{(i_t, j_t) | (i_t, j_t) \in \text{H2S}, 1 \le t \le T\}$, $\mathscr{A}_{ss} = \{(i_t, j_t) | (i_t, j_t) \in \text{S2S}, 1 \le t \le T\}$, $\mathscr{A}_{sg} = \{(i_t, j_t) | (i_t, j_t) \in \text{S2G}, 1 \le t \le T\}$, and $\mathscr{A}_{gd} = \{(i_t, j_t) | (i_t, j_t) \in \text{G2D}, 1 \le t \le T\}$ denote the connections of U2H, H2S, S2S, S2G, and G2D in the TEG, respectively. Lastly, $\mathscr{A}_s = \{(i_t, i_{t+1}) | i_t \in \mathscr{V}_h \cup \mathscr{V}_s, 1 \le t \le T - 1\}$ denotes the storage links of HAPs or LEO satellites from time slot t to the next time slot $t+1$ in the TEG.

Because G2D connections between ground stations and the DPC are high-speed optical fiber, the G2D capacity is deemed unlimited, and $\mathscr{A}$ is simplified as $\mathscr{A} = \mathscr{A}_{uh} \cup \mathscr{A}_{hs} \cup \mathscr{A}_{ss} \cup \mathscr{A}_{sg} \cup \mathscr{A}_s$. If the data from a user cannot complete transmission to a HAP in one time slot, the user data is decomposed into a set of small-sized data so that the decomposed data can complete transmission in the same time slot. As such, the user replicas in successive time slots send small-sized data from $i_t \in \mathscr{V}_u$ to the HAP $j_t \in \mathscr{V}_h$. An example data flow is depicted in Fig. 6.2: The data from user u_1^1 is collected by

Table 6.1 Key notations

Notations	Parameters
$\mathscr{S}$	Set of LEO satellite
$\mathscr{H}$	Set of HAP
$\mathscr{U}$	Set of user in the remote area
$\mathscr{G}$	Set of ground station
D	Data processing center
$\mathbb{G}_{\mathscr{T}}$	TEG, $\mathbb{G}_{\mathscr{T}} = \{\mathscr{V}, \mathscr{A}\}$
$\mathscr{T}$	Schedule horizon
T	Time slot numbers, $t \in T$
τ	Length of a time slot
$\mathscr{V}$	Vertices in TEG
$\mathscr{A}$	Links in TEG
$\mathscr{V}_s$	Vertices of LEO satellite in TEG
$\mathscr{V}_h$	Vertices of HAP in TEG
$\mathscr{V}_u$	Vertices of users in TEG
$\mathscr{V}_g$	Vertices of ground station in TEG
$\mathscr{V}_d$	Vertices of data processing center in TEG
i_t	Node in TEG, $i_t \in \mathscr{V}$
$\mathscr{A}_{uh}$	U2H links in TEG
$\mathscr{A}_{hs}$	H2S links in TEG
$\mathscr{A}_{ss}$	S2S links in TEG
$\mathscr{A}_{sg}$	S2G links in TEG
$\mathscr{A}_{gd}$	G2D links in TEG
$\mathscr{A}_s$	Storage links in TEG
(i_t, j_t)	Link in TEG, $(i_t, j_t) \in \mathscr{A}$
$c(i_t, j_t)$	Data rate of link $(i_t, j_t) \in \mathscr{A} \backslash \mathscr{A}_s$
$\mathscr{C}(i_t, j_t)$	Maximum containable data size of link (i_t, j_t)
$\mathscr{C}_{i_t}$	Storage capacity of i_t, $i_t \in \mathscr{V}_h \cup \mathscr{V}_s$
$E_{i^t}^{re,s}$	Receiver energy cost of LEO satellites
$E_{i^t}^{tr,s}$	Transmitter energy cost of LEO satellites
$E_{i^t}^{re,h}$	Receiver energy cost of HAPs
$E_{i^t}^{tr,h}$	Transmitter energy cost of HAPs
EB_s	Energy capacity of LEO satellites
EB_h	Energy capacity of HAPs
α_h	Battery maximum depth of discharge of HAPs
α_s	Battery maximum depth of discharge of LEO satellites
N	Maximum number of users accommodated by a HAP
σ_u	Data size of traffic flow from user u
f_u	Data flow of user u in TEG
$\mathscr{F}$	Data flow set of all users, $\mathscr{F} = \{f_u \mid 1 \leq u \leq \lvert\mathscr{U}\rvert\}$
ρ	Size of each unit flow
N_ρ^u	Number of unit flows corresponding to user u
Ω	Set of unit flows
$w(i_t, j_t)$	Weight of link (i_t, j_t) in TEG
$\Delta(i_t, j_t)$	Total amount of unit flow already loaded on link (i_t, j_t)

Table 6.1 (*conti.*)

	Variables
$x(i_t, j_t)$	Binary variable indicating whether link $(i_t, j_t) \in \mathscr{A}$ is active, $x(i_t, j_t) \in \{0, 1\}$
$f_u(i_t, j_t)$	Continuous variable denoting the task flow data of user u on link $(i_t, j_t) \in \mathscr{A}$, $f_u(i_t, j_t) \in [0, \sigma_u]$
$s_u(i_t)$	Continuous variable representing the traffic flow of user $u \in \mathscr{V}_u$ stored in node $i_t \in \mathscr{V}_h \cup \mathscr{V}_s$, $s_u(i_t) \in [0, \sigma_u]$

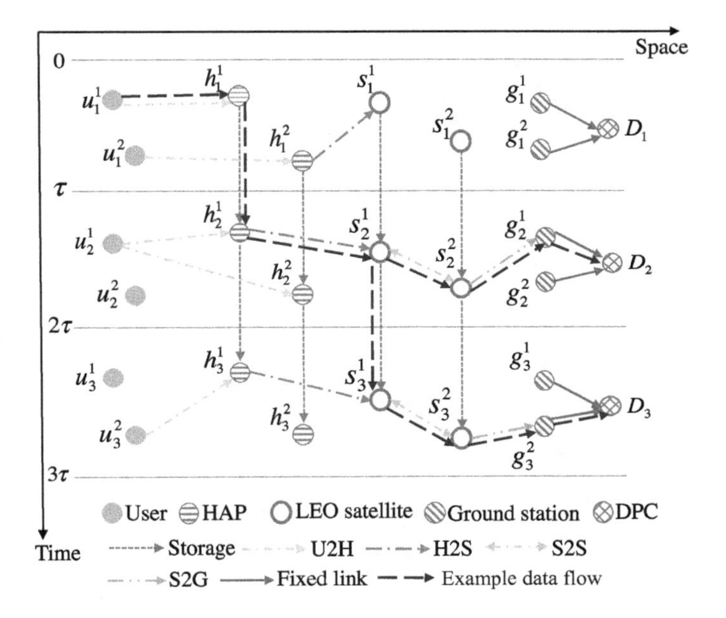

Figure 6.2 TEG of AAN

HAP h_1^1 in the first time slot and stored in the current HAP to the second time slot, h_2^1. Then, the data in h_2^1 is transmitted to LEO satellite s_2^1, and the data flow from s_2^1 is split into two parts: one part is transmitted to LEO satellite s_2^2, and another part is stored in s_2^1 to the third time slot, s_3^1. The data received in LEO satellite s_2^2 is transmitted to the ground station g_2^1 in the second time slot, while the data in LEO satellite s_3^1 is forwarded to LEO satellite s_3^2 and then to the ground station g_3^2 in the third time slot. Finally, the data received at g_2^1 and g_3^2 are transmitted to DPC D_2 and D_3, respectively.

Contact Model in TEG

Because the AAN is dynamic and the contact relationships among various nodes in the TEG change over time, let binary variable $x(i_t, j_t)$ indicate whether link $(i_t, j_t) \in \mathscr{A}$ is active. If $x(i_t, j_t) = 1$, link $(i_t, j_t) \in \mathscr{A}$ is active; otherwise, $x(i_t, j_t) = 0$:

$$x(i_t, j_t) = \begin{cases} 1, & \text{link } (i_t, j_t) \in \mathscr{A} \text{ is active,} \\ 0, & \text{otherwise.} \end{cases} \tag{6.1}$$

Due to the antenna limitation of terrestrial users, each user can connect to only one HAP in each time slot:

$$\sum_{(i_t, j_t) \in \mathscr{A}_{uh}} x(i_t, j_t) \leq 1, \quad \forall i_t \in \mathscr{V}_u, t \in T. \tag{6.2}$$

Similarly, we assume that each HAP can connect to only one LEO satellite in each time slot:

$$\sum_{(i_t, j_t) \in \mathscr{A}_{hs}} x(i_t, j_t) \leq 1, \quad \forall i_t \in \mathscr{V}_h, t \in T. \tag{6.3}$$

Suppose N is the maximum number of users that a HAP can accommodate in each time slot, so there exists

$$\sum_{(i_t, j_t) \in \mathscr{A}_{uh}} x(i_t, j_t) \leq N, \quad \forall j_t \in \mathscr{V}_h, t \in T, \tag{6.4}$$

which implies that there exists no collision if the total number of users trying to access to a HAP in one time slot is less than N.

In addition, assume that each LEO satellite can connect with only one HAP in each time slot, and each LEO satellite can only connect to one other LEO satellite in each time slot, so we have

$$\sum_{(i_t, j_t) \in \mathscr{A}_{hs}} x(i_t, j_t) \leq 1, \quad \forall j_t \in \mathscr{V}_s, t \in T, \tag{6.5}$$

and

$$\sum_{(i_t, j_t) \in \mathscr{A}_{ss}} x(i_t, j_t) \leq 1, \quad \forall i_t \in \mathscr{V}_s, t \in T. \tag{6.6}$$

Following [156] and [157], we assume that each LEO satellite can connect to only one ground station, and each ground station can only connect to one satellite in each time slot due to antenna restrictions:

$$\sum_{(i_t, j_t) \in \mathscr{A}_{sg}} x(i_t, j_t) \leq 1, \quad \forall i_t \in \mathscr{V}_s, t \in T, \tag{6.7}$$

and

$$\sum_{(i_t, j_t) \in \mathscr{A}_{sg}} x(i_t, j_t) \leq 1, \quad \forall j_t \in \mathscr{V}_g, t \in T. \tag{6.8}$$

Data Flow Model in the TEG

Because the data flow in the TEG can be split as in Fig. 6.2: Variable $f_u(i_t, j_k)$ is used to denoting the data flow (i.e., the data amount of user u on link $(i_t, j_k) \in \mathscr{A}$, which is composed of the data flow $f_u(i_t, j_t)$ on the transmission links and the data flow $f_u(i_t, i_{t+1})$ on the storage arcs). The set of flows $\mathscr{F} = \{f_u | 1 \leq u \leq |\mathscr{U}|\}$ includes the flows in the TEG for all users. Particularly, $f_u(i_t, i_{t+1})$ indicates the data amount stored in $i_t \in \mathscr{V}$ at the end of time slot t for f_u because there are no active links to transmit f_u in time slot t.

For a HAP $i_t \in \mathscr{V}_h$ in the TEG, the flow conservation constraint should be satisfied:

$$\sum_{(j_t, i_t) \in \mathscr{A}_{uh}} \sigma_u x(j_t, i_t) = \sum_{(i_t, j_t) \in \mathscr{A}_{hs}} f_u(i_t, j_t) + s_u(i_t),$$
$$\forall i_t \in \mathscr{V}_h, f_u \in \mathscr{F}, t \in T, \quad (6.9)$$

in which $s_u(i_t)$ indicates the data flow of user u stored in node $i_t \in \mathscr{V}_h \cup \mathscr{V}_s$ (i.e., the storage capability of HAPs and LEO satellites are employed). It is noted that $s_u(i_t) = f_u(i_t, i_{t+1})$.

Besides, the flow conservation constraint on LEO satellite $i_t \in \mathscr{V}_s$ in the TEG is

$$\sum_{(j_t, i_t) \in \mathscr{A}_{hs} \cup \mathscr{A}_{ss}} f_u(j_t, i_t) + s_u(i_{t-1})$$
$$= \sum_{(i_t, j_t) \in \mathscr{A}_{ss} \cup \mathscr{A}_{sg}} f_u(i_t, j_t) + s_u(i_t), \quad \forall i_t \in \mathscr{V}_s, f_u \in \mathscr{F}, t \in T, \quad (6.10)$$

in which $s_u(i_{t-1}) = 0$ when $t = 1$, and $s_u(i_T)$ signifies the data amount stored in $i_T \in \mathscr{V}_s$ at the end of time slot T. Moreover, because the storage capacity of HAPs and LEO satellites is limited, the total data flow stored in node $i_t \in \mathscr{V}_h \cup \mathscr{V}_s$ cannot violate the storage capacity:

$$\sum_{f_u \in \mathscr{F}} s_u(i_t) \leq \mathscr{C}_{i_t}, \quad \forall i_t \in \mathscr{V}_h \cup \mathscr{V}_s, t \in T, \quad (6.11)$$

where $\mathscr{C}_{i_t}$ denotes the storage capacity of HAPs ($i_t \in \mathscr{V}_h$) or LEO satellites ($i_t \in \mathscr{V}_s$).

6.1.3 Channel Model

According to [158, 159], the attainable data rate of link $(i_t, j_t) \in \mathscr{A} \backslash \mathscr{A}_s$ is

$$c(i_t, j_t) = \frac{P^{tr}_{i_t j_t} G^{tr}_{i_t j_t} G^{re}_{i_t j_t} L_{i_t j_t} L_l}{k_B T_s (E_b/N_0)_{req} M}, \quad \forall (i_t, j_t) \in \mathscr{A} \backslash \mathscr{A}_s, t \in T, \quad (6.12)$$

in which $L_{i_t j_t}(t) = (\frac{c}{4\pi \cdot d_{i_t j_t} \cdot \nu_{i_t j_t}})^2$ is the free space loss, $P^{tr}_{i_t j_t}$ is the transmission power (in watts) from node i_t to node node j_t, and $G^{tr}_{i_t j_t}$ and $G^{re}_{i_t j_t}$ are the transmission antenna gain of node i_t and the receiver antenna gain of node j_t, respectively. Here, L_l is the total line loss, k_B is the Boltzmann's constant, and T_s is the system noise temperature. Also, $(E_b/N_0)_{req}$ is the required ratio of received energy per bit to noise density, M is the link margin, c is the speed of light, $d_{i_t j_t}$ is the maximum slant range, and $\nu_{i_t j_t}$ is the center frequency. Therefore, the maximum data size that can be transmitted in a time slot is

$$\mathscr{C}(i_t, j_t) = c(i_t, j_t) \cdot \tau, \quad \forall (i_t, j_t) \in \mathscr{A} \backslash \mathscr{A}_s, t \in T. \quad (6.13)$$

Also recall that τ is the length of a time slot.

Because the communication capacity of link $(i_t, j_t) \in \mathscr{A}\backslash\mathscr{A}_s$ is limited, the following constraint should be satisfied:

$$\sum_{f_u \in \mathscr{F}} f_u(i_t, j_t) \leq x(i_t, j_t)\mathscr{C}(i_t, j_t), \quad \forall (i_t, j_t) \in \mathscr{A}\backslash\mathscr{A}_s, t \in T. \tag{6.14}$$

6.1.4 Energy Cost Model

Energy Cost of LEO Satellite

The energy consumed on the LEO satellite is mainly due to the data transceiver [159], and therefore the energy cost of the receiver (H2S, S2S) of LEO satellites can be expressed as

$$E_{i^t}^{re,s} = \sum_{f_u \in \mathscr{F}} \left(\sum_{j_t \in \mathscr{V}_h} \frac{P_{hs}^{re} f_u(j_t, i_t)}{\mathscr{C}(j_t, i_t)} + \sum_{j_t \in \mathscr{V}_s} \frac{P_{ss}^{re} f_u(j_t, i_t)}{\mathscr{C}(j_t, i_t)} \right) \tau,$$
$$\forall i_t \in \mathscr{V}_s, t \in T, \tag{6.15}$$

and the energy cost of the transmitter (S2S, S2G) is calculated as

$$E_{i^t}^{tr,s} = \sum_{f_u \in \mathscr{F}} \left(\sum_{j_t \in \mathscr{V}_s} \frac{P_{ss}^{tr} f_u(i_t, j_t)}{\mathscr{C}(i_t, j_t)} + \sum_{j_t \in \mathscr{V}_g} \frac{P_{sg}^{tr} f_u(i_t, j_t)}{\mathscr{C}(i_t, j_t)} \right) \tau,$$
$$\forall i_t \in \mathscr{V}_s, t \in T, \tag{6.16}$$

in which P_{hs}^{re}, P_{ss}^{re} are the receiver power (in watts) of H2S and S2S, respectively. Recall that $f_u(i_t, j_t)$ is the flow amount of user u on link (i_t, j_t). We denote $E_{i^t}^{o,s}$ the general operation energy cost of the LEO satellite, and the total energy consumption of the LEO satellite is

$$E_{i^t}^{co,s} = E_{i^t}^{re,s} + E_{i^t}^{tr,s} + E_{i^t}^{o,s}. \tag{6.17}$$

Energy Cost of the HAP

Similar to the energy cost of the LEO satellites, the energy consumption of the receiver (U2H) of the HAPs is

$$E_{i^t}^{re,h} = \sum_{f_u \in \mathscr{F}} \sum_{j_t \in \mathscr{V}_u} \frac{P_{uh}^{re} f_u(j_t, i_t)\tau}{\mathscr{C}(j_t, i_t)}, \quad \forall i_t \in \mathscr{V}_h, t \in T, \tag{6.18}$$

and the energy consumption of the transmitter of the HAPs is

$$E_{i^t}^{tr,h} = \sum_{f_u \in \mathscr{F}} \sum_{j_t \in \mathscr{V}_s} \frac{P_{hs}^{tr} f_u(i_t, j_t)\tau}{\mathscr{C}(i_t, j_t)}, \quad \forall i_t \in \mathscr{V}_h, t \in T, \tag{6.19}$$

where P_{uh}^{re} is the receiver power of U2H. The total energy cost of a HAP is

$$E_{i^t}^{co,h} = E_{i^t}^{re,h} + E_{i^t}^{tr,h} + E_{i^t}^{o,h}, \tag{6.20}$$

in which $E_{i^t}^{o,h}$ is the operational energy cost of the HAP.

It is noted that the energy cost model of the HAP and LEO satellite is related to the continuous variable $f_u(i_t, j_t)$. In addition, the energy capacity limitation of HAPs should be satisfied

$$E_{i^t}^{co,h} \leq \alpha_h \cdot EB_h, \quad \forall i_t \in \mathscr{V}_h, t \in T, \tag{6.21}$$

in which EB_h is the battery capacity of HAP h, and α_h is the HAP's battery maximum depth of discharge. Similarly, the energy capacity limitation of the LEO satellite also should be satisfied:

$$E_{i^t}^{co,s} \leq \alpha_s \cdot EB_s, \quad \forall i_t \in \mathscr{V}_s, t \in T, \tag{6.22}$$

where EB_s is the battery capacity of the LEO satellite, and α_s is the LEO satellite's battery maximum depth of discharge.

6.1.5 Problem Formulation

The original problem is formulated to maximize the total collected data at the DPC, and the constraints discussed earlier should be satisfied:

$$\mathscr{P}0: \max_{\boldsymbol{x}, \boldsymbol{f}, \boldsymbol{s}} \sum_{t \in T} \sum_{(i_t, j_t) \in \mathscr{A}_{sg}} \sum_{f_u \in \mathscr{F}} f_u(i_t, j_t) \tag{6.23}$$

$$\text{s.t.} \quad (6.2)-(6.11), (6.14), (6.21), (6.22),$$

$$x(i_t, j_t) \in \{0, 1\}, \quad \forall (i_t, j_t) \in \mathscr{A} \backslash \mathscr{A}_s, t \in T, \tag{6.24}$$

$$f_u(i_t, j_t) \in [0, \sigma_u], \quad \forall (i_t, j_t) \in \mathscr{A} \backslash \mathscr{A}_s, t \in T, \tag{6.25}$$

$$s_u(i_t) \in [0, \sigma_u], \quad \forall i_t \in \mathscr{V}_h \cup \mathscr{V}_s, t \in T, \tag{6.26}$$

where $\boldsymbol{x} = \{x(i_t, j_t), \forall (i_t, j_t) \in \mathscr{A} \backslash \mathscr{A}_s, t \in T\}$, $\boldsymbol{f} = \{f_u(i_t, j_t), \forall f_u \in \mathscr{F}, (i_t, j_t) \in \mathscr{A} \backslash \mathscr{A}_s, t \in T\}$, and $\boldsymbol{s} = \{s_u(i_t), \forall f_u \in \mathscr{F}, i_t \in \mathscr{V}_h \cup \mathscr{V}_s, t \in T\}$ are the variable vectors of the contact plan, data flow amount, and storage amount, respectively. In $\mathscr{P}0$, constraints (6.2) through (6.8) are the contact plan restrictions, constraints (6.9) and (6.10) are the flow conservation constraints in the TEG, (6.11) is the storage resource capacity restriction, and (6.14) is the communication resource capacity constraint. Constraints (6.21) and (6.22) are the energy capacity restrictions of the HAPs and LEO satellites, respectively. It is noted that $\mathscr{P}0$ is in the form of MILP, which is NP-hard to solve in general complexity theory [160, 161], and the brute-force enumeration has an exponential complexity due to the number of integer variables [160, 162].

6.2 Algorithm Design

As mentioned earlier, the original problem $\mathscr{P}0$ is a MILP problem and brute-force search is unacceptable due to the dimension curse. Note that the integer variable $\boldsymbol{x}$ in $\mathscr{P}0$ is the complicating variable, and if $\boldsymbol{x}$ is fixed, the residual problem is only related to the continuous variables $\boldsymbol{f}$ and $\boldsymbol{s}$, which are tractable by the optimization tools. Such a structure satisfies the conditions of the Benders decomposition, so we design the algorithm BDOA in Section 6.2.1. Moreover, to further alleviate the time complexity in a large-scale TEG, an acceleration algorithm is proposed in Section 6.2.2 to obtain the suboptimal solution with high efficiency.

6.2.1 Benders-Decomposition-Based Optimal Algorithm

Preliminaries of Benders Decomposition

Benders decomposition is a scheme used to tackle the complex problem in the form of MILP with complicating variables [163]. Instead of directly handling all the variables and constraints in the original problem, Benders decomposition can split the original problem into two smaller problems with fewer variables and constraints: a subproblem in the form of linear programming with only continuous variables, and a master problem in the form of a small-scale MILP with integer variables. The master problem and subproblem are solved iteratively to acquire the optimal solution for the original problem. Note that the original problem $\mathscr{P}0$ is in the form of MILP with two types of variables (i.e., the integer variables $\boldsymbol{x}$ and the continuous variables $\boldsymbol{f}$ and $\boldsymbol{s}$). Hence, $\mathscr{P}0$ meets the conditions of Benders decomposition. Specifically, the subproblem is used to update dual information and generate Benders cuts for the master problem. Benders cuts include the feasibility cuts and optimality cuts, depending on if the dual value of the subproblem is bounded. The master problem leverages the newly added cuts to update its solution and corresponding objective value. In turn, the integer solution of the master problem is updated as fixed values in the subproblem. In such a process, the lower bound (LB) and upper bound (UB) are renewed, and the gap between LB and UB shrinks. The iteration between the master problem and subproblem ceases when the UB equals the LB, and the optimal solution is obtained. The optimal solution can be regarded as the benchmark to evaluate the solution from other heuristic algorithms. Benders decomposition has wide applications in industries such as the electricity market [164], and it has also been used in communication networks such as network function virtualization and caching [165–167].

BDOA for $\mathscr{P}0$

As previously explained, via Benders decomposition, the original problem $\mathscr{P}0$ can be transformed into two smaller problems: a subproblem $\mathscr{P}1$ and a master problem $\mathscr{P}2$. In detail, the integer variable $\boldsymbol{x}$ is the complicating variable in $\mathscr{P}0$, and by fixing $\boldsymbol{x}$, $\mathscr{P}0$ is transformed into subproblem $\mathscr{P}1$,

$$\mathscr{P}1: \max_{\boldsymbol{f},\boldsymbol{s}} \sum_{t\in T} \sum_{(i_t,j_t)\in\mathscr{A}_{sg}} \sum_{f_u\in\mathscr{F}} f_u(i_t,j_t) \tag{6.27}$$

$$\text{s.t. } (6.10), (6.11), (6.21), (6.22),$$

$$\sum_{(i_t,j_t)\in\mathscr{A}_{hs}} f_u(i_t,j_t) + s_u(i_t) = \sum_{(j_t,i_t)\in\mathscr{A}_{uh}} \sigma_u x^*(j_t,i_t),$$

$$\forall i_t \in \mathscr{V}_h, f_u \in \mathscr{F}, t \in T, \tag{6.28}$$

$$\sum_{f_u \in \mathcal{F}} f_u(i_t, j_t) \le x^*(i_t, j_t)\mathcal{C}(i_t, j_t),$$

$$\forall (i_t, j_t) \in \mathcal{A} \backslash \mathcal{A}_s, t \in T, \tag{6.29}$$

$$f_u(i_t, j_t) \in [0, \sigma_u], \forall (i_t, j_t) \in \mathcal{A} \backslash \mathcal{A}_s, t \in T, \tag{6.30}$$

$$s_u(i_t) \in [0, \sigma_u], \forall i_t \in \mathcal{V}_h \cup \mathcal{V}_s, t \in T, \tag{6.31}$$

where $x^*(i_t, j_t)$ is the fixed value of $x(i_t, j_t)$. The dual problem of $\mathcal{P}1$ is solved to obtain the dual variables $\boldsymbol{\lambda} = \{\lambda^{(6.28)}_{i_t, f_u}, \lambda^{(6.29)}_{i_t j_t}\}$, in which $\lambda^{(6.28)}_{i_t, f_u}$ and $\lambda^{(6.29)}_{i_t j_t}$ are the dual variables related to the constraints (6.28) and (6.29) of $\mathcal{P}1$, respectively. Then, $\boldsymbol{\lambda}$ is used to generate Benders cuts for $\mathcal{P}2$. Benders cuts include two types: the feasibility cuts and optimality cuts. In detail, if the dual value $\boldsymbol{\lambda}$ is unbounded, the feasibility cut (6.32) is added to $\mathcal{P}2$:

$$\sum_{t \in T} \sum_{i_t \in \mathcal{V}_u} \sum_{f_u \in \mathcal{F}} \sum_{(j_t, i_t) \in \mathcal{A}_{uh}} \lambda^{(6.28)}_{i_t, f_u} \sigma_u x(j_t, i_t)$$

$$+ \sum_{t \in T} \sum_{(i_t, j_t) \in \mathcal{A} \backslash \mathcal{A}_s} \lambda^{(6.29)}_{i_t j_t} \mathcal{C}(i_t, j_t) x(i_t, j_t) \ge 0. \tag{6.32}$$

Otherwise, if the dual value $\boldsymbol{\lambda}$ is bounded, the optimality cut (6.33) is added to $\mathcal{P}2$:

$$\sum_{t \in T} \sum_{i_t \in \mathcal{V}_u} \sum_{f_u \in \mathcal{F}} \sum_{(j_t, i_t) \in \mathcal{A}_{uh}} \lambda^{(6.28)}_{i_t, f_u} \sigma_u x(j_t, i_t)$$

$$+ \sum_{t \in T} \sum_{(i_t, j_t) \in \mathcal{A} \backslash \mathcal{A}_s} \lambda^{(6.29)}_{i_t j_t} \mathcal{C}(i_t, j_t) x(i_t, j_t) \ge z \tag{6.33}$$

because the finite dual value $\boldsymbol{\lambda}$ provides a valid bound for the objective function z of the mater problem $\mathcal{P}2$.

The master problem $\mathcal{P}2$ is solved to update integer variables $\boldsymbol{x}$:

$$\mathcal{P}2: \max_{\boldsymbol{x}, z} z \tag{6.34}$$

s.t. (6.2) – (6.8),

feasibility cuts (6.32),

optimality cuts (6.33),

$$x(i_t, j_t) \in \{0, 1\}, \forall (i_t, j_t) \in \mathcal{A} \backslash \mathcal{A}_s, t \in T, \tag{6.35}$$

$$z > 0, \tag{6.36}$$

in which the objective function value z yields a lower bound in BDOA.

The specific BDOA is described in Algorithm 6.1. First, the integer variable is initialized as $\boldsymbol{x} = 0$, which is feasible for the original problem $\mathcal{P}0$. Here, UB and LB are initialized as UB $= +\infty$ and LB $= 0$, respectively. Then, $\mathcal{P}1$ and $\mathcal{P}2$ are iteratively solved until the UB equals the LB. In each iteration, the dual problem of $\mathcal{P}1$ is solved to obtain $\boldsymbol{\lambda}$. If $\boldsymbol{\lambda}$ is unbounded, the new feasibility cut (6.32) is added to $\mathcal{P}2$, as step 5 of Algorithm 6.1. Otherwise, the LB is updated as (6.37) of Algorithm 6.1 (step 7), and the newly generated optimality cut (6.33) is added to $\mathcal{P}2$. Note the previous generated cuts are still kept in $\mathcal{P}2$. Because $\mathcal{P}2$ is a small-scale MILP

Algorithm 6.1 BDOA

Input: TEG, network parameters, demands.
Output: $\boldsymbol{x}$, $\boldsymbol{f}$, $\boldsymbol{s}$, optimal solution.

1 *Initialization:* $\boldsymbol{x} = 0$, UB $= +\infty$, LB $= 0$;
2 **repeat**
3 Solve the dual problem of $\mathscr{P}1$ to obtain $\boldsymbol{\lambda}$;
4 **if** $\boldsymbol{\lambda}$ *is unbounded* **then**
5 Add the feasibility cut (6.32) to $\mathscr{P}2$.;
6 **else**
7 Update LB as

$$\text{LB} = \max\{\text{LB}, \sum_{t\in T}\sum_{i_t\in\mathscr{V}_u}\sum_{f_u\in\mathscr{F}}\sum_{(j_t,i_t)\in\mathscr{A}_{uh}}\lambda_{i_t,f_u}^{(6.28)}\sigma_u x^*(j_t,i_t) + \sum_{t\in T}\sum_{(i_t,j_t)\in\mathscr{A}\setminus\mathscr{A}_s}\lambda_{i_t j_t}^{(6.29)}\mathscr{C}(i_t,j_t)x^*(i_t,j_t) \tag{6.37}$$

Add the optimality cut (6.33) to $\mathscr{P}2$;
8 **end**
9 Solve the master problem $\mathscr{P}2$ via the branch-and-bound method to obtain $\boldsymbol{x}^*$ and z;
10 Update UB $= z$.
11 **until** *UB = LB.*

problem, it can be solved directly by the branch-and-bound algorithm to obtain $\boldsymbol{x}^*$ and z. Then, $\boldsymbol{x}^*$ is provided to $\mathscr{P}1$, and the UB is updated as UB $= z$ (step 10).

The proposed BDOA is guaranteed to converge to the optimal solution, and it is a centralized algorithm that is implemented at the network control center [168]. However, in some cases of large-scale TEG or demands, the slow convergence and time complexity may not be accepted. Hence, a more efficient and suboptimal algorithm will be more adaptive to such cases, which will be presented in detail later.

6.2.2 Acceleration Algorithm

In the event of a large-scale TEG, the time complexity of BDOA is still intractable, and an alternative acceleration method is designed further. Specifically, we resort to an acceleration algorithm (AA) in Algorithm 6.2, to obtain an approximate solution for the original problem $\mathscr{P}0$ with high efficiency.

Note that two tips are introduced in Algorithm 6.2: First, it terminates in finite steps when the gap between the UB and LB satisfies $\frac{\text{UB-LB}}{\text{UB}} \leq \varepsilon$, instead of when the UB equals to the LB, as in Algorithm 6.1. Second, with the increment of iteration in Algorithm 6.2, although the gap between the UB and LB is shrinking, the convergence speed is slowing down. As a consequence, a parameter k is introduced to count the iterations. When $k \geq K$, the iteration is terminated, and K is a predefined value.

Algorithm 6.2 Acceleration algorithm (AA)

Input: TEG, network parameters, demands.
Output: $\boldsymbol{x}$, $\boldsymbol{f}$, $\boldsymbol{s}$, solution for the original problem $\mathscr{P}0$.
1 *Initialization:* ϵ, K, $k = 0$, and $\boldsymbol{x} = 0$;
2 **repeat**
3 Solve the subproblem $\mathscr{P}1$ by Algorithm 6.3, and update LB as in Algorithm 6.1;
4 Solve the master problem $\mathscr{P}2$ by Algorithm 6.4, and update UB as in Algorithm 6.1;
5 $k = k + 1$;
6 **if** $k \geq K$ **then**
7 Go to step 10;
8 **end**
9 **until** $\frac{UB\text{-}LB}{UB} \leq \epsilon$;
10 **return** $\boldsymbol{x}$, $\boldsymbol{f}$, $\boldsymbol{s}$, and the solution of $\mathscr{P}0$.

In particular, in Algorithm 6.2, the parameters ϵ, K, and k and variable $\boldsymbol{x}$ are initialized before the iteration. Then, the subproblem $\mathscr{P}1$ is handled by Algorithm 6.3, and the LB is updated as in Algorithm 6.1. In addition, the master problem $\mathscr{P}2$ is tackled by Algorithm 6.4, and the UB is renewed as in Algorithm 6.1. Such an iteration is executed until $\frac{\text{UB-LB}}{\text{UB}} \leq \epsilon$. In each iteration, k is updated as $k = k + 1$. When $k \geq K$, Algorithm 6.2 ceases, and $\boldsymbol{x}$, $\boldsymbol{f}$, and $\boldsymbol{s}$ as well as the solution of $\mathscr{P}0$ are returned. Algorithm 6.3 and 6.4 will be elaborated on later.

Unit-Flow-Based Shortest-Path Algorithm for Subproblem $\mathscr{P}1$

Because the subproblem $\mathscr{P}1$ is a multipath routing problem, it can only be solved directly by the optimization tool rather than by the existing efficient shortest-path algorithm, and the complexity is not acceptable with large-scale networks. To further accelerate the solution for subproblem $\mathscr{P}1$, when $\boldsymbol{x}$ is acquired, the concept of unit flow [169] is introduced to approximately transform the subproblem $\mathscr{P}1$ into the single-path routing problem for multiple independent unit flows. Then, the classical shortest-path algorithm is applicable to the routing of unit flow without directly solving $\mathscr{P}1$.

In particular, the unit flow with a fixed data size cannot be further split when it is delivered to the TEG. Besides, the size of the unit flow will affect both the accuracy and computational complexity. We denote ρ as the data size of a unit flow, and the data σ_u from each user is split into N_ρ^u unit flows as $\sigma_u = \rho \cdot N_\rho^u$. Hence, the classical weighted shortest-path algorithm will be executed N_ρ^u times for each user to find the routing for the unit flows. Furthermore, note that the weight of links in the TEG is related to the link capacity, the unit flows allocated on the link, and the data size of the unit flow. As a result, the weight of link (i_t, j_t) in the TEG can be defined as

Algorithm 6.3 Unit-flow-based shortest-path algorithm for subproblem $\mathscr{P}1$

Input: TEG, σ_u, $\boldsymbol{x}$.
Output: $\boldsymbol{f}$, $\boldsymbol{s}$, $\boldsymbol{\lambda}$, and result of $\mathscr{P}1$.

1 *Initialization:* $\boldsymbol{f} = 0$, $\boldsymbol{s} = 0$, $\Delta(i_t, j_t) = 0$, $\Omega = \emptyset$;
2 **for** *each user u* **do**
3 Split σ_u into N_ρ^u unit flows ($N_\rho^u = \frac{\sigma_u}{\rho}$);
4 Add the unit flows to Ω;
5 **end**
6 **while** Ω *is not empty* **do**
7 Randomly select a unit flow from Ω;
8 Remove the impossible nodes and links before the start time of the current unit flow in the TEG;
9 Remove the links in the TEG with a link capacity that is less than ρ;
10 Use the weighted shortest path algorithm to find the shortest path with the minimum weight for the incumbent unit flow in the TEG;
11 **if** *failed* **then**
12 Discard the incumbent unit flow from Ω;
13 **else**
14 Allocate the shortest path to the current unit flow;
15 Update $\Delta(i_t, j_t) = \Delta(i_t, j_t) + \rho$ and weight $w(i_t, j_t)$ following (6.38), an[d] remove the incumbent unit flow from Ω;
16 **end**
17 **end**
18 Merge the paths for unit flows to obtain $\boldsymbol{f}$ and $\boldsymbol{s}$;
19 Calculate the dual variable $\boldsymbol{\lambda}$ of subproblem $\mathscr{P}1$;
20 **return** $\boldsymbol{f}$, $\boldsymbol{s}$, $\boldsymbol{\lambda}$, and result of $\mathscr{P}1$.

$$w(i_t, j_t) = \frac{\rho}{\mathscr{C}(i_t, j_t) - \Delta(i_t, j_t)}, \quad (6.\text{:}$$

in which $\Delta(i_t, j_t)$ indicates the total amount of unit flow already loaded on link $(i_t,$

The unit-flow-based shortest-path (UFSP) algorithm for subproblem $\mathscr{P}1$ is detai in Algorithm 6.3. First, $\boldsymbol{f}$, $\boldsymbol{s}$, and $\Delta(i_t, j_t)$ are initialized as 0, and Ω is initialized as Then, in regard to each user, the data size σ_u is split to into N_ρ^u unit flows ($N_\rho^u =$ and all unit flows are to the set Ω. The loop in Algorithm 6.3 is executed until Ω empty. Step 7 in Algorithm 6.3 guarantees fairness for all users by randomly selecti a unit flow from Ω because the data of each user is decomposed into a couple of u flows in Ω. To simplify the TEG topology, the nodes and links before the start time the current unit flow are removed in step 8, and the links with link capacities that less than ρ are also removed in step 9. In step 10, the weighted shortest-path algorit is employed to find the shortest path with the minimum weight in the TEG for incumbent unit flow. If no possible path exits, the incumbent unit flow is discard from Ω in step 12. Otherwise, the shortest path is allocated to the current unit flow. Then, $\Delta(i_t, j_t)$ is updated as

$$\Delta(i_t, j_t) = \Delta(i_t, j_t) + \rho, \tag{6.39}$$

$w(i_t, j_t)$ is updated following (6.38), and the incumbent unit flow is removed from Ω. Finally, when Ω is empty, the paths for unit flows are merged to obtain $\boldsymbol{f}$ and $\boldsymbol{s}$, the dual value $\boldsymbol{\lambda}$ is calculated, and the final solution for $\mathscr{P}1$ is obtained.

Approximation Algorithm for Master Problem $\mathscr{P}2$

Because it is not necessary to acquire the optimal solution for the master problem in every iteration to achieve global convergence, a suboptimal solution is useful [163]. Accordingly, we design the approximation algorithm for master problem $\mathscr{P}2$ in Algorithm 6.4. First, the integer variable $\boldsymbol{x}$ ($x(i_t, j_t) \in \{0, 1\}$) is relaxed to the continuous variable ($x(i_t, j_t) \in [0, 1]$), and the relaxed master problem $\mathscr{P}2$ is solved in step 1 to obtain $\boldsymbol{x}$. Then, we check if $\boldsymbol{x}$ is an integer or not: If $\boldsymbol{x}$ is an integer, it is the solution of $\mathscr{P}2$. Otherwise, the maximum $\boldsymbol{x}$ is rounded to 1 and other $\boldsymbol{x}$ is rounded as 0 in step 5. In effect, the rounded solution may violate the contact constraints, so the feasibility of $\boldsymbol{x}$ is checked in step 6. If $\boldsymbol{x}$ is feasible, $\boldsymbol{x}$ and the corresponding z are returned. If there exist $x(i_t, j_t)$ in $\boldsymbol{x}$ that is unfeasible, the unfeasible $x(i_t, j_t)$ is set as 0, and the residual feasible $\boldsymbol{x}$ as well as the suboptimal solution z of $\mathscr{P}2$ are returned.

Algorithm 6.4 Approximation algorithm for master problem $\mathscr{P}2$

Input: Dual value $\boldsymbol{\lambda}$ of subproblem $\mathscr{P}1$, and other parameters in master problem $\mathscr{P}2$.

Output: $\boldsymbol{x}$ and z.

```
1  Solve the relaxed master problem P2 to obtain x;
2  if x is integer then
3  |   Go to step 13;
4  else
5  |   Round the maximum x to 1 and other x as 0;
6  |   Check the feasibility of x;
7  |   if x is feasible then
8  |   |   Go to step 13;
9  |   else
10 |   |   Set the unfeasible x(i_t, j_t) in x to 0.
11 |   end
12 end
13 return x and solution z of P2.
```

6.3 Performance Evaluation

In this section, a series of simulations are conducted for the AAN composed of LEO satellites, HAPs, and users to verify the effectiveness and efficiency of the proposed algorithms as well as to analyze the impacts of different parameters on the performance of the AAN. The connections among LEO satellites as well as LEO satellites and HAPs in the AAN are built by the Satellite Tool Kit (STK), and the algorithm implementation is based on MATLAB. The optimization tools CVX and MOSEK are employed for the algorithm design.

6.3.1 Parameter Settings

In detail, 16 LEO satellites with a height of 780 km are distributed in 4 orbits, composing a 4×4 walker constellation, and the LEO satellites have a period of 100 min. There are four HAPs with a height of 100 min distributed in a square area of 100 km $\times$ 100 km. Two ground stations are located at (40°N, 116°E) and (39.5°N, 76°E), respectively. Terrestrial users are randomly distributed in the area covered by the HAPs, and the data size of the users is randomly generated between [10 Mbit and 200 Mbit]. The time horizon is set as $\mathscr{T} = 100$ min, and the time slot length is $\tau = 200$ s. Besides, following [158, 170–172], the energy-related parameters are set to $P_{uh}^{tr} = 1$ W, $P_{hs}^{tr} = 10$ W, $P_{ss}^{tr} = 20$ W, $P_{sg}^{tr} = 20$ W, $\alpha_h = 80\%$, and $\alpha_s = 80\%$. The channel corresponding parameters are set as $G_{uh}^{tr} G_{uh}^{re} = 41$ dB, $G_{hs}^{tr} G_{hs}^{re} =$ 42 dB, $G_{ss}^{tr} G_{ss}^{re} = 52$ dB, $G_{sg}^{tr} G_{sg}^{re} = 42$ dB, $L_l = -23$ dB, $k_B = 1.38 \times 10^{-23}$ J/K, $T_s = 1{,}000$ K, $\nu_{uh} = 3.4$ GHz, $\nu_{hs} = 48$ GHz, $\nu_{ss} = 2.2$ GHz, $\nu_{sg} = 20$ GHz, and $M = 5$ dB. Furthermore, other parameters are set as $N = 100$, $K = 20$, and $\epsilon = 0.05$.

6.3.2 Numerical Results

To evaluate the optimality of algorithm BDOA and the performance of algorithm AA, we compare the proposed algorithms with the optimal solution of $\mathscr{P}0$ in Fig. 6.3. The optimal solution of $\mathscr{P}0$ is obtained by exhaustive search. A related algorithm CPCG from [173] is also compared with the proposed algorithms. The CPCG algorithm adopts the conflict graph to obtain the contact plan in each time slot, and the results in different time slots are acquired step by step. Therein, the energy capacity of the HAPs and LEO satellites are set as $EB_h = 30$ KJ and $EB_s = 30$ KJ, respectively. The storage capacity of the HAPs and LEO satellites are set as $\mathscr{C}_h = 20$ Gbit and $\mathscr{C}_s = 20$ Gbit. The size of unit flow is $\rho = 2$ Mbit. As shown in Fig. 6.3a, the execution time of the optimal solution exponentially grows with the number of users, whereas the algorithm BDOA can rapidly obtain the final results, and the algorithm AA is more efficient than the compared algorithm CPCG. Meanwhile, in Fig. 6.3b, it is observed that the total received data at the DPC by the BDOA algorithm acquires the same optimal result compared with the optimal solution, and the result from AA is suboptimal but with high efficiency. Both BDOA and AA perform better than CPCG in regard to the total

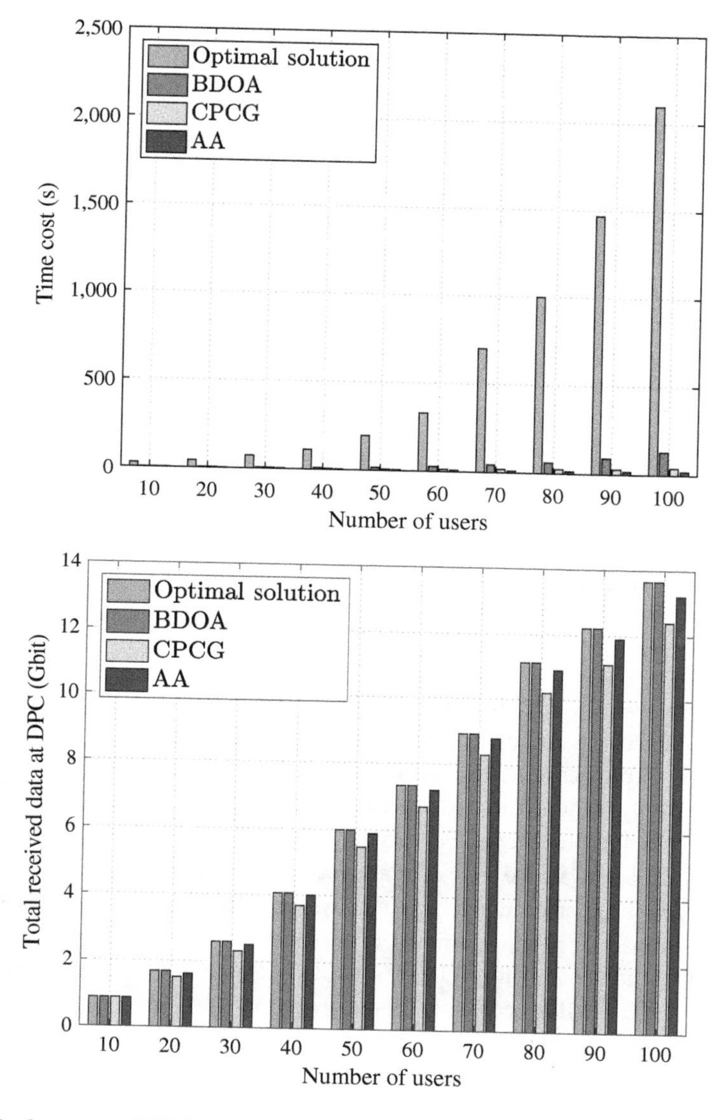

Figure 6.3 Performance of BDOA and AA ($EB_h = 30$ KJ, $EB_s = 30$ KJ, $\mathscr{C}_h = 20$ Gbit, $\mathscr{C}_s = 20$ Gbit, $\rho = 2$ Mbit): (a) time complexity and (b) optimization results

received data at the DPC, which is the objective of the original problem. From Fig. 6.3, it is observed that there always exist a tradeoff between the time complexity and optimal solution.

In Fig. 6.4, the convergence of the designed algorithms BDOA and AA is depicted by the UB and LB. It is observed that the gap between the UB and LB of BDOA narrows with the iterations until it converges. Compared with BDOA, the converge speed between the UB and LB of AA is faster, and the iteration ceases when $K = 20$.

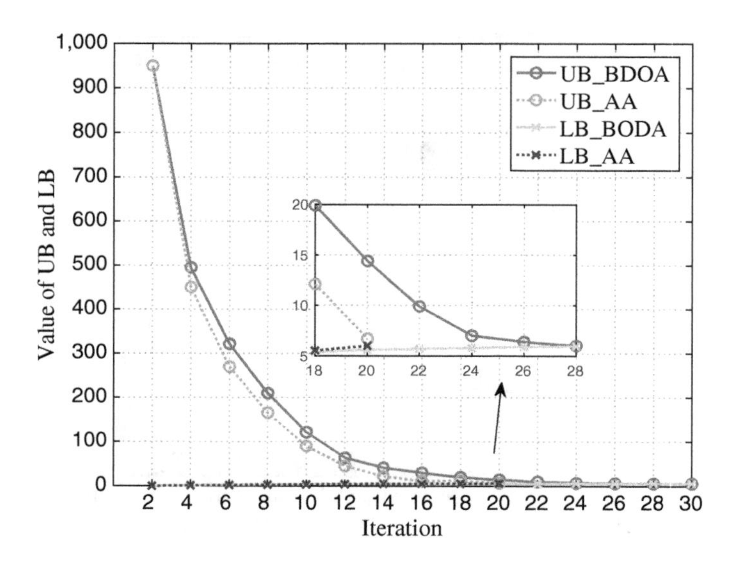

Figure 6.4 Convergence of BDOA and AA ($EB_h = 30$ KJ, $EB_s = 30$ KJ, $\mathscr{C}_h = 20$ Gbit, $\mathscr{C}_s = 20$ Gbit, $\rho = 2$ Mbit, $|\mathscr{U}| = 50$)

However, the final UB and LB of BDOA cannot acquire the complete convergence due to the approximation property of AA.

To investigate the impact of energy capacity of the HAPs and LEO satellites, we study the total data received at the DPC with different number of users in Figs. 6.5 and 6.6. It is observed in Fig. 6.5 that the total received data at the DPC increases and levels off when all the user data are successfully transmitted back to the DPC. In terms of the LEO satellite energy capacity, the results in Fig. 6.6 show similar trends, and the impact is more significant than that of the HAP energy capacity, because the high-speed movement of LEO satellites and intermittent connections between different LEO satellites consume more energy than HAPs.

We further study the impacts of the storage capacity of HAPs and LEO satellites on the total data acquired at the DPC in Figs. 6.7 and 6.8, respectively. As shown in Fig. 6.7, when HAP storage capacity is less than 20 Gbit, the data amount received at the DPC is not satisfactory. Such trends are expected because the data transmitted in the TEG is limited by the HAP's storage capacity. As the HAP's storage capacity increases, the data amount received at the DPC considerably improves, especially when $\mathscr{C}_h > 20$ Gbit. Similarly, with respect to the storage capacity of LEO satellites, note that Fig. 6.8 shows a similar trend, but the impacts from the storage capacity of LEO satellites are larger. It can be explained that the data received at LEO satellites may be stored more often due to the intermittent connections of satellite-to-satellite links as well as satellite-to-ground links, and the storage resources of LEO satellite are utilized more frequently.

To evaluate the effect of the size of unit flow of Algorithm 6.3, in Fig. 6.9, the total received data at the DPC as well as the time cost versus the unit flow ρ are

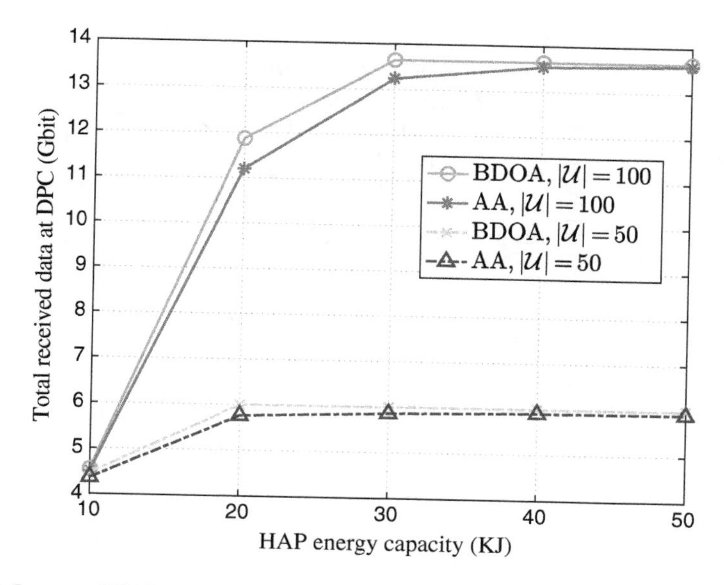

Figure 6.5 Impact of HAP energy capacity on network performance ($EB_s = 30$ KJ, $\mathscr{C}_h = 20$ Gbit, $\mathscr{C}_s = 20$ Gbit, $\rho = 2$ Mbit)

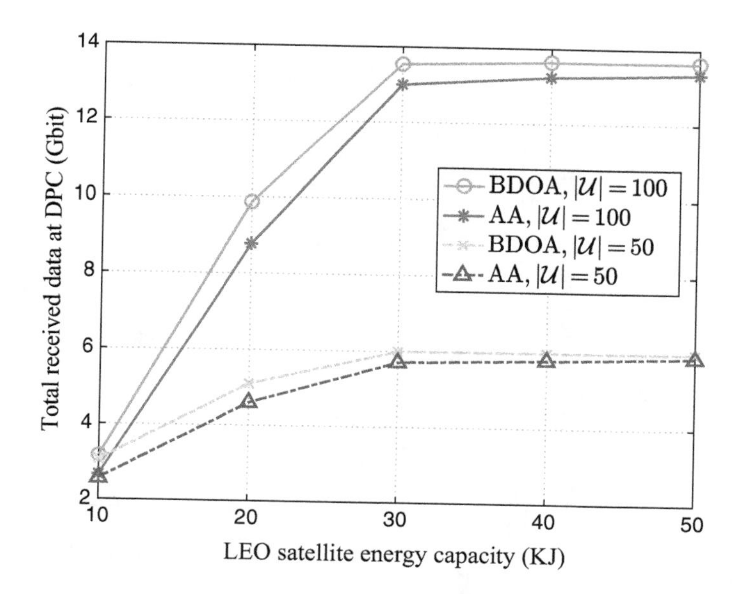

Figure 6.6 Impact of LEO satellite energy capacity on network performance ($EB_h = 30$ KJ, $\mathscr{C}_h = 20$ Gbit, $\mathscr{C}_s = 20$ Gbit, $\rho = 2$ Mbit)

investigated. For example, $\rho = 1$ Mbit indicates the data flow of each user will split into a couple of unit flows with a data size of 1 Mbit. From Fig. 6.9, we can observe that with the increment of ρ, the total data received at the DPC decreases, because it is more difficult to discover an end-to-end path for the unit flow with a large data

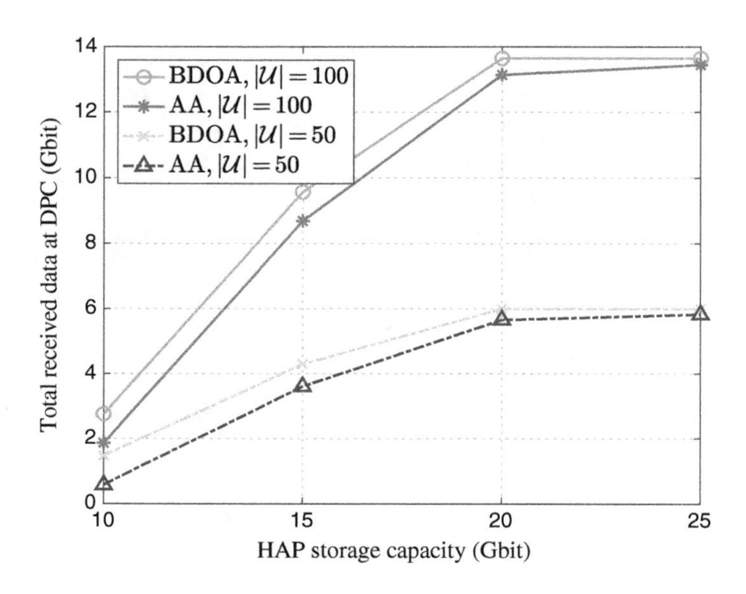

Figure 6.7 Impact of HAP storage capacity on network performance ($EB_h = 30$ KJ, $EB_s = 30$ KJ, $\mathscr{C}_s = 20$ Gbit, $\rho = 2$ Mbit)

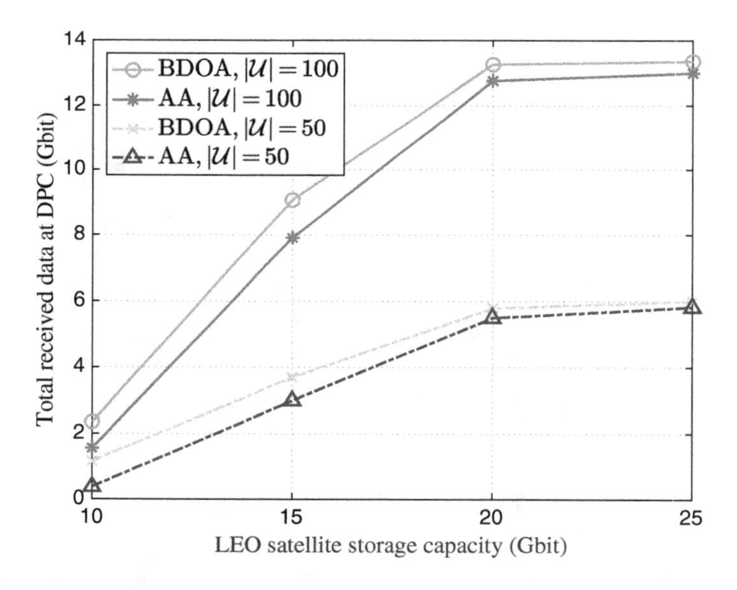

Figure 6.8 Impact of LEO satellite storage capacity on network performance ($EB_h = 30$ KJ, $EB_s = 30$ KJ, $\mathscr{C}_h = 20$ Gbit, $\rho = 2$ Mbit)

size due to resource limitations. In addition, the time cost also has a descending trend because the total number of data flows to be handled decreases with the increment of data size ρ. It is observed that there exist a tradeoff between the total received data at the DPC and the time cost with respect to ρ.

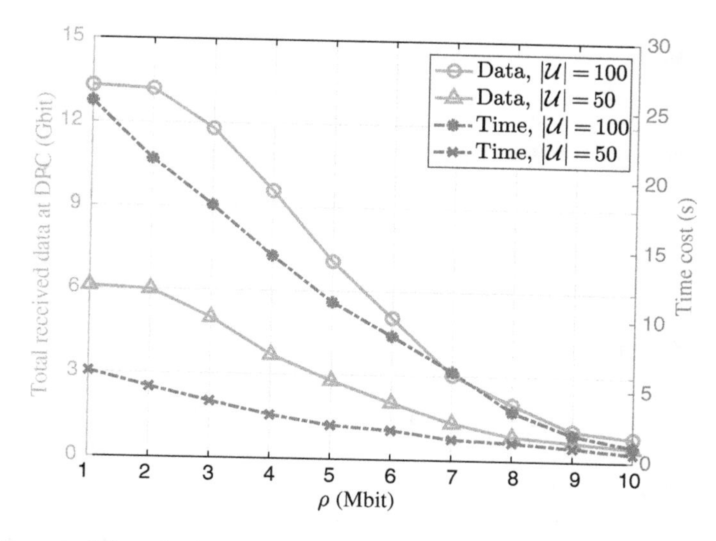

Figure 6.9 Impact of size of unit flow ρ on network performance ($EB_h = 30$ KJ, $EB_s = 30$ KJ, $\mathscr{C}_h = 20$ Gbit, $\mathscr{C}_s = 20$ Gbit)

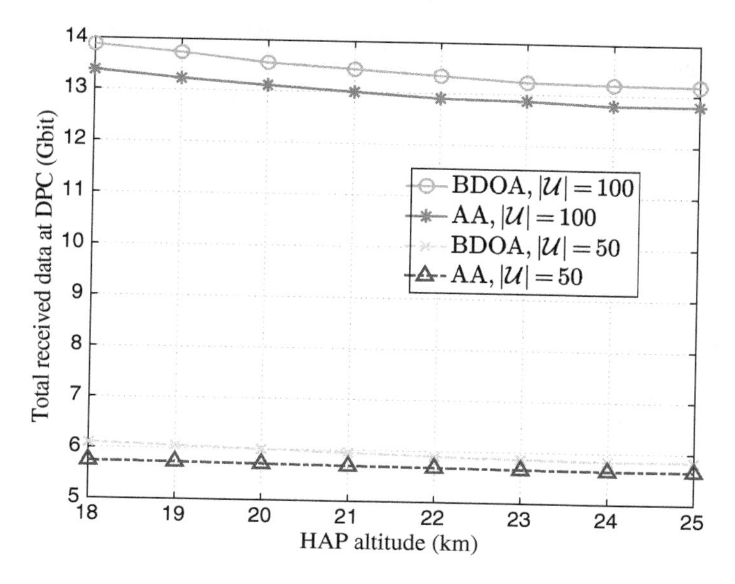

Figure 6.10 Impact of HAP altitude on network performance ($EB_h = 30$ KJ, $EB_s = 30$ KJ, $\mathscr{C}_h = 20$ Gbit, $\mathscr{C}_s = 20$ Gbit, $\rho = 2$ Mbit)

In addition, we investigate the impact of HAP altitude on network performance because it affects the transmission distance between users and HAPs as well as the distance between HAPs and LEO satellites. Specifically, in Fig. 6.10, the HAP altitude varies from 18 km to 25 km [174], and it is observed that total data received at the DPC decreases as the HAP altitude increases. In other words, the data successfully transmitted from users to HAPs decreases because the HAP's height affects the distance between users and HAPs, and the small transmission power of the user can only support limited data transmission. Besides, the distance between HAPs and LEO satellites is much larger than the distance from users to HAPs, and the decrease in the distance between HAPs and LEO satellites only slightly affects network performance.

6.4 Conclusion

In this chapter, we investigated cooperative HAPs and LEO satellites in an AAN for terrestrial data collection and transmission. We employed the TEG to precisely represent the resources of the AAN. Based on the TEG, we focused on maximizing the total data received at the DPC in a time horizon. Because the formulated problem is in the form of MILP and is NP-hard, we designed the algorithm BDOA to obtain the optimal solution by separating the original problem into a smaller master problem and a subproblem. To further alleviate the time complexity or a large-scale TEG, we presented the acceleration algorithm to acquire the suboptimal solution with high efficiency. The acceleration algorithm consists of an approximation algorithm for the master problem and a unit-flow-based algorithm for the subproblem. Simulation and numerical results verified the efficiency of the proposed algorithms, and the performance with respect to various parameters was also analyzed, which will promote the network design for the AAN.

7 HAP-Reserved Communications in Space-Air-Ground Integrated Networks

In recent years, traditional terrestrial communications have realized significant achievements in providing convenient communication services for ground users. However, for economic or technological reasons, there is still a need for high-quality communication services in challenging areas, such as mountains, deserts, oceans, and rural areas. Motivated by this challenge, air and space communications have been leveraged to complement ground communications and provide more comprehensive coverage [138]. Compare to the ground communication system, air and space communication systems have better coverage but suffer from limited capacity and long propagation latency. To overcome these shortcomings, some researchers have proposed the space-air-ground integrated network (SAGIN) to complement and ground, air, and space communications at different altitudes and achieve more flexible end-to-end services [10]. By integrating ground, air, and space segments, transmissions from ground users to different destinations can be initiated via various paths. As a result, different segments can provide differential quality transmissions to meet distinguished services, which is essential for next-generation communication networks. Currently, extensive work has investigated SAGIN in the areas of machine learning [175], nonorthogonal multiple access (NOMA) [176], edge computing [177], reconfigurable intelligent surface (RIS) [178], and spatial resource allocation [179]. However, research on the effective cooperation among ground, air, and space communications is still in its infant stage, especially for transmission access and control at the medium access control (MAC) layer of SAGIN.

In this chapter, we mainly focus on integrating ground, air, and space communications by implementing a transmission control strategy, as highlighted in Fig. 7.1. To complement the terrestrial communication system, we propose a HAP-reserved ground-air-space (GAS) transmission scheme to support ground-to-space (G2S) link transmissions, thereby achieving flexible and differentiated services, efficient spectrum utilization, energy savings, and coverage enhancement. In the proposed GAS transmission scheme, HAPs can be reserved by ground users to support their GAS link transmissions, where a GAS link includes a ground-to-air link and an air-to-space link. Based on the proposed GAS and G2S transmission schemes, we explore a transmission control strategy that enables ground users to adopt the GAS link transmission with a certain probability. Furthermore, we formulate an optimization problem to maximize the throughput, thereby deriving the optimal probability and optimal HAP for each

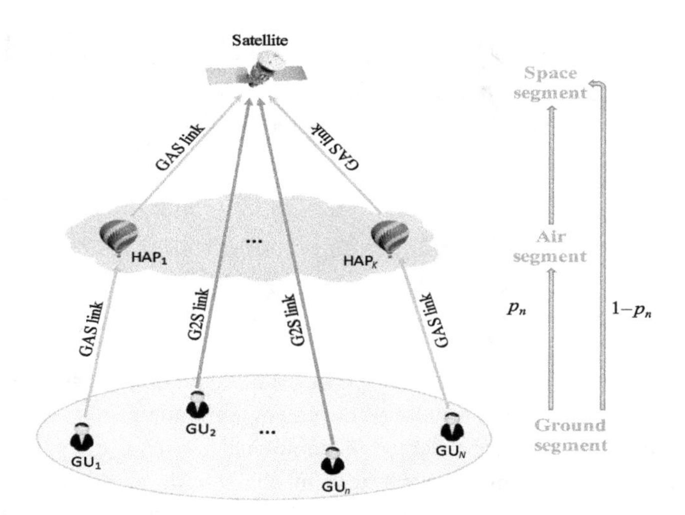

Figure 7.1 Uplink ground-space communication system, where GAS links proposed to complement G2S links. In particular, a transmission control strategy is presented, where the ground user n initiates the GAS link transmission with probability ρ_n and the G2S link transmission is initiated with a probability of $1-\rho_n$

ground user. Finally, numerical results demonstrate the advantage of the proposed transmission control strategy.

The rest of this chapter is organized as follows. In Section 7.1, we describe the system model. Then, we formulate a transmission strategy optimization problem and give the problem solution in Section 7.2. Section 7.3 analyzes the overall throughput performance and presents numerical results. Section 7.4 concludes the chapter.

7.1 System Model

The system is presented in Fig. 7.1, where three-segment networks are constructed by the ground, air, and space systems. The ground system includes N ground users (i.e., $\mathscr{N}=\{\mathrm{GU}_1, \mathrm{GU}_2, \ldots, \mathrm{GU}_N\}$). The air system employs K total HAPs (e.g., airships or balloons) as carriers (i.e., $\mathscr{K}=\{\mathrm{HAP}_1, \mathrm{HAP}_2, \ldots, \mathrm{HAP}_K\}$). As a relay, each HAP serves one ground user at a time. The space segment that comprises one narrow-band satellite can connect with the ground and air segments. As shown in Fig. 7.1, the HAPs deployed in the air segment provide the potential assistance for wireless communications between the ground users and the satellite. In the proposed system, the ground user decides to communicate with the satellite via the HAP or not, which means that the ground user $n, n \in \mathscr{N}$, may use the G2S link or the GAS link with a probability of ρ_n or $1-\rho_n$ accordingly. By performing negotiation, the ground user is aware of whether HAPs are available and selects the optimal HAP for its GAS link transmission if available, thereby impacting the system performance. Without loss of generality, we assume that the number of HAPs is limited, and the system cannot provide services for all ground users. In addition, the total system bandwidth is B, and the bandwidth for the G2S links and the GAS links is $\omega_1 B$ and $\omega_2 B$, respectively. Here, ω_1 and ω_2 are the weights of the allocated bandwidth, and $\omega_1 + \omega_2 \leq 1$ holds.

We define $\rho_n, \rho_n \in [0,1]$ as the transmission probability of GU$_n$ via a HAP, and $1 - \rho_n$ denotes the transmission probability of GU$_n$ without the assistance of HAPs. Specifically, we have

$$\rho_n = \begin{cases} 0, & \text{GU}_n \text{ only use G2S link,} \\ 1, & \text{GU}_n \text{ only use GAS link,} \\ (0,1), & \text{GU}_n \text{ uses GAS link with a probability.} \end{cases} \tag{7.1}$$

In addition, let $\mathbf{U}_n = \{u_{n1}, u_{n2}, \ldots, u_{nK}\}, \forall n \in \mathcal{N}$, denote the K-dimensional link decision vector of GU$_n$, where $u_{nk} \in \{0,1\}$ indicates that GU$_n$ communicates with the satellite via HAP k or not. Then, we have

$$u_{nk} = \begin{cases} 1, & \text{if GU}_n \text{ transmits via the HAP } k, \\ 0, & \text{otherwise.} \end{cases} \tag{7.2}$$

Moreover, we consider that GU$_n$ either transmits without the assistance of HAPs or transmits only via a HAP. Then, we have

$$\sum_{k=0}^{K} u_{nk} = \begin{cases} 1, & \text{if GU}_n \text{ transmits via a HAP,} \\ 0, & \text{otherwise.} \end{cases} \tag{7.3}$$

Therefore, the transmission strategy of GU$_n$ is illustrated as

$$\mathcal{D}_n = \{\rho_n, \mathbf{U}_n\}. \tag{7.4}$$

Therefore, the transmission control decision for ground users can be given by $\mathbb{D} = \{\mathcal{D}_1, \ldots, \mathcal{D}_N\}, \ \forall n \in \mathcal{N}$.

7.1.1 G2S Link Transmission Model

Channel Access Scheme

For the G2S link transmission, the ground users communicate with the satellite using an orthogonal frequency-division multiple access (OFDMA) based carrier-sense multiple access/collision avoidance (CSMA/CA) scheme. Suppose that there are M available subfrequency bands (i.e., $\mathcal{M} = \{1, 2, \ldots, M\}$), and the bandwidth of each subfrequency is $\omega_1 B/M$. Based on the CSMA/CA scheme, a ground user competes for transmission access on the M subfrequency band and then transmits to the satellite via the G2S link.

Propagation Model

The G2S link transmission is a direct link transmission. According to the pathloss propagation model, the G2S link rate for GU_n is denoted by

$$R_n^{G2S} = \frac{\omega_1 B}{M} \log_2\left(1 + \frac{E_n^{G2S} 10^{-L_n^{G2S}/10}}{\sigma_0^2}\right), \tag{7.5}$$

where E_n^{G2S} denotes the transmit power at the GU_n using the G2S link, L_n^{G2S} is the pathloss from GU_n to the satellite, and σ_0^2 is Gaussian noise power of the G2S link.

7.1.2 GAS Link Transmission Model

Channel Access Scheme

For the proposed GAS transmission scheme, the ground users communicate with the satellite using a time-division multiple access (TDMA) based reservation scheme, where the time is divided into a series of frames. Each frame consists of two periods: the ground negotiation period with a length of t_h and the reserved GAS transmission period with a length of t_r. Thus, the length of one time frame, denoted as T, is equal to $t_h + t_r$. During the ground negotiation period, following the specified licensed assisted access (LAA) listen-before-talk (LBT) scheme in 3GPP release 14, the ground users compete for the reservation privilege on the unlicensed band to reserve the HAP and slots. Based on the 3GPP standard protocol, LAA defines the spectrum sharing on the unlicensed band. Specifically, LAA adopts the LBT scheme, which allows the licensed users to leverage the unlicensed band when the channel is sensed to be idle, thus improving data rates and providing better services. During the reserved GAS transmission period, the ground users communicate with the satellite via the reserved HAP within the reserved slots. Compared with the G2S link transmission the proposed GAS transmission scheme can reduce the power consumption of the ground user.

Propagation Model

The GAS link transmission can be considered an amplify-and-forward transmission. Each GAS link transmission occupies $\omega_2 B$ bandwidths because only one GAS link transmission can be allowed to exit at a time. According to the pathloss propagation model, the GAS link rate via the HAP_k for GU_n is denoted by

$$R_{nk}^{GAS} = \omega_2 B \log_2\left(1 + f\left(r_{nk}, r_{ks}\right)\right), \tag{7.6}$$

where

$$f\left(r_{nk}, r_{ks}\right) = \frac{r_{nk} r_{ks}}{r_{nk} + r_{ks} + 1}. \tag{7.7}$$

In (7.7), r_{nk} and r_{ks} denote the signal-to-noise ratio (SNR) at the HAP$_k$ and the satellite, respectively. They are expressed as

$$r_{nk} = \frac{E_n^{GAS} 10^{-L_{nk}^{GAS}/10}}{\sigma_1^2}, \tag{7.8}$$

and

$$r_{ks} = \frac{E_k^{GAS} 10^{-L_{ks}^{GAS}/10}}{\sigma_2^2}, \tag{7.9}$$

where E_n^{GAS} and E_k^{GAS} are the transmit power at the GU$_n$ and HAP$_k$, respectively. The pathloss from GU$_n$ to HAP$_k$ and from HAP$_k$ to the satellite are L_{nk}^{GAS} and L_{ks}^{GAS}, respectively. And σ_1^2 and σ_2^2 are Gaussian noise power from GU$_n$ to HAP$_k$ and from HAP$_k$ to the satellite, respectively.

7.2 Problem Formulation and Solution

7.2.1 Problem Formulation

Having defined the system model and proposed the transmission control strategy in Section 7.1, we formulate the optimization problem to maximize the overall throughput of all ground users as follows:

$$\begin{aligned}
\mathscr{P}1 : \max_{\mathbb{D}} & \left(\sum_{m=1}^{M}\sum_{n=1}^{N} \alpha_n (1-\rho_n) R_n^{G2S} + \sum_{n=1}^{N}\sum_{k=1}^{K} \beta_n \rho_n u_{nk} R_{nk}^{GAS} \right) \\
s.t.\ & \text{C1: } 0 \le \rho_n \le 1, \quad \forall n, \\
& \text{C2: } u_{nk} \in \{0,1\}, \quad \forall n, \exists k, \\
& \text{C3: } \sum_{k=1}^{K} u_{nk} \in \{0,1\}, \quad \forall n, \exists k, \\
& \text{C4: } \sum_{n=1}^{N}\sum_{k=1}^{K} u_{nk} = N_s, \quad \forall n, \exists k, \\
& \text{C5: } 0 < \alpha_n < 1, 0 < \beta_n < 1, \quad \forall n,
\end{aligned} \tag{7.10}$$

where α_n and β_n are defined as the channel utilization of the G2S transmission scheme and the GAS transmission scheme, respectively. Constraint 1 (C1) limits the transmission probability of GAS link of GU$_n$. Constraints 2 and 3 are the constrains of u_{nk}, and C4 constrains the number of ground users that adopt the GAS link transmission. The number of ground users that successfully reserve the HAP is denoted by N_s, which is affected by the competition process during the period of t_h. Lastly, C5 shows the feasibility constraints of α_n and β_n.

In problem $\mathscr{P}1$, the overall throughput includes the throughput of the G2S and GAS links. Let N_1 be the number of ground users that use the OFDMA-based G2S link transmission to connect the satellite. Using [180] as a reference as N_1 ground users compete for access at a subchannel, the successful transmission probability (p_s), the idle probability (p_e), and the failed transmission probability (p_c) can be expressed as

$$\begin{cases} p_s = N_1 \tau (1-\tau)^{N_1-1}, \\ p_e = (1-\tau)^{N_1}, \\ p_c = 1 - p_s - p_e, \end{cases} \tag{7.11}$$

where τ is the stationary probability that a ground user transmits in a random slot, which is denoted by

$$\tau = \frac{2(1-2p)}{(1-2p)(W+1) + pW(1-(2p)^l)}. \tag{7.12}$$

In (7.12), $W \in [W_{\min}, W_{\max}]$ is the contention window and l is the backoff stage. The minimum and maximum contention windows are denoted by $W_{\min}$ and $W_{\max}$, respectively. And p is the collision probability, which is expressed as

$$p = 1 - (1-\tau)^{N_1-1}. \tag{7.13}$$

According to [180], we can obtain the channel utilization of each subchannel as

$$\varphi = \frac{p_s t_s}{p_e \delta + p_s t_s + p_c t_c}, \tag{7.14}$$

where $t_s = RTS + CTS + t_p + 2SIFS + DIFS + 2\delta$ and $t_c = RTS + DIFS + \delta$ are the time length that the channel undergoes a successful transmission and failed transmission, respectively. The time length of a payload is denoted by t_p. Lastly, we have the duration of request-to-send (RTS), clear-to-send (CTS), short interframe space (SISF), and DCF interframe space (DISF).

For the proposed GAS transmission scheme, because GU_n can use the G2S link transmission on M subchannels with a probability of $1 - \rho_n$, the total throughput of GAS link transmissions is denoted by

$$\mathbb{S}^{G2S} = \sum_{m=1}^{M} \sum_{n=1}^{N} \frac{p_s t_s}{p_e \delta + p_s t_s + p_c t_c} (1-\rho_n) R_n^{G2S}. \tag{7.15}$$

Furthermore, when N_2 ground users adopt the TDMA-based GAS transmission scheme to connect with the satellite, the total throughput of GAS links within a frame is expressed as

$$\mathbb{S}^{GAS} = \sum_{n=1}^{N} \rho_n S_{nk}^{GAS}, \tag{7.16}$$

where S_{nk}^{GAS} denotes the throughput of a ground user that adopts the GAS link transmission, which is denoted by

$$S_{nk}^{GAS} = \sum_{k=1}^{K} \frac{t_p r_n}{T} u_{nk} R_{nk}^{GAS}. \tag{7.17}$$

In (7.17), r_n denotes the reservation step (i.e., the number of data packets that GU_n can transmit in one frame). Also, $\frac{t_p r_n}{T}$ is the channel utilization when GU_n adopts the GAS link transmission. Due to the limitations of the ground negotiation period and the reserved GAS transmission period, the number of successful reserved ground users should meet the condition that $\sum_{n=1}^{N}\sum_{k=1}^{K} u_{nk} = N_s = \frac{t_h \zeta_s}{t_s'}$. Here, $t_s' = RTS + CTS + SIFS + DIFS$ is the required negotiation time for GU_n, and ζ_s is the probability of successful negotiation for GU_n, which can be given by

$$\zeta_s = N_2 q \epsilon (1-\epsilon)^{N_2 q - 1}, \tag{7.18}$$

where $\epsilon = \frac{2(1-2\rho)q}{q[(W+1)(1-2\rho)+\rho W(1-(2\rho)^l)]+2(1-q)(1-\rho)(1-2\rho)}$ represents the stationary probability that GU_n transmits a data packet in a random slot during the ground negotiation period. Here, $\rho = 1-(1-\epsilon)^{N_2 q - 1}$ is the collision probability of GU_n, and $q = \frac{\eta}{\gamma+\eta}$ is the stationary probability that GU_n stays in the ground negotiation period. Also, γ and η denote the arrival rate and the service rate, respectively.

Then, the total throughput $\mathbb{S}^{GAS}$ can be expressed as

$$\mathbb{S}^{GAS} = \sum_{n=1}^{N}\sum_{k=1}^{K} \frac{t_p r_n}{T} \rho_n u_{nk} R_{nk}^{GAS}. \tag{7.19}$$

7.2.2 Problem Solution

According to the throughput analysis in (7.15) and (7.19) for each scheme, we have $\alpha_n = \frac{p_s t_s}{p_e \delta + p_s t_s + p_c t_c}$ and $\beta_n = \frac{t_p r_n}{T}$ for GU_n, where $p_s = N_1 \tau (1-\tau)^{N_1 - 1}$, $p_e = (1-\tau)^{N_1}$, $p_c = 1 - p_s - p_e$, $r_n = \frac{t_r}{N_s t_p} = \frac{t_r t_s'}{t_p t_h \zeta_s}$, and $\zeta_s = N_2 q \epsilon (1-\epsilon)^{N_2 q - 1}$. Thus, the optimization problem $\mathscr{P}1$ is rewritten as problem $\mathscr{P}2$, which is given by

$$\mathscr{P}2: \max_{\mathbb{D}} \left(\sum_{m=1}^{M}\sum_{n=1}^{N} \frac{p_s t_s}{p_e \delta + p_s t_s + p_c t_c} (1-\rho_n) R_n^{G2S} + \sum_{n=1}^{N} \frac{t_p r_n}{T} \rho_n \sum_{k=1}^{K} u_{nk} R_{nk}^{GAS} \right)$$

$$s.t.\ \text{C1–C5},$$

$$\text{C6: } N_1 = N - \sum_{n=1}^{N} \rho_n,\ N_2 = \sum_{n=1}^{N} \rho_n,\ \forall n. \tag{7.20}$$

To simplify problem $\mathscr{P}2$, we assume that $\rho_n = \rho, r_n = r, \forall n \in \mathcal{N}$, and then we have $N_1 = N(1-\rho)$ and $N_2 = N\rho$. Because p_s, p_e, and p_c are related to the value of N_1, and ζ_s is related to the value of N_2, it is observed that α_n and β_n are the functions of ρ. Let $\alpha_n = \alpha(\rho)$ and $\beta_n = \beta(\rho)$. Then, we have

$$\mathscr{P}3: \max_{\mathbb{D}} \left(\alpha(\rho)(1-\rho) \sum_{m=1}^{M}\sum_{n=1}^{N} R_n^{G2S} + \beta(\rho)\rho \sum_{n=1}^{N}\sum_{k=1}^{K} u_{nk} R_{nk}^{GAS} \right)$$

$$\text{s.t. C2–C4,}$$
$$\text{C7: } 0 \leq \rho \leq 1, \ \forall n,$$
$$\text{C8: } 0 < \alpha(\rho) < 1, \ 0 < \beta(\rho) < 1, \ \forall n,$$
$$\text{C9: } N_2 = N\rho, \ N_1 = N - N\rho. \tag{7.21}$$

Problem $\mathscr{P}3$ is a mixed-integer nonlinear programming (MINLP) problem. To solve $\mathscr{P}3$, we can decompose it into two subproblems: GAS link transmission probability and HAP selection. These two subproblems can be optimized iteratively in an alternating manner with one being fixed in each iteration until both reach convergence.

For a given HAP selection, $\{\mathbf{U}_1, \ldots, \mathbf{U}_N\}$, problem $\mathscr{P}3$ can be transformed as

$$\mathscr{P}3.1: \max_{\rho} \left(M\alpha(\rho)(1-\rho)\sum_{n=1}^{N} R_n^{G2S} + \beta(\rho)\rho \sum_{n=1}^{N}\sum_{k=1}^{K} u_{nk} R_{nk}^{GAS} \right)$$
$$\text{s.t.} \quad \text{C7–C9}, \tag{7.22}$$

which is convex, and the optimal ρ can be solved by existed math tools.

Then, given the GAS link transmission probability, ρ, for the ground users, problem $\mathscr{P}3$ can be transformed as

$$\mathscr{P}3.2: \max_{\{\mathbf{U}_1,\ldots,\mathbf{U}_N\}} \beta(\rho)\rho \sum_{n=1}^{N}\sum_{k=1}^{K} u_{nk} R_{nk}^{GAS}$$
$$\text{s.t.} \quad \text{C2–C4}. \tag{7.23}$$

Problem $\mathscr{P}3.2$ is integer linear programming, and the optimal $\{\mathbf{U}_1, \ldots, \mathbf{U}_N\}$ for GU can be solved by using the exhaustive method. The computing complexity is $\mathscr{O}(2^{NK}$ when the exhaustive method is adopted to solve the HAP selection subproblem. O this basis, it is observed that the exhaustive method can be adaptive to the scenario wit only a few number of ground users and HAPs. When a large number of ground user and HAPs exist in the network, the complexity of the exhaustive method become large because it exponentially increases with the number of ground users and HAP However, an optimal solution can be obtained through the alternating iteration metho only when the number of iterations is large enough. To address these challenge we can use a deep-learning method to solve the HAP selection problem mo efficiently.

According to the optimization solution of $\mathbb{D}$, the implementation algorithm of t GAS transmission scheme is presented in Algorithm 7.1, where each ground us makes a transmission decision to communicate with the satellite via the GAS link G2S link and then reserves the HAP$_k$ to assist its GAS link transmissions if the GA transmission scheme is selected.

Algorithm 7.1 HAP-reserved algorithm

```
1  Input: 𝔻.
2  for n= 1 to N do
3      if ρ_n = 0 then
4          GU_n transmits data to the satellite with G2S link;
5      end
6      else if ρ_n = 1 then
7          for k= 1 to K do
8              if u_nk = 1 then
9                  GU_n reserves HAP_k in the period of t_h;
10             end
11         end
12         GU_n transmits data to the satellite via HAP_k with GAS link in the period
           of t_r ;
13     end
14     else if 0 < ρ_n < 1 then
15         GU_n uses G2S link with probability 1−ρ_n;
16         GU_n uses GAS link with probability ρ_n;
17         Repeats line 7 to line 12;
18     end
19 end
```

7.3 Performance Analysis and Evaluation

7.3.1 Performance Analysis

We present three cases to show the overall throughput of the ground system when ground users adopt a GAS link with probability ρ_n (i.e., $0 < \rho_n < 1$, $\rho_n = 0$, and $\rho_n = 1$).

Case Given $0 < \rho_n < 1, \forall n \in \mathcal{N}, N_1 = N - \sum_{n=1}^{N} \rho_n$, and $N_2 = \sum_{n=1}^{N} \rho_n$, the overall throughput that ground users communicate with the satellite via HAP with probability ρ_n, $\mathbb{S}^{sum}$, is expressed as

$$\begin{aligned}\mathbb{S}^{sum} &= \mathbb{S}^{G2S} + \mathbb{S}^{GAS} \\ &= \omega_1 B \sum_{n=1}^{N} \frac{t_s p_s (1-\rho_n) \log_2 \left(1 + \frac{E_n^{G2S} 10^{-L_n^{G2S}/10}}{\sigma^2}\right)}{p_e \delta + p_s t_s + p_c t_c} \\ &\quad + \omega_2 B \sum_{n=1}^{N} \frac{t_r t_s'}{T t_h \zeta_s} \rho_n \sum_{k=1}^{K} u_{nk} \log_2 (1 + f(r_{nk}, r_{ks})).\end{aligned} \tag{7.24}$$

Proof When $\rho_n \in (0,1), \forall n \in \mathcal{N}$, it means that GU$_n$ transmits via the GAS link with a certain probability. In other words, N_1 ground users adopt the OFDMA-based G2S link transmission, and N_2 ground users adopt the GAS link transmission. Based on (7.5) and (7.6), by using $N_1 = N - \sum_{n=1}^{N} \rho_n$ in (7.15), and by using $N_2 = \sum_{n=1}^{N} \rho_n$ and $r_n = \frac{t_r}{N_s t_p}$ in (7.19), we can obtain the overall throughput as shown in (7.24). □

Case Given $\rho_n = 0, \forall n \in \mathcal{N}$, the overall throughput that ground users communicate with the satellite without the assistance of HAP, $\mathbb{S}^{G2S}$, is expressed as

$$\begin{aligned}\mathbb{S}^{G2S} &= \sum_{m=1}^{M} \frac{p'_s t_s}{p'_e \delta + p'_s t_s + p'_c t_c} \sum_{n=1}^{N} R_n^{G2S} \\ &= \frac{\omega_1 B p'_s t_s}{p'_e \delta + p'_s t_s + p'_c t_c} \sum_{n=1}^{N} \log_2 \left(1 + \frac{E_n^{G2S} 10^{-L_n^{G2S}/10}}{\sigma^2}\right).\end{aligned} \tag{7.25}$$

Proof When $\rho_n = 0, \forall n \in \mathcal{N}$, it means that all the ground users communicate with the satellite by adopting the G2S link transmission (i.e., $N_1 = N$, and $N_2 = 0$). According to (7.15), the overall throughput of all ground users can be calculated as in (7.25), where p'_s, p'_e, p'_c, τ', and p' can be calculated using N instead of N_1 in (7.11) through (7.13). □

Case Given $\rho_n = 1, \forall n \in \mathcal{N}$, the overall throughput that ground users communicate with the satellite via HAP, $\mathbb{S}^{GAS}$, is expressed as

$$\begin{aligned}\mathbb{S}^{GAS} &= \sum_{n=1}^{N} \frac{t_p r_n}{T} \sum_{k=1}^{K} u_{nk} R_{nk}^{GAS} \\ &= \frac{\omega_2 B t_r t'_s}{T t_h \zeta'_s} \sum_{n=1}^{N} \sum_{k=1}^{K} u_{nk} \log_2 (1 + f(r_{nk}, r_{ks})).\end{aligned} \tag{7.26}$$

Proof When $\rho_n = 1, \forall n \in \mathcal{N}$, it means that all the ground users communicate with the satellite by adopting the GAS link transmission (i.e., $N_1 = 0$, and $N_2 = N$). According to (7.16), the overall throughput of all ground users can be calculated as in (7.26), where ζ'_s, ρ', q', and ϵ' can be calculated using N instead of N_2 in (7.18). □

Based on the analysis of these three cases, we conclude that the performance of the GAS transmission scheme is better than the performance of the G2S transmission scheme when the number of ground users is small and there is an existing rational range of ρ to achieve better system performance.

7.3.2 Performance Evaluation

In the following, we evaluate the performance of the proposed transmission control strategy. The key parameters are $N = 100, M = 5, T = 200$ ms, $t_h = 10$ ms, and $t_p = 0.5$ ms. The packet size of RTS and CTS are 24 bytes and 16 bytes, respectively.

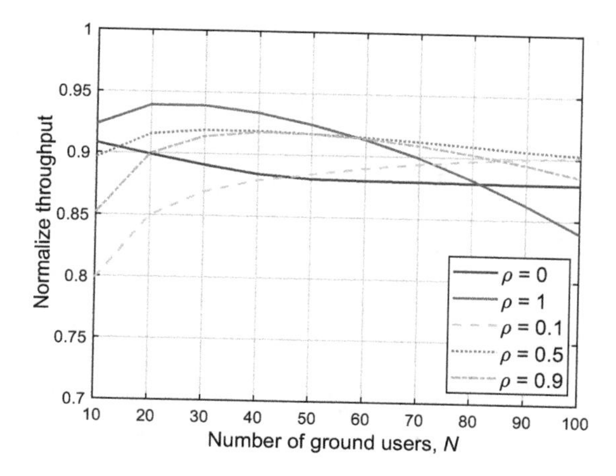

Figure 7.2 Normalized throughput versus N

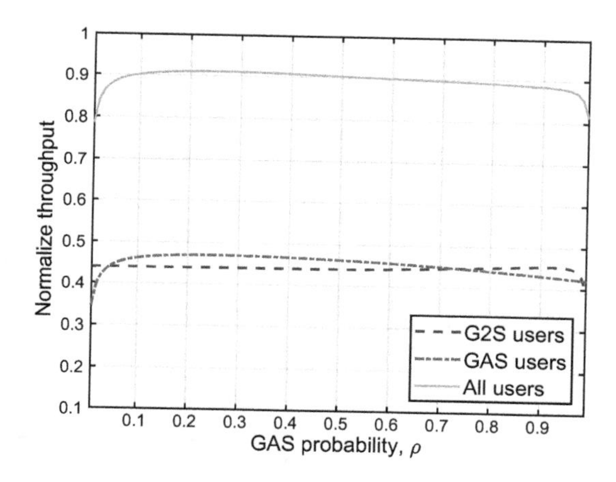

Figure 7.3 Normalized throughput versus ρ

Figure 7.2 shows that the normalized throughput changes with the number of ground users (N) under different GAS link probabilities (ρ). When $\rho = 0$, the ground users adopt G2S link to transmit, and the throughput of ground communication slightly declines due to the competition collision. When $\rho = 1$, each ground user selects a GAS link to communicate with the satellite. The throughput of ground communication first increases and then decreases. When $0 < \rho < 1$, each ground user uses a GAS link with probability ρ and uses a G2S link with probability $1 - \rho$. We can observe from Fig. 7.2 that when ρ is 0.1, 0.5, and 0.9, the throughput of ground communications in the three cases have similar trends: first increase and then decrease. Therefore, the proposed transmission control strategy can balance G2S and GAS. Figure 7.3 shows that the normalized throughput changes with the GAS link probability (ρ), where $0 < \rho < 1$. As shown in Fig. 7.3, the throughput of G2S users and GAS users first climbs and then decreases as ρ increases. Compare to the throughput of G2S users,

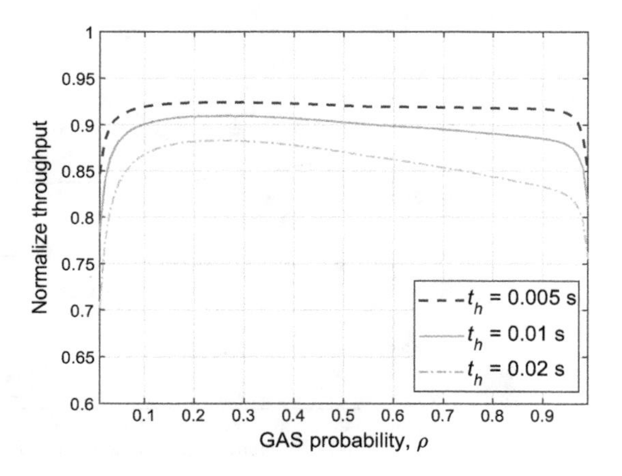

Figure 7.4 Normalized throughput versus ρ

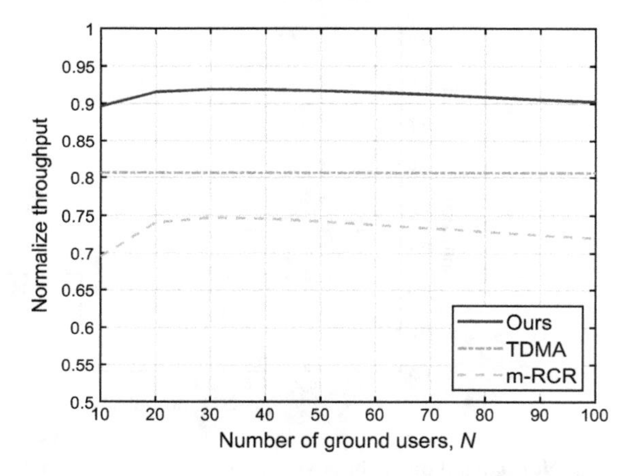

Figure 7.5 Normalized throughput versus N

the throughput of GAS users reaches the top earlier due to the low competition for GAS links and the high competition for G2S links when ρ is low.

Figure 7.4 shows that the normalized throughput changes with the probability (ρ) under the different ground negotiation periods (t_h). It can be seen that an optimal ρ exists in the three cases to maximize the system throughput. It also can be observed that a lower t_h leads to a better throughput due to the low negotiation overhead. Figure 7.5 shows that the normalized throughput changes with the number of ground users (N) in different schemes. Compared with the existed TDMA scheme and m-RCR scheme proposed in [181], the proposed strategy that can be adaptive to the GAS and G2S link outperforms the other two schemes because the spectrum efficiency is improved.

7.4 Conclusion

In this chapter, the HAP-reserved GAS transmission scheme was proposed to complement G2S link transmissions and strengthen conventional terrestrial communications. We investigated a transmission control strategy and explored a rational range of the GAS link transmission probability to maximize the system throughput. The numerical results show that the GAS transmission scheme can beat the G2S transmission scheme when the number of ground users is lower than 80 and achieve better system performance when ρ is in the range of $[0.1, 0.7]$.

Part IV

Satellite Communication Networks

8 Ultra-dense LEO Satellite Networks

Compared with MEO and GEO satellites, LEO satellite constellations have the advantages of low propagation delay and high link quality. Compared with traditional terrestrial networks, they are also superior in terms of seamless coverage and high-capacity backhaul support. Benefiting from high altitudes, broad operating spectrum, and an ultra-dense topology, LEO satellite networks can support a massive number of users with their more flexible access technique, which is less dependent on real environments. However, to further exploit the potential of LEO satellites, it remains to be explored how a LEO constellation influences the capacity of the network and how to deploy it to satisfy the demands of terrestrial users.

In this chapter, we first introduce the network architecture of satellite communications in Section 8.1. The application of satellite formation flying enabled by LEO satellites is presented in Section 8.2. A new satellite-terrestrial transmission technique enabled by reconfigurable holographic surfaces is also discussed in Section 8.3. The chapter is concluded by Section 8.4.

8.1 System Architecture

In this section, we first introduce the general network architecture where the LEO satellite networks can work in conjunction with terrestrial networks. Three different types of global communication services supported by such a network architecture are then illustrated.

8.1.1 General Network Architecture

Before discussing the network architectures, we first present two access modes of the satellite networks over different operating spectrum bands. As shown in Fig. 8.1, both the C-band (below 6 GHz) and high-frequency band (above 6 GHz), which require different access terminals, are considered.

1. *C-Band*: User devices can directly transmit to the satellite due to the strong diffraction capability of low-frequency signals. However, the data rate may be limited by the narrow frequency band.

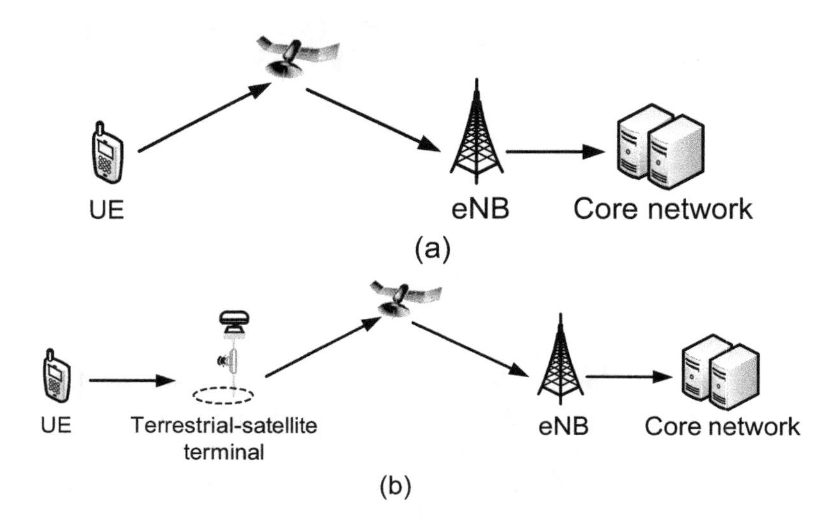

Figure 8.1 Two access modes for 5G satellite access networks. Satellite below 8 GHz (a) and above 6 GHz (b)

2. *High-Frequency Band*: User devices do not support direct communications, and each user needs to access the network via a terrestrial-satellite terminal (TST) as an access point (AP). This is because the user device is usually equipped with a limited number of antennas, which cannot support the direct connection between the user device and the satellite in a highly sensitive propagation environment. Therefore, to access the network, the user device first connects with the TST via the C-band spectrum, and then the TST uploads to the satellite via the high-frequency band.

In both cases, the satellite forwards the received data to the base station (BS) or the Earth gateway station (EGS), which is connected to the core network.

8.1.2 Global Communication Services via LEO-Based Satellite Access Networks

Based on the described network architecture, we now introduce the use cases and divide them into three types based on the 3GPP's classifications [182]. Each type depicts the network from a different perspective of service demand, but the corresponding use cases are not exclusive.

Service Continuity

As mentioned earlier, the ultra-dense topology of satellites guarantees seamless network access for users. Unlike traditional terrestrial networks where the BSs are deployed in populated areas, the deployment of a LEO satellite constellation is geographically egalitarian. Thus, network access is available to users outside the radio coverage of a terrestrial network. Based on the combination of both population-centric and geography-centric topologies in the integrated network, service continuity addresses the users who are not be well served while moving between terrestrial networks. Typical users include moving land vehicles (cars, trains, etc.) and airborne and maritime platforms (airplanes, ocean vessels, etc.).

Service Ubiquity

Service ubiquity is concerned with the satellite networks being able to provide services to unserved or underserved users suffering from a lack of terrestrial infrastructure or radio resources. Some use cases include the Internet of Things (IoT) for agriculture and environmental measurement [182], safety-critical emergency networks, and home access in remote areas and hot spots. Specifically, for IoT applications, the below 6 GHz access mode can be implemented in remote areas with little interference over the C-band to other users. The above 6 GHz access mode can be designed to work in populated areas such that cochannel interference to traditional terrestrial users can be avoided.

Service Scalability

Unlike traditional networks, the integrated networks enable large coverage corresponding to thousands of cells, thereby supporting both broadcast and multicast services via the satellites and user-specific services via the terrestrial small cells. This is especially efficient for offloading traffic from terrestrial networks by either multicasting during busy hours or broadcasting non-time-sensitive data when the propagation delay is tolerable.

8.2 LEO Satellite Formation Flying: Capacity Analysis

To further enable ultra-dense LEO satellite systems to accomplish various aerospace and earth science missions such as navigation, remote sensing, and Earth observation, there has been interest in applying the satellite *formation-flying* technique [183] to ultra-dense LEO satellite systems. Specifically, a fleet of small LEO satellites is placed into nearby orbits, forming a satellite cluster supported by intersatellite coordination, to achieve a common mission where the cooperation among LEO satellites enables communication and information processing with real-time communication capabilities. Based on the ultra-dense LEO satellite topology, satellites in the formation can cooperate as a large antenna array that can bring extra diversity gain [184], thereby overcoming the problem that only power gain can be obtained from traditional satellite LoS channels due to highly correlated satellite antennas. Therefore, the ultra-dense LEO satellite formation-flying system can provide much more reliable, efficient, and flexible data service to terrestrial terminal stations at a lower cost than traditional satellite communication networks.

Most works on LEO-based multiterminal satellite systems either focus on capacity optimization [185, 186] and beamforming design [187] or discuss the satellite formation-flying control [188]. In [185, 189], a geometrically optimized antenna arrangement scheme was proposed to improve the LoS channel capacity in multiterminal satellite systems. In [186], a collaborative scheme allowing LEO satellites to offload data among themselves utilizing intersatellite links was developed to maximize capacity in satellite-terminal links. In [187], the joint design of linear precoding and ground-based beamforming was proposed to mitigate the interference among users and improve the throughput in multibeam broadband satellite systems.

In [188], a predictive smooth variable structure filter based on a variable structure control concept was presented for attitude synchronization during satellite formation flying.

However, in most existing works the influence of satellite distribution and formation size have not been considered in ultra-dense LEO satellite formation design, although these are two crucial factors to improving the capacity performance. Existing theoretical results on capacity analysis have also not revealed such influence directly. Here, we consider evaluating the capacity performance in a fully connected ultra-dense LEO-based multiterminal satellite formation-flying system, where each terminal station connects to all satellites within its visibility region in one snapshot. Based on the theoretical capacity, we aim to investigate how satellite distribution and formation size influence capacity performance.

To achieve this goal, in this section we first analyze the upper bound of the channel capacity in the general case where each LEO satellite is equipped with multiple antennas. A special case of single-antenna satellites is considered where both the closed-form upper bound and lower bound of the channel capacity are derived. Based on the derived capacity, we then analyze the influence of satellite distribution (i.e., the visual range and altitude) and satellite formation size (i.e., the number of satellites in the formation) on the channel capacity. Simulation results are presented to verify our results.

8.2.1 System Model

As shown in Fig. 8.2, we consider a downlink ultra-dense LEO-based multiterminal satellite formation-flying system consisting of a set of satellites flying in a formation used as transmitters and a group of terminal stations used as receivers. Each LEO satellite and terminal station is equipped with K_t and K_r antennas, respectively. Benefiting from the dense satellite constellation, a group of LEO satellites are deployed in different orbits and remain close to each other, forming a satellite cluster supported by intersatellite coordination to extend coverage [190]. Specifically, the formation-flying LEO satellites send navigational measurements to the ground control center, based on which the center can issue instructions to maneuver satellites into appropriate positions in the formation. The LEO satellites in the formation can maintain a specific distance and orientation relative to each other at specified altitudes [183]. Therefore, by arranging the satellites in a planar or vertical manner, we consider the case where the set of satellites covers the same area of interest simultaneously, and each terminal station will connect to all these satellites.

Terrestrial-Satellite Topology

We denote the set of satellites flying in a formation as $\mathscr{M} = \{1, 2, \ldots, M\}$. We assume that the orbit of a satellite is circular, and the radii of satellite orbits are denoted as $\mathscr{R} = \{R_1, R_2, \ldots, R_M\}$. The Cartesian coordinate of satellite m can then be given by

$$\mathbf{a}_m^s = (R_m \cos\phi_m \cos\theta_m, R_m \cos\phi_m \sin\theta_m, R_m \sin\phi_m), \tag{8.1}$$

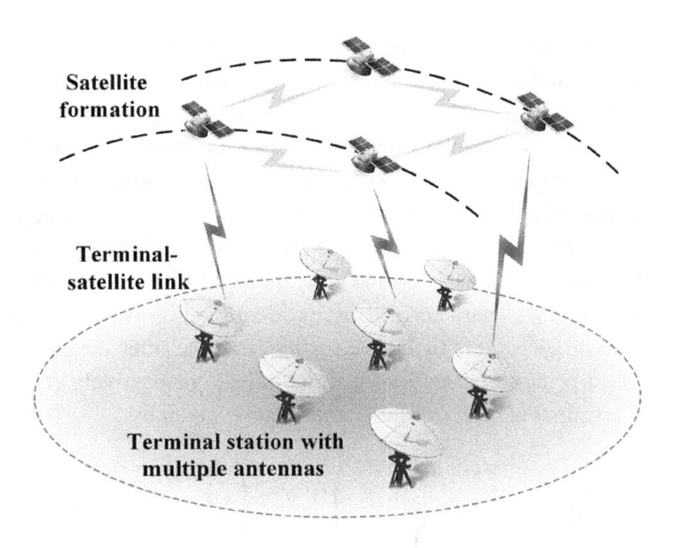

Figure 8.2 Ultra-dense LEO-based multiterminal satellite formation-flying system

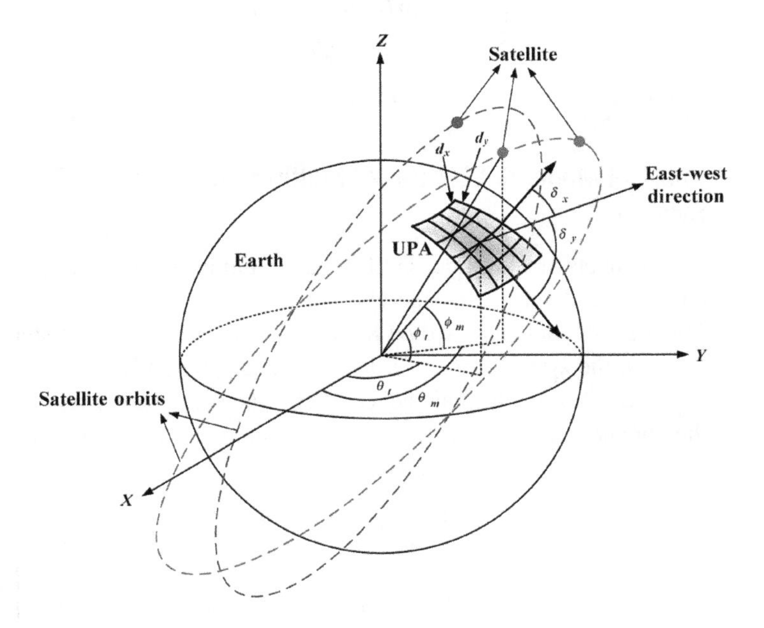

Figure 8.3 Terrestrial-satellite topology

where ϕ_m and θ_m denote the latitude and the longitude of the satellite orbit,[1] respectively, as illustrated in Fig. 8.3.

Note that the distance between any two terminal stations is much smaller than the propagation distance between each terminal station and the satellites. Therefore, from the perspective of each satellite, the distribution of terminal stations can be approximated by a uniform planar array (UPA) within the area of interest. For convenience, we assume that the terminal stations are arranged in a rectangular UPA, such that they are located on lines going in two orthogonal principle directions, forming a lattice structure. As shown in Fig. 8.3, the orientation of the UPA is characterized by the angles between its two orthogonal principal directions and the east-west direction δ_x and δ_y. The latitude and the longitude of the center of the UPA are denoted as ϕ_t and θ_t. We denote the number of terminal stations in the first and second principle directions as N and L, respectively. The position of a terminal station in the lattice can be characterized by its identification (ID) number in the first and second principle directions (n,l). We assume that the inter-terminal distances in the first and second principle directions are d_x and d_y, respectively. The Cartesian coordinate of the terminal station (n,l) $\mathbf{a}_{n,l}^t$ can then be given by

$$\begin{aligned}\mathbf{a}_{n,l}^t = R_e \begin{pmatrix} \cos\phi_t \cos\theta_t \\ \cos\phi_t \sin\theta_t \\ \sin\phi_t \end{pmatrix} - d_{x,n} \begin{pmatrix} \sin\phi_t \cos\theta_t \sin\delta_x + \sin\theta_t \cos\delta_x \\ \sin\phi_t \sin\theta_t \sin\delta_x - \cos\theta_t \cos\delta_x \\ -\cos\phi_t \sin\delta_x \end{pmatrix} \\ - d_{y,l} \begin{pmatrix} \sin\phi_t \cos\theta_t \sin\delta_y + \sin\theta_t \cos\delta_y \\ \sin\phi_t \sin\theta_t \sin\delta_y - \cos\theta_t \cos\delta_y \\ -\cos\phi_t \sin\delta_y \end{pmatrix},\end{aligned} \tag{8.2}$$

where $d_{x,n} = d_x(n - \frac{N+1}{2})$ and $d_{y,l} = d_y(l - \frac{L+1}{2})$.

8.2.2 Channel Model of LEO-Based Multiterminal Satellite Formation-Flying System

The channels of a ultra-dense LEO-based multiterminal satellite formation-flying system can be described by the channel transfer matrix $\boldsymbol{H} \in \mathbb{C}^{NLK_r \times MK_t}$ consisting of MK_t transmit antennas and NLK_r receive antennas. We denote the channel coefficient between the k_tth antenna of satellite m and k_rth antenna of terminal station (n,l) as $h_{nl,m}^{k_r,k_t}$. By merging the channel coefficient between each satellite and terminal station, the channel transfer matrix $\boldsymbol{H}$ in the system can be written in the form of a *block matrix*, which can then be given by

$$\boldsymbol{H} = \begin{pmatrix} H_{11,1} & \cdots & H_{11,m} & \cdots & H_{11,M} \\ \vdots & \ddots & \vdots & \ddots & \vdots \\ H_{1L,1} & \cdots & H_{1L,m} & \cdots & H_{1L,M} \\ \vdots & \ddots & \vdots & \ddots & \vdots \\ H_{nl,1} & \cdots & H_{nl,m} & \cdots & H_{nl,M} \\ \vdots & \ddots & \vdots & \ddots & \vdots \\ H_{N1,1} & \cdots & H_{N1,m} & \cdots & H_{N1,M} \\ \vdots & \ddots & \vdots & \ddots & \vdots \\ H_{NL,1} & \cdots & H_{NL,m} & \cdots & H_{NL,M} \end{pmatrix}, \tag{8.3}$$

where

$$H_{nl,m} = \begin{pmatrix} h_{nl,m}^{1,1} & h_{nl,m}^{1,2} & \cdots & h_{nl,m}^{1,K_t} \\ h_{nl,m}^{2,1} & h_{nl,m}^{2,2} & \cdots & h_{nl,m}^{2,K_t} \\ \vdots & \vdots & \ddots & \vdots \\ h_{nl,m}^{K_r,1} & h_{nl,m}^{K_r,2} & \cdots & h_{nl,m}^{K_r,K_t} \end{pmatrix}. \tag{8.4}$$

We consider the case where the LoS links dominate the communications between multiple terminal stations and satellites given frequency-flat fading due to the high-frequency bands [191]. Based on free-space propagation, the LoS channel coefficient $h_{nl,m}^{k_r,k_t}$ in equivalent baseband notation can then be given by [189]

$$h_{nl,m}^{k_r,k_t} = a_{nl,m}^{k_r,k_t} \exp\left\{-j\frac{2\pi f_c}{c_0} d_{nl,m}^{k_r,k_t}\right\}, \tag{8.5}$$

where $d_{nl,m}^{k_r,k_t}$ is the distance between the k_tth antenna of satellite m and k_rth antenna of terminal station (n,l). Carrier frequency is denoted by f_c, and the speed of light in free space by c_0. The notion $a_{nl,m}^{k_r,k_t}$ denotes the free-space propagation gain, which can be calculated by $a_{nl,m}^{k_r,k_t} = c_0 \cdot e^{j\varphi_0}/(4\pi f_c d_{nl,m}^{k_r,k_t})$, where φ_0 is the carrier phase angle at the time of observation.[2]

8.2.3 Channel Capacity of LEO-Based Multiterminal Satellite Formation-Flying System

We denote the signal vectors at the transmitter and the receiver as $\boldsymbol{x} \in \mathbb{C}^{MK_t}$ and $\boldsymbol{y} \in \mathbb{C}^{NLK_r}$, respectively. The additive white Gaussian noise (AWGN) is denoted as $\boldsymbol{w} \in \mathbb{C}^{NLK_r}$ with the covariance matrix $\mathbb{E}[\boldsymbol{w}\boldsymbol{w}^{\mathrm{H}}] = \sigma^2 \boldsymbol{I}_{NLK_r}$, where $\mathbb{E}[\cdot]$ is the expectation operation and $(\cdot)^{\mathrm{H}}$ denotes the conjugate, complex transpose. Therefore, the signal received by terminal stations from LEO satellites can be given by

$$\boldsymbol{y} = \boldsymbol{H}\boldsymbol{x} + \boldsymbol{w}. \tag{8.6}$$

Unlike traditional multiple input, multiple output (MIMO) systems, it is hard to obtain the spatial diversity gain for satellite systems only from multiple antennas of the same satellite due to the highly correlated antenna coefficients. Due to the dense topology of satellites, high channel capacity can be obtained from spatial diversity. We assume that each LEO satellite has perfect channel state information (CSI) for all terminal stations, each terminal station only has perfect CSI for its own channel, and the transmit symbols are realizations of uncorrelated independent, identically distributed (i.i.d.) Gaussian random variables. Because the SNR of the satellite-terminal station link is high due to the large antenna gain, the optimal channel capacity can be achieved by equal-power allocation [192], which is calculated by [193]

$$C = \log_2[\det(\boldsymbol{I}_{MK_t} + \rho \cdot \boldsymbol{H}^{\mathrm{H}}\boldsymbol{H})], \tag{8.7}$$

where ρ denotes all the gains of the satellite-terminal station link budget except the LoS pathloss of the channel, while the LoS pathloss is included in the channel transfer matrix $\boldsymbol{H}$.

8.2.4 Capacity Analysis of LEO-Based Multiterminal Satellite Formation-Flying

We first derive the upper bound of the channel capacity of downlink LEO-based multiterminal satellite formation-flying systems. We then apply the derived bounds to the special case of single-antenna satellites to give the specific forms of upper and lower bounds of the channel capacity.

Upper Bound of the Channel Capacity

To derive the upper bound of the channel capacity, we first reformulate the channel capacity into the determinant of a matrix with columns of the same magnitude. Following that, we give the upper bound of the determinant by estimating the eigenvalues of the reformulated matrix, based on which upper bound of the channel capacity is derived.

Considering that the distance between different satellites and the terminal stations varies, to balance the scale of the columns of $\boldsymbol{H}$, we decompose $\boldsymbol{H}$ into $\boldsymbol{H} = \tilde{\boldsymbol{H}}\boldsymbol{P}$, where $\boldsymbol{P}$ is a diagonal matrix. Moreover, $\boldsymbol{P}$ can be written as a block matrix, which can be given by

$$\boldsymbol{P} = \begin{pmatrix} a_1 \boldsymbol{I}_{K_t} & & & \\ & a_2 \boldsymbol{I}_{K_t} & & \\ & & \ddots & \\ & & & a_M \boldsymbol{I}_{K_t} \end{pmatrix}, \tag{8.8}$$

where $a_m = c_0/(4\pi f_c d_m)$, and d_m is the distance between satellite m and the center of the UPA. Because the Cartesian coordinate of the center of the UPA can be given by

$$\boldsymbol{a}^t = (R_e \cos\phi_t \cos\theta_t, R_e \cos\phi_t \sin\theta_t, R_e \sin\phi_t), \tag{8.9}$$

we have

$$d_m^2 = \|\boldsymbol{a}_m^s - \boldsymbol{a}^t\|^2 = R_m^2 + R_e^2 - 2R_m R_e A_m, \tag{8.10}$$

where

$$A_m = \cos(\phi_m - \phi_t)\cos(\theta_m - \theta_t) + 2\sin\phi_m \sin\phi_t \sin^2\frac{\theta_m - \theta_t}{2}. \tag{8.11}$$

The channel capacity of LEO-based multiterminal satellite formation-flying system given by (8.7) can then be rewritten as

$$\begin{aligned} C_{\text{multiple}} &= \log_2\{\det(\boldsymbol{P}^{\mathrm{H}}[(\boldsymbol{P}\boldsymbol{P}^{\mathrm{H}})^{-1} + \rho\tilde{\boldsymbol{H}}^{\mathrm{H}}\tilde{\boldsymbol{H}}]\boldsymbol{P})\} \\ &= \log_2\{\det(\boldsymbol{P}^{\mathrm{H}}) \cdot \det[(\boldsymbol{P}\boldsymbol{P}^{\mathrm{H}})^{-1} + \rho\tilde{\boldsymbol{H}}^{\mathrm{H}}\tilde{\boldsymbol{H}}] \cdot \det(\boldsymbol{P})\} \\ &= \log_2\left\{\prod_{m=1}^{M}(a_m^2)^{K_t} \cdot \det[(\boldsymbol{P}\boldsymbol{P}^{\mathrm{H}})^{-1} + \rho\tilde{\boldsymbol{H}}^{\mathrm{H}}\tilde{\boldsymbol{H}}]\right\}. \end{aligned} \tag{8.12}$$

Note that the rank of $\tilde{\boldsymbol{H}}^H \tilde{\boldsymbol{H}}$ is determined by $\min(MK_t, NLK_r)$. We derive the upper bound of the channel capacity in two cases: $MK_t \leq NLK_r$ and $MK_t > NLK_r$.

In this case of $MK_t \leq NLK_r$, we first give a lemma to illustrate that the determinant of a matrix can be upper bounded by its trace.

LEMMA 8.1 *Define a matrix* $\boldsymbol{A}$ *and denote its eigenvalues as* $\lambda_1, \lambda_2, \ldots, \lambda_{N_a}$. *By utilizing the inequality of arithmetic and geometric means, the upper bound of* $\det \boldsymbol{A}$ *can then be obtained by*

$$\det \boldsymbol{A} = \prod_{n_a=1}^{N_a} \lambda_{n_a} \leq \left(\frac{\sum_{n_a=1}^{N_a} \lambda_{n_a}}{N_a} \right)^{N_a} = \left(\frac{\operatorname{tr} \boldsymbol{A}}{N_a} \right)^{N_a}. \tag{8.13}$$

Because the trace of $(\boldsymbol{P}\boldsymbol{P}^{\mathrm{H}})^{-1} + \rho \tilde{\boldsymbol{H}}^{\mathrm{H}} \tilde{\boldsymbol{H}}$ *can be given by*

$$\operatorname{tr}\left[(\boldsymbol{P}\boldsymbol{P}^H)^{-1} + \rho \tilde{\boldsymbol{H}}^H \tilde{\boldsymbol{H}} \right] = K_t \sum_{m=1}^{M} a_m^{-2} + \rho \sum_{m=1}^{M} \sum_{k_t=1}^{K_t} \left[a_m^{-2} \sum_{n=1}^{N} \sum_{l=1}^{L} \sum_{k_r=1}^{K_r} (a_{nl,m}^{k_r,k_t})^2 \right], \tag{8.14}$$

substituting (8.14) *with* (8.12), *the upper bound of the channel capacity in the case of* $MK_t \leq NLK_r$ *can be obtained by*

$$C_{\text{multiple}} \leq K_t \log_2 \left(\prod_{m=1}^{M} a_m^2 \right) + MK_t \log_2 \left(\frac{\operatorname{tr}[(\boldsymbol{P}\boldsymbol{P}^H)^{-1} + \rho \tilde{\boldsymbol{H}}^H \tilde{\boldsymbol{H}}]}{MK_t} \right). \tag{8.15}$$

In this case of $MK_t > NLK_r$, note that the magnitude of ρ is larger than that of a_m^{-2}, indicating that the magnitude of nonzero eigenvalues of $\rho \tilde{\boldsymbol{H}}^H \tilde{\boldsymbol{H}}$ is larger than that of $(\boldsymbol{P}\boldsymbol{P}^H)^{-1}$.

Therefore, $(\boldsymbol{P}^H)^{-1}$ can be seen as a small perturbation to $\rho \tilde{\boldsymbol{H}}^H \tilde{\boldsymbol{H}}$, and thus, we give the following lemma estimating the perturbation of the eigenvalues of $\rho \tilde{\boldsymbol{H}}^H \tilde{\boldsymbol{H}}$ to derive the upper bound of $\det[(\boldsymbol{P}\boldsymbol{P}^H)^{-1} + \rho \tilde{\boldsymbol{H}}^H \tilde{\boldsymbol{H}}]$.

LEMMA 8.2 (Weyl's inequality in matrix theory [194]) *Suppose* A *and* B *are two* $n-by-n$ *Hermitian matrices, where* A *has eigenvalues* $\lambda_1 \geq \cdots \geq \lambda_n$ *and* $A+B$ *has eigenvalues* $\mu_1 \geq \cdots \geq \mu_n$. *Then,* $|\mu_i - \lambda_i| \leq \|B\|_2$, $\forall 1 \leq i \leq n$, *where* $\|\cdot\|_2$ *denotes the spectral norm of the matrix.*

We denote the eigenvalues of $\tilde{\boldsymbol{H}}^{\mathrm{H}} \tilde{\boldsymbol{H}}$ and $(\boldsymbol{P}\boldsymbol{P}^{\mathrm{H}})^{-1} + \rho \tilde{\boldsymbol{H}}^{\mathrm{H}} \tilde{\boldsymbol{H}}$as $\lambda_1, \ldots, \lambda_{MK_t}$ and $\mu_1, \ldots, \mu_{MK_t}$, respectively, which are arranged in descending order. Because $MK_t > NLK_r$, the rank of $\tilde{\boldsymbol{H}}^{\mathrm{H}} \tilde{\boldsymbol{H}}$ is NLK_r, indicating that $\lambda_{NLK_r+1} = \cdots = \lambda_{MK_t} = 0$. Note that $\|(\boldsymbol{P}\boldsymbol{P}^{\mathrm{H}})^{-1}\|_2 = \max_{1 \leq m \leq M} \{a_m^{-2}\}$, by applying Weyl's inequality given in Lemma 8.2 to $\rho \tilde{\boldsymbol{H}}^{\mathrm{H}} \tilde{\boldsymbol{H}}$ and $(\boldsymbol{P}\boldsymbol{P}^{\mathrm{H}})^{-1}$, we have

$$\begin{aligned} \mu_i &\leq \rho \lambda_i + \max_{1 \leq m \leq M} \{a_m^{-2}\}, \quad \forall\, 1 \leq i \leq NLK_r, \\ \mu_i &\leq \max_{1 \leq m \leq M} \{a_m^{-2}\}, \quad \forall\, NLK_r + 1 \leq i \leq MK_t. \end{aligned} \tag{8.16}$$

Therefore, the upper bound of $\det[(\boldsymbol{P}\boldsymbol{P}^{\mathrm{H}})^{-1}+\rho\tilde{\boldsymbol{H}}^{\mathrm{H}}\tilde{\boldsymbol{H}}]$ can be given by

$$\begin{aligned}\det\left[(\boldsymbol{P}\boldsymbol{P}^{\mathrm{H}})^{-1}+\rho\tilde{\boldsymbol{H}}^{\mathrm{H}}\tilde{\boldsymbol{H}}\right] &= \prod_{i=1}^{MK_t}\mu_i \\ &\leq \prod_{i=1}^{NLK_r}\left(\rho\lambda_i+\max_{1\leq m\leq M}\{a_m^{-2}\}\right)\cdot\left(\max_{1\leq m\leq M}\{a_m^{-2}\}\right)^{MK_t-NLK_r} \\ &\overset{(a)}{\leq}\left[\frac{NLK_r\max_{1\leq m\leq M}\{a_m^{-2}\}+\rho\mathrm{tr}(\tilde{\boldsymbol{H}}^{\mathrm{H}}\tilde{\boldsymbol{H}})}{NLK_r}\right]^{NLK_r}\cdot\left(\max_{1\leq m\leq M}\{a_m^{-2}\}\right)^{MK_t-NLK_r},\end{aligned} \tag{8.17}$$

where in (a) the inequality of arithmetic and geometric means is utilized. Because the trace of $\tilde{\boldsymbol{H}}^{\mathrm{H}}\tilde{\boldsymbol{H}}$ satisfies

$$\mathrm{tr}(\tilde{\boldsymbol{H}}^{\mathrm{H}}\tilde{\boldsymbol{H}})=\sum_{m=1}^{M}\sum_{k_t=1}^{K_t}\left[a_m^{-2}\sum_{n=1}^{N}\sum_{l=1}^{L}\sum_{k_r=1}^{K_r}\left(a_{nl,m}^{k_r,k_t}\right)^2\right], \tag{8.18}$$

combining (8.12) and (8.17), the upper bound of the channel capacity in the case of $MK_t > NLK_r$ can be obtained by

$$\begin{aligned}C_{\text{multiple}} \leq\, & NLK_r\log_2\left\{\max_{1\leq m\leq M}\{a_m^{-2}\}+\frac{\rho}{NLK_r}\sum_{m=1}^{M}\sum_{k_t=1}^{K_t}\left[a_m^{-2}\sum_{n=1}^{N}\sum_{l=1}^{L}\sum_{k_r=1}^{K_r}(a_{nl,m}^{k_r,k_t})^2\right]\right\} \\ &+(MK_t-NLK_r)\log_2\left(\max_{1\leq m\leq M}\{a_m^{-2}\}\right)+K_t\log_2\left(\prod_{m=1}^{M}a_m^2\right).\end{aligned} \tag{8.19}$$

Combining the upper bound of the channel capacity obtained in the two cases given in (8.15) and (8.19), the following proposition holds.

PROPOSITION 8.1 *The upper bound of the channel capacity in downlink LEO-based multiterminal satellite formation-flying systems can be obtained by*

$$C_{\text{multiple}} \leq \begin{cases} K_t\log_2\left(\prod_{m=1}^{M}a_m^2\right)+MK_t\log_2\left(\frac{\mathrm{tr}\left[(\boldsymbol{P}\boldsymbol{P}^{\mathrm{H}})^{-1}+\rho\tilde{\boldsymbol{H}}^{\mathrm{H}}\tilde{\boldsymbol{H}}\right]}{MK_t}\right), & MK_t\leq NLK_r, \quad (a) \\ NLK_r\log_2\left\{\max_{1\leq m\leq M}\{a_m^{-2}\}+\frac{\rho}{NLK_r}\sum_{m=1}^{M}\sum_{k_t=1}^{K_t}\left[a_m^{-2}\sum_{n=1}^{N}\sum_{l=1}^{L}\sum_{k_r=1}^{K_r}\left(a_{nl,m}^{k_r,k_t}\right)^2\right]\right\} & \\ +K_t\log_2\left(\prod_{m=1}^{M}a_m^2\right)+(MK_t-NLK_r)\log_2\left(\max_{1\leq m\leq M}\{a_m^{-2}\}\right), & MK_t>NLK_r, \quad (b), \end{cases} \tag{8.20}$$

where $\mathrm{tr}[(\boldsymbol{P}\boldsymbol{P}^{\mathrm{H}})^{-1}+\rho\tilde{\boldsymbol{H}}^{\mathrm{H}}\tilde{\boldsymbol{H}}]$ *is given in* (8.14).

Special Case: LEO-Based Multiterminal Systems with Multiple Single-Antenna Satellites

To reveal more insights about the system capacity, we consider the special case where each formation-flying LEO satellite is equipped with a single antenna. We utilize the UPA structure to model the uniform distribution of the terminal stations.

Remark 8.1 The single-antenna satellite model is common in the performance analysis of satellite networks [189, 195, 196]. It has also been adopted for the cube satellite antenna design to avoid the use of complicated deployment mechanisms [197].

Note that it is not accurate enough to analyze the capacity properties (such as monotonicity) only based on the upper bound derived in the last section. Here we derive both closed-form upper bound and lower bound of the channel capacity in the single-antenna satellite case, which gives a good approximation of the actual value. We will explain in detail later that the insights obtained in the single-antenna satellite case can also be extended to the general multiantenna satellite case.

The key idea is to recast capacity into the determinant of a positive definite Hermitian matrix. Because each satellite is equipped with a single antenna, and the distance among different antennas of the same terminal station is much smaller than the propagation distance between the terminal station and the satellites (i.e., $d_{nl,m}^{k_r} = d_{nl,m}$), the distance between satellite m and terminal station (n,l) can be given by

$$\begin{aligned} d_{nl,m}^2 &= \|\mathbf{a}_m^s - \mathbf{a}_{n,l}^t\|^2 \\ &= R_m^2 + R_e^2 - 2R_m R_e A_m - 2R_m d_{x,n} B_m \\ &\quad - 2R_m d_{y,l} C_m + d_{x,n}^2 + d_{y,l}^2 + \mathcal{O}(d_{x,n} \cdot d_{y,l}), \end{aligned} \tag{8.21}$$

where

$$\begin{aligned} B_m &= \sin(\phi_m - \phi_t)\sin\delta_x \cos(\theta_m - \theta_t) + \cos\phi_m \cos\delta_x \\ &\quad \cdot \sin(\theta_m - \theta_t) + 2\sin\phi_m \cos\phi_t \sin\delta_x \sin^2\frac{\theta_m - \theta_t}{2}, \\ C_m &= \sin(\phi_m - \phi_t)\sin\delta_y \cos(\theta_m - \theta_t) + \cos\phi_m \cos\delta_y \\ &\quad \cdot \sin(\theta_m - \theta_t) + 2\sin\phi_m \cos\phi_t \sin\delta_y \sin^2\frac{\theta_m - \theta_t}{2}. \end{aligned} \tag{8.22}$$

Note that A_m is defined in (8.11), and $\mathcal{O}(\cdot)$ means that the item is of the same order as $d_{x,n} \cdot d_{y,l}$, which can be neglected. The distance $d_{nl,m}$ can then be rewritten as

$$d_{nl,m} = Z_m - U_m d_{x,n} - V_m d_{y,l}, \tag{8.23}$$

where $Z_m = \sqrt{R_m^2 + R_e^2 - 2R_m R_e A_m}$, $U_m = B_m R_m / Z_m$, $V_m = C_m R_m / Z_m$. Moreover, we apply the approximation $a_{nl,m}^{k_r} \approx a_m = c_0/(4\pi f_c d_m)$, where d_m denotes the distance between satellite m and the center of the UPA. Therefore, the element $[\boldsymbol{H}]_{nl,m}^{k_r,k_t}$ of the channel transfer matrix can be rewritten as

$$H_{nl,m}^{k_r,k_t} = H_{nl,m} = a_m \exp\left\{-j\frac{2\pi f_c}{c_0} d_{nl,m}\right\} = \tilde{a}_m u_m^{n-1} v_m^{l-1}, \tag{8.24}$$

where $\tilde{a}_m = a_m \exp\{-j\frac{2\pi f_c}{c_0}(Z_m - U_m d_{x,1} - V_m d_{y,1})\}$, $u_m = \exp(-j\frac{2\pi f_c}{c_0} U_m d_x)$, and $v_m = \exp(-j\frac{2\pi f_c}{c_0} V_m d_y)$. $\boldsymbol{H}$ is a nonorthogonal LOS channel transfer matrix, and thus, $\boldsymbol{H}^{\mathrm{H}}\boldsymbol{H}$ can be expressed as $\boldsymbol{H}^{\mathrm{H}}\boldsymbol{H} = K_r \boldsymbol{D}^{\mathrm{H}}\boldsymbol{Q}^{\mathrm{H}}\boldsymbol{Q}\boldsymbol{D}$, where $\boldsymbol{D} = \mathrm{diag}\{\tilde{a}_1, \tilde{a}_2, \ldots, \tilde{a}_M\}$, $\boldsymbol{Q} \in \mathbb{C}^{NL\times M}$, and the $((n-1)L+l, m)$th element of $\boldsymbol{Q}$ (i.e., $[\boldsymbol{Q}]_{nl,m}$) is given by $u_m^{n-1} v_m^{l-1}$. Therefore, the channel capacity given in (8.7) is equivalent to

$$C_{\text{single}} = \log_2[\det(\boldsymbol{I}_M + \rho K_r \cdot \boldsymbol{D}^{\mathrm{H}}\boldsymbol{Q}^{\mathrm{H}}\boldsymbol{Q}\boldsymbol{D})]. \tag{8.25}$$

Upper Bound of the Channel Capacity

As shown in (8.25), the channel capacity can be calculated by $\det(\boldsymbol{I}_M + \rho K_t \cdot \boldsymbol{D}^{\mathrm{H}}\boldsymbol{Q}^{\mathrm{H}}\boldsymbol{Q}\boldsymbol{D})$, which can be rewritten as

$$\det\left[\rho K_r \boldsymbol{D}^{\mathrm{H}}\left(\frac{1}{\rho K_r}\mathrm{diag}\left\{\frac{1}{a_1^2}, \ldots, \frac{1}{a_M^2}\right\} + \boldsymbol{Q}^{\mathrm{H}}\boldsymbol{Q}\right)\boldsymbol{D}\right]. \tag{8.26}$$

The channel capacity can be upper bounded by the trace of $(\frac{1}{\rho K_r}\mathrm{diag}\{\frac{1}{a_1^2}, \ldots, \frac{1}{a_M^2}\} + \boldsymbol{Q}^{\mathrm{H}}\boldsymbol{Q})$. Because $[\boldsymbol{Q}^{\mathrm{H}}\boldsymbol{Q}]_{mm} = NL$, the following proposition can then be derived from Lemma 8.1.

PROPOSITION 8.2 *The upper bound of the channel capacity in downlink LEO-based multi-terminal systems with multiple single-antenna satellites can be obtained by*

$$C_{\text{single}} \le \log_2\left(\prod_{m=1}^{M} a_m^2\right) + \log_2\left(\frac{\rho K_r MNL + \sum_m \frac{1}{a_m^2}}{M}\right)^M. \tag{8.27}$$

Remark 8.2 The upper bound of the channel capacity in this special single-antenna satellite given in Proposition 8.2 can also be obtained directly by setting $K_t = 1$ in the case of $MK_t \le NLK_r$ and applying the approximation $a_{nl,m}^{k_r} \approx a_m$.

Lower Bound of the Channel Capacity

To derive the lower bound of the channel capacity, we first illustrate that the channel capacity can be lower bounded by the determinant of $\boldsymbol{Q}^{\mathrm{H}}\boldsymbol{Q}$ (i.e., Proposition 8.3), which is determined by the eigenvalues of $\boldsymbol{Q}^{\mathrm{H}}\boldsymbol{Q}$. Following that, we analyze the range of the eigenvalues of $\boldsymbol{Q}^{\mathrm{H}}\boldsymbol{Q}$ (i.e., Proposition 8.4), and the lower bound of the channel capacity can be finally derived.

LEMMA 8.3 *If $M < L$ and $v_1, \ldots, v_M$ are distinct (or $M < N$ and $u_1, \ldots, u_M$ are distinct correspondingly), then $\boldsymbol{Q}$ has full column rank, and $\boldsymbol{Q}^{\mathrm{H}}\boldsymbol{Q}$ is a positive definite Hermitian matrix.*

Proof Without loss of generality, we assume that $M < L$, and $v_1, \ldots, v_M$ are distinct. The first M rows of $\boldsymbol{Q}$ then form a Vandermonde matrix of order M with determinant $\prod_{1<j<i\le M}(v_i - v_j) \ne 0$. Therefore, $\boldsymbol{Q}$ has a nonzero $M \times M$ subdeterminant, indicating that $\mathrm{rank}(\boldsymbol{Q}) = M$ and thus $\boldsymbol{Q}$ has full column rank. Therefore, $\boldsymbol{Q}^{\mathrm{H}}\boldsymbol{Q}$ is a positive definite Hermitian matrix [198]. □

The following proposition can then be derived from Lemma 8.3.

PROPOSITION 8.3

$$C_{\text{single}} > \log_2\left[(\rho K_r)^M \prod_{m=1}^{M} a_m^2 \det(\boldsymbol{Q}^{\mathrm{H}}\boldsymbol{Q})\right]. \tag{8.28}$$

Proof Because $\boldsymbol{Q}^{\mathrm{H}}\boldsymbol{Q}$ is a positive definite Hermitian matrix, $\boldsymbol{D}^{\mathrm{H}}\boldsymbol{Q}^{\mathrm{H}}\boldsymbol{Q}\boldsymbol{D}$ is also positive definite and has positive eigenvalues. We denote the eigenvalues of $\boldsymbol{D}^{\mathrm{H}}\boldsymbol{Q}^{\mathrm{H}}\boldsymbol{Q}\boldsymbol{D}$ as $\tilde{\eta}_1, \tilde{\eta}_2, \ldots, \tilde{\eta}_M$, where $\prod_{m=1}^{M}\tilde{\eta}_m = \det(\boldsymbol{D}^{\mathrm{H}}\boldsymbol{Q}^{\mathrm{H}}\boldsymbol{Q}\boldsymbol{D})$. Therefore, $\det(\boldsymbol{I}_{MK_t} + \rho\boldsymbol{H}^{\mathrm{H}}\boldsymbol{H})$ can be given by

$$\begin{aligned}\det(\boldsymbol{I}_{MK_t} + \rho\boldsymbol{H}^{\mathrm{H}}\boldsymbol{H}) &= \prod_{m=1}^{M}(1+\rho K_r\tilde{\eta}_m) > \prod_{m=1}^{M}\rho K_t\tilde{\eta}_m \\ &= (\rho K_r)^M \prod_{m=1}^{M} a_m^2 \det(\boldsymbol{Q}^{\mathrm{H}}\boldsymbol{Q}).\end{aligned} \tag{8.29}$$

□

Therefore, we can obtain the lower bound of C_{single} by analyzing the eigenvalues of $\boldsymbol{Q}^{\mathrm{H}}\boldsymbol{Q}$. We first define

$$q_{ij} = (1+\bar{u}_i u_j + \cdots + \bar{u}_i^{\,N-1}u_j^{N-1})(1+\bar{v}_i v_j + \cdots + \bar{v}_i^{\,L-1}v_j^{L-1}). \tag{8.30}$$

Note that $q_{11} = q_{22} = \cdots = q_{MM} = NL$, and $\boldsymbol{Q}^{\mathrm{H}}\boldsymbol{Q}$ can then be given by

$$\boldsymbol{Q}^{\mathrm{H}}\boldsymbol{Q} = \begin{bmatrix} NL & q_{12} & \cdots & q_{1M} \\ q_{21} & NL & \cdots & q_{2M} \\ \vdots & \vdots & & \vdots \\ q_{M1} & q_{M2} & \cdots & NL \end{bmatrix}. \tag{8.31}$$

LEMMA 8.4 (Gershgorin circle theorem [199]) *Suppose* $\boldsymbol{A} = [a_{i,j}] \in \mathbb{C}^{M\times M}$. *Then, each eigenvalue of* $\boldsymbol{A}$ *lies in* $\cup_{i=1}^{M} D_i$, *where* D_i *is the disk centered at* $a_{i,i}$ *with radius* $r_i = \sum_{j=1, j\neq i}^{M} |a_{i,j}|$ *(i.e.,* $D_i = \{z \| z - a_{i,i}| < r_i, z \in \mathbb{C}\}$*).*

Applying Lemma 8.4 gives the following proposition.

PROPOSITION 8.4 *The eigenvalues of* $\boldsymbol{Q}^{\mathrm{H}}\boldsymbol{Q}$ *fall within* $[NL-\Delta,\ NL+\Delta]$, *where*

$$\Delta = \max_{1\le i\le M}\left\{\sum_{j=1, j\neq i}^{M} |q_{ij}|\right\}. \tag{8.32}$$

We set the eigenvalues of $\boldsymbol{Q}^{\mathrm{H}}\boldsymbol{Q}$ as $\eta_1, \ldots, \eta_M \in [NL-\Delta, NL+\Delta]$. Because $\sum_{m=1}^{M}\eta_m = \mathrm{tr}(\boldsymbol{Q}^{\mathrm{H}}\boldsymbol{Q}) = NLM$, we have $\det(\boldsymbol{Q}^{\mathrm{H}}\boldsymbol{Q}) = \prod_{m=1}^{M}\eta_m \ge (N^2L^2 - \Delta^2)^{\frac{M}{2}}$ [200], and thus, the following proposition follows from Proposition 8.3.

PROPOSITION 8.5 *The lower bound of the channel capacity in downlink LEO-based multi-terminal systems with multiple single-antenna satellites can be obtained by*

$$C_{\text{single}} > \log_2\left\{(\rho K_r)^M \prod_{m=1}^{M} a_m^2 \left[(NL)^2 - \Delta^2\right]^{\frac{M}{2}}\right\}. \tag{8.33}$$

8.2.5 Influence of Satellite Distribution and Formation Size on Capacity

In this section, based on the capacity bounds derived in the previous section, we analyze the influence of satellite distribution (including the visual range and formation altitude) and formation size on the channel capacity of downlink LEO-based systems with multiple single-antenna satellites. The applicability of the theoretical result in the single-antenna satellite case to the general multiantenna satellite case will be verified in Section 8.2.6.

Note that the satellites flying in the formation can be seen as a large "virtual satellite" given a specific formation configuration [184]. Both the single-antenna satellite case and the multiantenna satellite case can be seen as a large "virtual satellite" with multiple antennas, indicating the similarities between the two cases. Specifically, the channel matrices in the two cases share a similar structure due to the LoS-dominated channels. Therefore, the capacity in the two cases is similar in their functional relation with respect to satellite distribution and satellite formation size. The theoretical results of the single-antenna satellite case can be reasonably generalized to the multiantenna satellite case.

Influence of the Satellite Distribution

Given a fixed number of satellites, we first illustrate the influence of the satellite distribution on the number of accessible terminal stations, (N, L). We then analyze the influence of satellite distribution on the upper and lower bounds of the channel capacity separately, based on which the influence of satellite distribution on the channel capacity is given.

Influence on (N, L)

We assume the latitude and the longitude of the satellites are uniformly distributed with a center (ϕ_t, θ_t) (i.e., $\phi_m \sim U[\phi_t - \Delta\phi, \phi_t + \Delta\phi]$, $\theta_m \sim U[\theta_t - \Delta\theta, \theta_t + \Delta\theta]$), and the altitude of the satellite formation is R_f. Therefore, the length of the subsatellite region (i.e., the region directly below the satellites) along the warp and the weft are $R_e \cos\phi_t \Delta\theta$ and $R_e \Delta\phi$, respectively. Without loss of generality, we assume that the length of the accessible terrestrial region outside the subsatellite region along the two principle directions of the UPA is proportional to the satellite formation altitude, as shown in Fig. 8.4. Therefore, we have $N \propto R_f\tau + R_e \cos\phi_t \Delta\theta$ and $L \propto R_f\tau + R_e\Delta\phi$, where τ is a parameter relevant to the value of the visual range and the satellite formation altitude.[3] The influence of the visual range where satellites are visible to the terrestrial terminal station in one snapshot (i.e., $\Delta\phi$ and $\Delta\theta$) and the satellite formation altitude (i.e., R_f) on the channel capacity will be analyzed separately.

Influence of the Satellite Visual Range

For the given satellite formation altitude, we first analyze the influence of the visual range where satellites in the formation are visible to the terrestrial terminal station in one snapshot (i.e., $\Delta\phi$ and $\Delta\theta$ on the upper bound of the channel capacity C_{single})

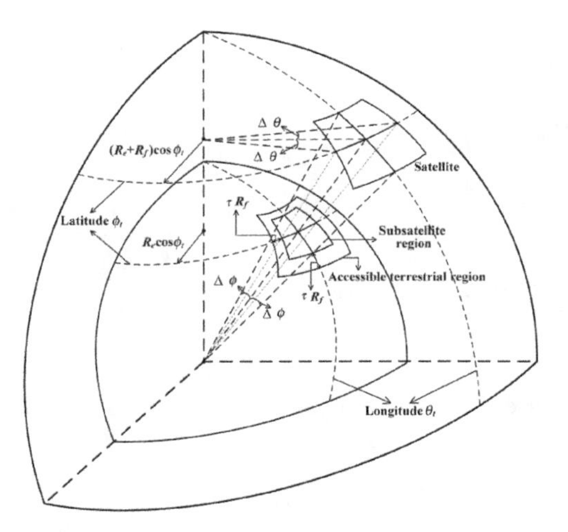

Figure 8.4 Visual range and formation altitude of satellites

Because $a_m = c_0/(4\pi f_c d_m)$, the upper bound of C_{single} shown in (8.27) can be rewritten as

$$C_{\text{single}} \leq M \log_2 \left[1 + \frac{\left(\frac{c_0}{4\pi f_c}\right)^2 \rho K_r NL}{\mathbb{E}(d_m^2)} \right]. \tag{8.34}$$

To analyze the influence of $\Delta\phi$ and $\Delta\theta$ on the channel capacity, we give the closed-form expression of $\mathbb{E}(d_m^2)$. Following that, we illustrate the monotonicity of C_{single} by combining the monotonicity of NL and $\mathbb{E}(d_m^2)$ with respect to $\Delta\phi$ and $\Delta\theta$.

LEMMA 8.5 *$\mathbb{E}(d_m^2)$ can be given by*

$$\mathbb{E}(d_m^2) = [2R_e^2 + 2R_e R_f] \left[1 - \sin^2\phi_t \frac{\sin\Delta\phi}{\Delta\phi} - \cos^2\phi_t \frac{\sin\Delta\phi}{\Delta\phi} \frac{\sin\Delta\theta}{\Delta\theta} \right] + R_f^2, \tag{8.35}$$

and it increases with $\Delta\phi$ and $\Delta\theta$.

Proof Because $R_m = R_e + R_f$, $\phi_m \sim U[\phi_t - \Delta\phi, \phi_t + \Delta\phi]$, $\theta_m \sim U[\theta_t - \Delta\theta, \theta_t + \Delta\theta]$, $\mathbb{E}(d_m^2)$ can be given by

$$\mathbb{E}(d_m^2) = \int_{\phi_t - \Delta\phi}^{\phi_t + \Delta\phi} \int_{\theta_t - \Delta\theta}^{\theta_t + \Delta\theta} \frac{(R_e + R_f)^2 + R_e^2 - 2(R_e + R_f)R_e A_m}{4\Delta\phi\Delta\theta} d\theta_m d\phi_m. \tag{8.36}$$

After substituting A_m given in (8.11), $\mathbb{E}(d_m^2)$ can be rewritten as

$$\begin{aligned}\mathbb{E}(d_m^2) &= \int_{-\Delta\phi}^{\Delta\phi}\int_{-\Delta\theta}^{\Delta\theta}\frac{1}{4\Delta\phi\Delta\theta}\Big\{(R_e+R_f)^2+R_e^2-2(R_e+R_f)R_e \\ &\quad\cdot\left[\cos\theta_m\cos\phi_m+2\sin(\phi_m+\phi_t)\sin\phi_t\sin^2\frac{\theta_m}{2}\right]\Big\}d\theta_m d\phi_m \\ &= [2R_e^2+2R_eR_f]\left[1-\sin^2\phi_t\frac{\sin\Delta\phi}{\Delta\phi}-\cos^2\phi_t\frac{\sin\Delta\phi}{\Delta\phi}\frac{\sin\Delta\theta}{\Delta\theta}\right]+R_f^2.\end{aligned} \tag{8.37}$$

Because $\frac{\sin x}{x}$ is monotonically decreasing with x, $\mathbb{E}(d_m^2)$ is increasing with $\Delta\phi$ and $\Delta\theta$. □

Note that $N \propto R_f\tau + R_e\cos\phi_t\Delta\theta$ and $L \propto R_f\tau + R_e\Delta\phi$, so the following lemma stands.

LEMMA 8.6 *The upper bound of C_{single} first increases and then decreases as $\Delta\phi$ and $\Delta\theta$ increase.*

Proof By utilizing the approximation $\frac{\sin x}{x} \approx 1-\frac{1}{6}x^2$, $\mathbb{E}(d_m^2)$ can be approximated by $\mathbb{E}(d_m^2) \approx \frac{1}{6}(\Delta\phi^2+\cos^2\phi_t\Delta\theta^2)[2R_e^2+2R_eR_f]+R_f^2$. Therefore, we only need to consider $f(\Delta\phi,\Delta\theta) \overset{def}{=} \frac{(R_f\tau+R_e\cos\phi_t\Delta\theta)(R_f\tau+R_e\Delta\phi)}{\Delta\phi^2+\cos^2\phi_t\Delta\theta^2+C_R}$, where $C_R = \frac{3R_f^2}{R_e^2+R_eR_f}$ is a constant. Note that τ first remains constant and then decreases linearly with respect to $\Delta\phi$ and $\Delta\theta$, and the molecule of $f(\Delta\phi,\Delta\theta)$ is a linear function of both $\Delta\theta$ and $\Delta\phi$. Because the denominator is a quadratic function of both $\Delta\theta$ and $\Delta\phi$, it can be shown by taking the derivative directly that $f(\Delta\phi,\Delta\theta)$ first increases and then decreases with respect to $\Delta\phi$ and $\Delta\theta$. Therefore, the upper bound of the channel capacity first increases and then decreases with respect to $\Delta\phi$ and $\Delta\theta$. □

We now analyze the influence of $(\Delta\phi,\Delta\theta)$ on the lower bound of the channel capacity C_{single}. Because $a_m = c_0/(4\pi f_c d_m)$, the lower bound of C_{single} shown in (8.33) can be rewritten as

$$C_{\text{single}} \geq M\log_2\left[\frac{(\frac{c_0}{4\pi f_c})^2\rho K_r\sqrt{N^2L^2-\Delta^2}}{\mathbb{E}(d_m^2)}\right]. \tag{8.38}$$

Applying Lemma 8.5, the following lemma stands.

LEMMA 8.7 *The lower bound of C_{single} first increases and then decreases as $\Delta\phi$ and $\Delta\theta$ increase.*

Because the $\sqrt{(NL)^2-\Delta^2}$ item can be approximated by a translation of NL, the proof of Lemma 8.7 is similar to that of Lemma 8.6 and is thus omitted. Combining Lemma 8.6 and Lemma 8.7 gives the following proposition.

PROPOSITION 8.6 *The channel capacity C_{single} first increases and then decreases as $\Delta\phi$ and $\Delta\theta$ increase.*

Influence of Satellite Formation Altitude

We denote the east-west span and south-north span of the satellite system as L_θ and L_ϕ, respectively, which are fixed due to satellite formation flying. For the given L_θ and L_ϕ, the latitude and longitude spans of the satellites are relevant to the satellite formation altitude and can be given by $\Delta\phi = \arcsin\frac{L_\phi}{2(R_e+R_f)}$ and $\Delta\theta = \arcsin\frac{L_\theta}{2(R_e+R_f)\cos\phi_t}$, respectively. Similar to the analysis on the influence of satellite visual range, by applying the upper bound and lower bound of the channel capacity C_{single} given by (8.34), (8.38), and Lemma 8.5, the following lemma stands.

LEMMA 8.8 *The upper bound and lower bound of C_{single} first increases and then decreases as the altitude of satellite formation increases.*

Proof We only give a proof of the monotonicity of the upper bound. The proof of the lower bound is similar because the $\sqrt{(NL)^2-\Delta^2}$ item can be approximated by a translation of NL. By utilizing the approximation $\frac{\sin x}{x}\approx 1$, $\frac{x}{\sin x}\approx 1$ to N and L, we have $N \propto h\tau + \frac{L_\theta R_e}{2(R_e+R_f)}\cdot\frac{\Delta\theta}{\sin\Delta\theta} \approx h\tau + \frac{L_\theta R_e}{2(R_e+R_f)}$, $L \propto h\tau + \frac{L_\phi R_e}{2(R_e+R_f)}\cdot\frac{\Delta\theta}{\sin\Delta\theta} \approx h\tau + \frac{L_\phi R_e}{2(R_e+R_f)}$. By utilizing the second-order approximation $\frac{\sin x}{x}\approx 1-\frac{1}{6}\sin^2 x$, $\frac{x}{\sin x}\approx 1+\frac{1}{6}\sin^2 x$ to $\mathbb{E}(d_m^2)$ given by Lemma 8.5, $\mathbb{E}(d_m^2)$ can be approximated by $\mathbb{E}(d_m^2)\approx \frac{L_\phi^2+L_\theta^2 R_e}{12(R_e+R_f)}+R_f^2$. Because the altitude of LEO satellites is no more than 2,000 km, we have $\frac{R_e}{R_e+R_f} = \frac{1}{1+\frac{R_f}{R_e}} \approx 1-\frac{2R_f}{5R_e}$. NL and $\mathbb{E}(d_m^2)$ can then be given by $NL \propto [(\tau-\frac{2L_\theta}{5R_e})R_f+\frac{L_\theta}{2}]\cdot[(\tau-\frac{2L_\phi}{5R_e})R_f+\frac{L_\phi}{2}]$, $\mathbb{E}(d_m^2)\approx(1+\frac{L_\phi^2+L_\theta^2}{12R_e^2})R_f^2-\frac{L_\phi^2+L_\theta^2}{12R_e}R_f+\frac{L_\phi^2+L_\theta^2}{12}$. Similar to the proof of Lemma 8.6, NL and $\mathbb{E}(d_m^2)$ can both be approximated by quadratic functions with respect to R_f, it can be shown by taking the derivative directly that $\frac{NL}{\mathbb{E}(d_m^2)}$ has a maximum point in [0, 2,000 km]. Therefore, the upper bound of channel capacity first increases and then decreases with respect to the satellite formation altitude. □

Therefore, the following proposition can be obtained.

PROPOSITION 8.7 *The channel capacity C_{single} first increases and then decreases as the altitude of satellite formation increases.*

Influence of the Satellite Formation Size

Given the satellite distribution (i.e., fixed values of $\Delta\phi$, $\Delta\theta$, and R_f), the channel correlation increases with the satellite formation such that the influence of the satellite formation size on the channel capacity cannot be revealed directly. In the following, we analyze the influence of the satellite formation size (i.e., the number of satellites in the formation) on the capacity bounds separately.

Influence on the Upper Bound

The upper bound of C_{single} given by (8.27) can be rewritten as

$$C_{\text{single}} \leq M\left[\frac{\sum_{m=1}^{M}\log_2 a_m^2}{M}+\log_2\left(\rho K_r NL+\frac{\sum_{m=1}^{M}\frac{1}{a_m^2}}{M}\right)\right]. \tag{8.39}$$

Because $\{a_1^2, a_2^2, \ldots, a_m^2\}$ are i.i.d. random variables, we have

$$\begin{aligned}&\mathbb{E}\left[\frac{\sum_{m=1}^{M}\log_2 a_m^2}{M}\right]=\mathbb{E}\left[\log_2 a_m^2\right],\\&\mathbb{E}\left[\log_2\left(\rho K_r NL+\frac{\sum_{m=1}^{M}\frac{1}{a_m^2}}{M}\right)\right]\approx\log_2\left(\rho K_r NL+\mathbb{E}\left[\frac{1}{a_m^2}\right]\right),\end{aligned} \tag{8.40}$$

which are independent of M. Therefore, the upper bound of C_{single} increases linearly with the size of the satellite formation.

Remark 8.3 In the general multiantenna satellite case, when $MK_t \leq NLK_r$, similar to the analysis earlier, it can be proved that the upper bound of C_{multiple} given by (8.20a) increases linearly with satellite formation size. When $MK_t > NLK_r$, the first item given by (8.20b) increases logarithmically with M, while the second and the third items increase linearly with M. Because the first term is dominant in the upper bound of the channel capacity, the capacity increases almost logarithmically with respect to M.

Influence on the Lower Bound

To analyze the influence of satellite formation size on the lower bound of the channel capacity shown in (8.33), we first consider the main influencing factors of Δ as shown in (8.42) and derive their distribution (i.e., Lemmas 8.9 and 8.10). Following that, we give an approximate expression of Δ (i.e., (8.52)), where the coefficient is determined by the special case of $M = 2$. The influence of satellite formation size on the channel capacity is finally derived by utilizing the approximate expression of Δ.

Note that $q_{i,j}$ defined in (8.30) can be rewritten as

$$\begin{aligned}|q_{i,j}| &= \left|\frac{1-\bar{u}_i^N u_j^N}{1-\bar{u}_i u_j}\,\frac{1-\bar{v}_i^N v_j^N}{1-\bar{v}_i v_j}\right| = \left|\frac{\sin(\frac{N}{2}\angle\frac{u_j}{u_i})}{\sin(\frac{1}{2}\angle\frac{u_j}{u_i})}\,\frac{\sin(\frac{L}{2}\angle\frac{v_j}{v_i})}{\sin(\frac{1}{2}\angle\frac{v_j}{v_i})}\right|\\&= \left|\frac{2\sin(N\arcsin\frac{|u_i-u_j|}{2})}{|u_i-u_j|}\cdot\frac{2\sin(L\arcsin\frac{|v_i-v_j|}{2})}{|v_i-v_j|}\right|\\&\leq \frac{2}{|u_i-u_j|}\cdot\frac{2}{|v_i-v_j|},\end{aligned} \tag{8.41}$$

where $\angle(\cdot)$ denotes the angle of a complex number's polar representation, and u_i and v_i are defined in (8.24). Because the function $\frac{\sin(Nx)}{\sin(x)}$ increases rapidly as $x \to 0$, $|q_{i,j}|$ is much larger than other off-diagonal elements in $\boldsymbol{Q}^{\mathrm{H}}\boldsymbol{Q}$ if $|u_i - u_j|$ or $|v_i - v_j|$ is close to zero. Therefore, $\Delta = \max_{1\leq i\leq M}\{\sum_{j=1,\, j\neq i}^{M}|q_{ij}|\}$ is mainly determined by

$$\Delta u = \min_{1\leq i<j\leq M}|u_i-u_j|, \quad \Delta v = \min_{1\leq i<j\leq M}|v_i-v_j|. \tag{8.42}$$

Because Δu and Δv have same probability distribution and generally $\arg\min_{1\le i<j\le M} |u_i - u_j| \ne \arg\min_{1\le i<j\le M} |v_i - v_j|$, we assume that $\Delta \propto \frac{2}{\mathbb{E}(\Delta u)} = \frac{2}{\mathbb{E}(\Delta v)}$. To analyze the distribution and expectation of Δu and Δv, we first give the following lemma about the property of $\{u_m\}_{m=1}^M$.

LEMMA 8.9 *$\{u_m\}_{m=1}^M$ are M points approximately uniformly distributed on an unit circle.*

Proof Because $u_m = \exp(-j2\pi \frac{f_c U_m d_x}{c_0}) = \exp(-j2\pi \{\frac{f_c U_m d_x}{c_0}\})$, where $\{\cdot\}$ denotes the fractional part of a real number. Because U_m is a continuous random variable, and the magnitude of d_x is relatively larger than that of $\frac{f_c U_m}{c_0}$, by [201] we obtain that the distribution of $\{\frac{f_c U_m d_x}{c_0}\}$ can be approximated by a uniform distribution on $[0, 1]$, and Lemma 8.9 is then proved. □

Therefore, Δu can be interpreted as the shortest distance among M uniformly distributed points on a unit circle. Note that the shortest distance must be taken between two adjacent points. Without loss of generality, we assume $0 \le \{\frac{u_1 d_x f_c}{c_0}\} < \{\frac{u_2 d_x f_c}{c_0}\} < \cdots < \{\frac{u_M d_x f_c}{c_0}\}$, and

$$\begin{aligned} z = \min\Bigg\{ & \left\{\frac{u_2 d_x f_c}{c_0}\right\} - \left\{\frac{u_1 d_x f_c}{c_0}\right\}, \ldots, \left\{\frac{u_M d_x f_c}{c_0}\right\} - \left\{\frac{u_{M-1} d_x f_c}{c_0}\right\}, \\ & 1 - \left\{\frac{u_M d_x f_c}{c_0}\right\} + \left\{\frac{u_1 d_x f_c}{c_0}\right\}\Bigg\}, \end{aligned} \tag{8.43}$$

where $\{\cdot\}$ denotes the fractional part. We then have $\Delta u = 2\sin \pi z$.

LEMMA 8.10 *The probability density function (PDF) of z is $f_z(z) = M(M-1)(1 - Mz)^{M-2}, 0 < z < \frac{1}{M}$.*

Proof For simplicity we denote $\{\frac{u_m d_x f_c}{c_0}\}$ as $T_m, m = 1, \ldots, M$ and $T_{(m)}$ as their order statistics. Therefore, z can be rewritten as $z = \min\{T_{(2)} - T_{(1)}, \ldots, T_{(M)} - T_{(M-1)}, 1 - T_{(M)} + T_{(1)}\}$. Because $T_m \sim i.i.d. U[0, 1]$, the joint PDF of $T_{(1)}, \ldots, T_{(M)}$ can be given by [202]

$$\begin{aligned} & f_{T_{(1)},\ldots,T_{(M)}}(t_{(1)}, \ldots, t_{(M)}) = M!, \\ & 0 < t_{(1)} < \cdots < t_{(M)} < 1. \end{aligned} \tag{8.46}$$

By utilizing the substitution $\{S_1 = T_{(1)}, S_m = T_{(m)} - T_{(m-1)},\ m = 2, \ldots, M\}$, the joint PDF of $S_1, \ldots, S_M$ is $f_{S_1,\ldots,S_M}(s_1, \ldots, s_M) = M!,\ s_1, \ldots, s_M > 0,\ s_1 + \cdots + s_M < 1$. By integrating y_1, the joint PDF of $S_2, \ldots, S_M$ can be given by

$$\begin{aligned} & f_{S_1,\ldots,S_M}(s_1, \ldots, s_M) = (1 - s_2 - \cdots - s_M)M!, \\ & s_2, \ldots, s_M > 0,\ s_2 + \cdots + s_M < 1. \end{aligned} \tag{8.47}$$

Therefore, we can compute the probability $P(z > z_0)$ as

$$\begin{aligned} P(z > z_0) &= P(S_2 > z_0, \ldots, S_M > z_0, S_2 + \cdots + S_M < 1 - z_0) \\ &= \int_{s_2,\ldots,s_M > z_0,\ s_2+\cdots+s_M < 1-z_0} (1 - s_2 - \cdots - s_M)M!\ ds_2 \cdots ds_M \end{aligned}$$

$$\overset{(a)}{=} \int_{\tilde{s}_2,\ldots,\tilde{s}_M>0,\ \tilde{s}_2+\cdots+\tilde{s}_M<1-Mz_0} [1-(M-1)z_0-\tilde{s}_2-\cdots-\tilde{s}_M]M!\, d\tilde{s}_2\cdots d\tilde{s}_M$$
$$\overset{(b)}{=} \int_0^{1-Mz_0} (1-(M-1)z_0-s)\frac{s^{M-2}}{(M-2)!}ds = (1-Mz_0)^{M-1}, \tag{8.48}$$

where in (a) we utilized the substitution $\tilde{s}_m = s_m - z_0$ and in (b) we used the polyhedron $s < \tilde{s}_2+\cdots+\tilde{s}_M < s+ds$ as the volume element in integration. Therefore, the cumulative distribution function (CDF) of Z is $F_z(z_0) = 1-P(z>z_0) = 1-(1-Mz_0)^{M-1}$, and thus, $f_z(z_0) = M(M-1)(1-Mz_0)^{M-2}$. □

Therefore, the expectation of z can be obtained by

$$\mathbb{E}(z) = \int_0^{\frac{1}{M}} M(M-1)(1-Mz)^{M-2}dz = \frac{1}{M^2}, \tag{8.49}$$

and thus the expectation of Δu and Δv can be approximated by $\mathbb{E}(\Delta u) = \mathbb{E}(\Delta v) \approx 2\sin(\pi\mathbb{E}(z)) = 2\sin\frac{\pi}{M^2}$. Because $\Delta \propto \frac{2}{\mathbb{E}(\Delta u)} = \frac{2}{\mathbb{E}(\Delta v)}$, we set $\Delta = k/\sin\frac{\pi}{M^2}$. To determine the coefficient k, we consider the case where $M=2$ is given by the following lemma.

LEMMA 8.11 *For $M=2$, the expectation of Δ can be obtained by*

$$\mathbb{E}(\Delta) = 4\int_0^1 u\left|\frac{\sin(N\pi u)}{\sin(\pi u)}\right|du\cdot\int_0^1 v\left|\frac{\sin(L\pi v)}{\sin(\pi v)}\right|dv. \tag{8.50}$$

Proof For $M=2$, $\Delta = \left|\frac{\sin[N\pi(\{\frac{f_c u_1 d_x}{c_0}\}-\{\frac{f_c u_2 d_x}{c_0}\})]}{\sin[\pi(\{\frac{f_c u_1 d_x}{c_0}\}-\{\frac{f_c u_2 d_x}{c_0}\})]}\right|\cdot\left|\frac{\sin[L\pi(\{\frac{f_c v_1 d_x}{c_0}\}-\{\frac{f_c v_2 d_x}{c_0}\})]}{\sin[\pi(\{\frac{f_c v_1 d_x}{c_0}\}-\{\frac{f_c v_2 d_x}{c_0}\})]}\right|$, where $\{\cdot\}$ denotes the fraction part of a real number. For simplicity, we set $\tilde{u}_1 = \{\frac{f_c u_1 d_x}{c_0}\}$ and $\tilde{u}_2 = \{\frac{f_c u_2 d_x}{c_0}\}$. According to Lemma 8.9, $\tilde{u}_1, \tilde{u}_2$ i.i.d. $\sim U[0,1]$, and thus the PDF of $\tilde{u} \triangleq \tilde{u}_1 - \tilde{u}_2$ can be given by $f_{\tilde{u}}(\tilde{u}) = 1-|\tilde{u}|, |\tilde{u}|<1$. Therefore, we have

$$\mathbb{E}\left(\left|\frac{\sin\left[N\pi\left(\left\{\frac{f_c u_1 d_x}{c_0}\right\}-\left\{\frac{f_c u_2 d_x}{c_0}\right\}\right)\right]}{\sin\left[\pi\left(\left\{\frac{f_c u_1 d_x}{c_0}\right\}-\left\{\frac{f_c u_2 d_x}{c_0}\right\}\right)\right]}\right|\right) = 2\int_0^1 \tilde{u}\left|\frac{\sin(N\pi\tilde{u})}{\sin(\pi\tilde{u})}\right|du. \tag{8.51}$$

Similarly, $\mathbb{E}(|\frac{\sin[L\pi(\{\frac{f_c v_1 d_x}{c_0}\}-\{\frac{f_c v_2 d_x}{c_0}\})]}{\sin[\pi(\{\frac{f_c v_1 d_x}{c_0}\}-\{\frac{f_c v_2 d_x}{c_0}\})]}|) = 2\int_0^1 \tilde{v}\left|\frac{\sin(L\pi\tilde{v})}{\sin(\pi\tilde{v})}\right|dv$, and Lemma 8.11 is then obtained. □

Therefore, by utilizing Lemma 8.11 to calculate the coefficient k, Δ can be given by

$$\Delta = \frac{2\sqrt{2}}{\sin\frac{\pi}{M^2}}\int_0^1 u\left|\frac{\sin(N\pi u)}{\sin(\pi u)}\right|du\cdot\int_0^1 v\left|\frac{\sin(L\pi v)}{\sin(\pi v)}\right|dv. \tag{8.52}$$

Remark 8.4 The numerical results of the real value and approximate value of Δ for $(N,L) = (20,20)$ are given in Table 8.1. It can be seen that the approximation approaches the real value as the size of satellite formation M increases.

Because the integral given in (8.52) can be computed numerically for given N and L, the numerical approximation of Δ can be obtained, based on which the following proposition holds.

Table 8.1 Real value and approximate value of Δ

M	4	6	8	10	12
Real value	29.6	51.4	81.8	120.5	148.4
Approximate value	17.6	39.4	70.9	109.3	157.4

PROPOSITION 8.8 *When $2 \leq M \leq 20, NL > 1.6M^2$, the lower bound of the channel capacity increases with the size of satellite formation.*

Proof According to the lower bound of channel capacity given in (8.38), we only need to prove that for $2 \leq M \leq 19, NL > 1.6M^2$, we have

$$\left(N^2L^2 - \Delta_M^2\right)^M < \left(N^2L^2 - \Delta_{M+1}^2\right)^{M+1} \cdot \left[\frac{\left(\frac{c_0}{4\pi f_c}\right)^2 \rho K_r}{\mathbb{E}(d_m^2)}\right]^2, \tag{8.53}$$

which can be equivalently written as

$$\left(1 + \frac{\Delta_{M+1}^2 - \Delta_M^2}{N^2L^2 - \Delta_{M+1}^2}\right)^M < \left(N^2L^2 - \Delta_{M+1}^2\right) \cdot \left[\frac{\left(\frac{c_0}{4\pi f_c}\right)^2 \rho K_r}{\mathbb{E}(d_m^2)}\right]^2. \tag{8.54}$$

Because $NL > 1.6M^2$, we only need to prove

$$\left(1 + \frac{\Delta_{M+1}^2 - \Delta_M^2}{1.6^2M^4 - \Delta_{M+1}^2}\right)^M < \left(1.6^2M^4 - \Delta_{M+1}^2\right) \cdot \left[\frac{\left(\frac{c_0}{4\pi f_c}\right)^2 \rho K_r}{\mathbb{E}(d_m^2)}\right]^2, \tag{8.55}$$

which can be verified numerically for $2 \leq M \leq 19$. □

Therefore, by combining the analysis on the upper bound and the lower bound of the channel capacity, the following proposition stands.

PROPOSITION 8.9 *The channel capacity C_{single} increases as the satellite formation size increases.*

Gap between the Channel Capacity Bounds

We now illustrate that the upper bound and lower bound of the channel capacity given in Propositions 8.2 and 8.5 are close to each other. Specifically, the gap between the upper bound and the lower bound can be rewritten as $C_{\text{gap}} = M \log_2[\frac{\mathbb{E}(\frac{1}{a_m^2}) + \rho K_r NL}{\rho K_r \sqrt{N^2L^2 - \Delta^2}}]$. Note that the magnitude of $\mathbb{E}(\frac{1}{a_m^2})$ is about 10^{18}, which is far less than the magnitude of ρ and can then be ignored. Therefore, the gap can be obtained by

$$C_{\text{gap}} = \frac{M}{2} \log_2 \left(1 + \frac{\Delta^2}{N^2L^2 - \Delta^2}\right). \tag{8.56}$$

Table 8.2 Gap between the upper and lower bounds of the channel capacity

(N, L)	(10, 10)	(20, 20)	(30, 30)	(40, 40)	(50, 50)
Gap (bits/s/Hz)	8.510	0.560	0.143	0.096	0.026

Table 8.3 Simulation parameters

Parameters	Values
Carrier frequency (Ka-band) f_C (GHz)	20
Average total received SNR ρ (dB)	22.4
Number of terminal stations (N, L) in single-antenna satellite case	(20, 20)
Number of transmit antennas in each BS K_r	2
Interterminal distance (d_x, d_y) (km)	(2, 2)
Number of LEO satellites M	10
Number of receive antennas in each LEO satellite K_t	20
Formation altitude of LEO satellites (km)	900
LEO satellite distribution $(\Delta\phi, \Delta\theta)$	$(15°, 15°)$
Noise density for Ka-band communications σ_l^2 (dBm/Hz)	−203

The numerical results of the gap for $M = 10$ and different (N, L) are given in Table 8.2.

Because the magnitude of the channel capacity is 10^2 bits/s/Hz, which will be verified in Fig. 8.6a, the gap shows that the derived upper bound and lower bound of the channel capacity are close to each other, indicating that they give good approximations of the channel capacity. Moreover, because the derived upper bound and lower bound of the channel capacity are close to each other, the rationality of Propositions 8.7 and 8.9 is ensured, where we claim that the monotonicity of the channel capacity is the same as the monotonicity of its upper bound and lower bound.

8.2.6 Performance Evaluation

In this section, we present the simulation results of the channel capacity in the downlink ultra-dense LEO-based multiterminal satellite formation-flying system to validate the theoretical analysis. The simulation parameters are based on prior work [203] and 3GPP specifications [204], as given in Table 8.3.

Figure 8.5 depicts the total channel capacity in the special single-antenna satellite case and the general multiantenna satellite case versus the visual range (i.e., $\Delta\phi$ and $\Delta\theta$) where satellites are visible to the terminal station in one snapshot. It can be seen that the channel capacity first increases and then decreases with $\Delta\phi$ and $\Delta\theta$ in both cases. This is because, as $\Delta\phi$ and $\Delta\theta$ become larger, the number of accessible terminal stations increases initially, leading to an increase in the capacity. However, when $\Delta\phi$ and $\Delta\theta$ continue to grow, severe pathloss will result in the decline of the capacity. Therefore, there exists an optimal visual range of satellites maximizing the channel capacity of the system, which also verifies the theoretical analysis in Section 8.2.5.

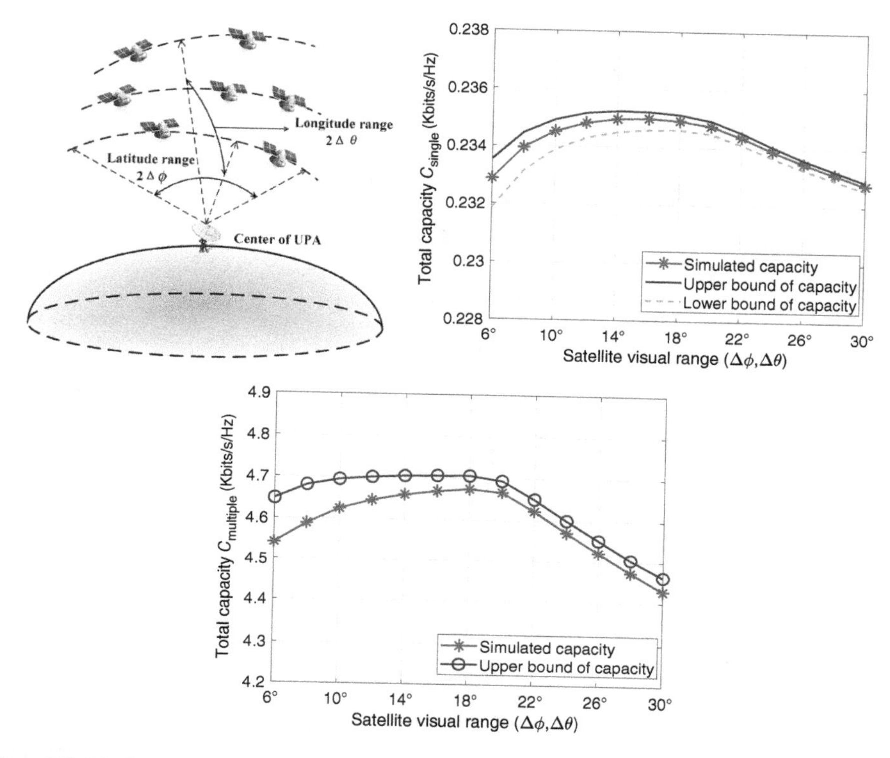

Figure 8.5 Total capacity in two cases versus the visual range where satellites are visible to the terminal station in one snapshot ($M = 10, K_r = 2, K_t = 20, R_f = 900$ km): (a) visual range, (b) single-antenna satellite case, and (c) multiantenna satellite case

Figure 8.6a and 8.6b illustrate the total channel capacity in the special single-antenna satellite case and general multiantenna satellite case versus the satellite formation altitude, respectively. Similar to Fig. 8.5, when the satellite formation altitude increases, the channel capacity first increases with the number of accessible terminal stations and then decreases due to severe pathloss. Therefore, there exists an optimal formation altitude of satellites maximizing the channel capacity of the system, which also verifies the theoretical analysis in Section 8.2.5.

Figure 8.7a and 8.7b show the total channel capacity in the special single-antenna satellite case and the general multiantenna satellite case versus the size of satellite formation, M. It can be seen that the capacity first increases almost linearly with M in the high SNR condition when the number of transmit antennas (i.e., MK_t is less than the number of transmit antennas, NLK_r in both cases), which also verifies the theoretical analysis in Section 8.2.5. However, when MK_t becomes larger than NLK_r with the increment of M in the general multiantenna satellite case,[4] the total channel capacity increases approximately logarithmically with M. This is because the channel capacity increases linearly with the degrees of freedom of channels in the high SNR condition. When $MK_t \leq NLK_r$, the rank of channel transfer matrix

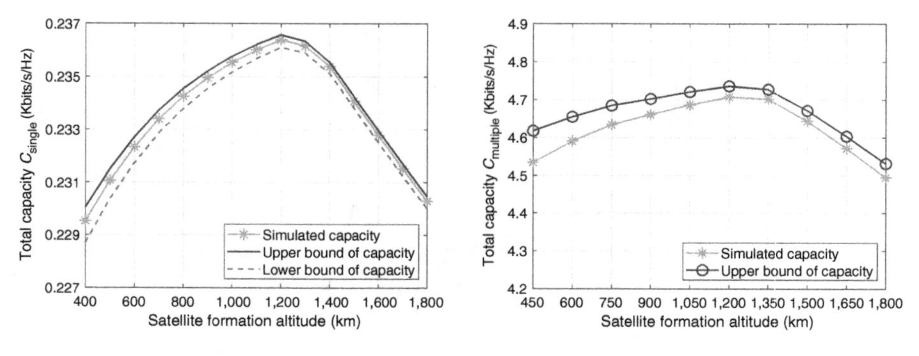

Figure 8.6 Total capacity in two cases versus the altitude of satellite formation ($M = 10, K_r = 2, K_t = 20, \Delta\phi = 15^\circ, \Delta\theta = 15^\circ$): (a) single-antenna satellite case and (b) multiantenna satellite case

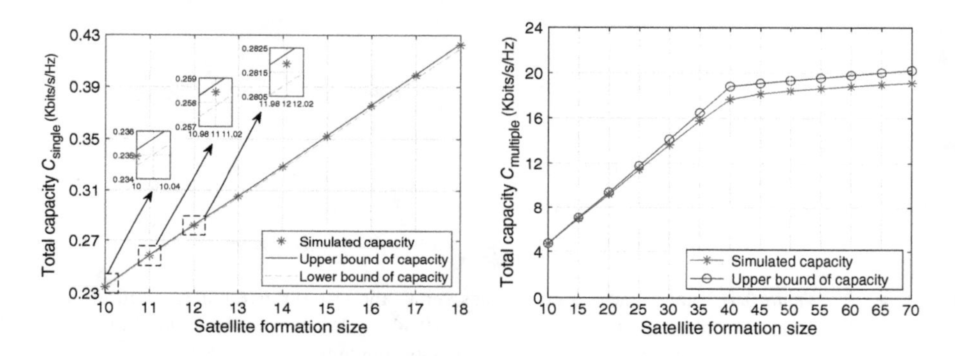

Figure 8.7 Total capacity in two cases versus the size of satellite formation ($N = 20, L = 20, K_r = 2, K_t = 20, R_f = 900$ km, $\Delta\phi = 15^\circ, \Delta\theta = 15^\circ$): (a) single-antenna satellite case and (b) multiantenna satellite case

$\boldsymbol{H}$ is MK_t. In contrast, when $MK_t > NLK_r$, the rank of channel transfer matrix $\boldsymbol{H}$ is NLK_r. In this case, the increase of the satellite formation size only contributes to the power gain rather than causing a space multiplexing gain. This implies that a satellite formation of moderate size is enough to achieve sufficient total channel capacity without unnecessary cost.

Figure 8.8 illustrates the total channel capacity in the general multiantenna satellite case versus the number of each satellite's antennas K_t. It shows that the total channel capacity first increases almost linearly with K_t in the high SNR condition when MK_t is less than NLK_r and then increases approximately logarithmically with K_t when MK_t becomes larger than NLK_r with the increment of K_t. The main reason is similar to that of Fig. 8.6; that is, the increment of K_t first contributes to the space multiplexing gain and then only contributes to the power gain. Note that the implementation difficulty also increases in practice as the number of antennas per satellite grows. A tradeoff can then be achieved between the total channel capacity and the number of antennas per satellite.

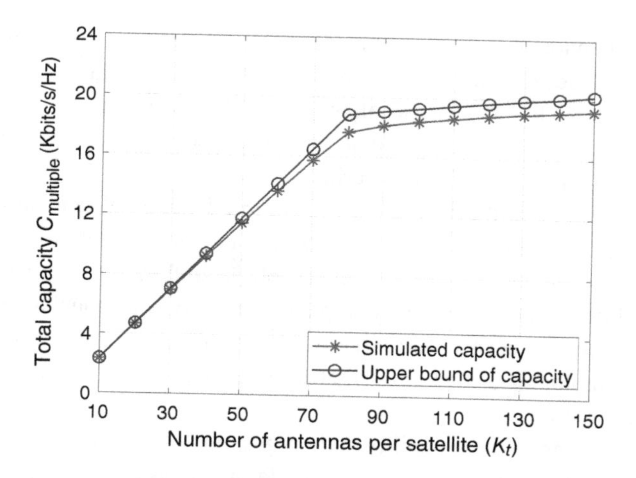

Figure 8.8 Total capacity in multiantenna satellite case versus the number of antennas per satellite ($M = 10, N = 20, L = 20, K_r = 2, R_f = 900$ km, $\Delta\phi = 15°, \Delta\theta = 15°$)

Most importantly, Figs. 8.5–8.8 all show that our closed-form lower and upper capacity bounds depicted by (8.33) and (8.27) in the single-antenna satellite case and the upper bound depicted by (8.20) in the multiantenna satellite case give good approximations of the capacity. The theoretical result in the single-antenna satellite case is also applicable to the general multiantenna satellite case.

8.2.7 Conclusion

In this section, we considered a downlink ultra-dense LEO-based multiterminal satellite formation-flying system and analyzed its performance capacity. The upper bound of the channel capacity in the general case where each satellite is equipped with multiple antennas was analyzed. Considering the special case of single-antenna satellites, both closed-form lower bound and upper bound of the channel capacity were also derived and were verified to give an accurate theoretical approximation of the capacity in the simulations. The influence of satellite distribution (i.e., the visual range and formation altitude) and formation size on the channel capacity was discussed separately. It was proved that there exist an optimal visual range and a formation altitude of LEO satellites to achieve the maximum total channel capacity of the downlink ultra-dense LEO-based multiterminal satellite formation-flying system.

8.3 Reconfigurable Holographic Surface-Aided LEO Satellite Communications

Increasingly, traditional terrestrial networks cannot meet the data transmission demands created by numerous mobile devices and applications due to scarce spectrum resources and limited coverage area [205]. Operating over high-frequency bands with wide bandwidth, ultra-dense LEO satellite communication networks [2] are being developed with the promise of providing high-speed data services and global connectivity for terrestrial users [203]. Compared with terrestrial networks, LEO satellite networks necessitate more stringent requirements on antenna technologies in terms of accurate beam steering and high antenna gain due to the high mobility of the satellites and the severe pathloss [206]. Traditional antennas integrated with UTs for satellite communications such as dish antennas and phased arrays either require heavy mechanical components or costly phase shifters [207]. Such bulky structures or high fabrication costs make the implementation of traditional antennas in practical systems prohibitive. Hence, LEO satellite communications call for novel cost-efficient antenna technologies with lightweight structures to overcome the limitations of traditional antennas.

Owing to the tunability and programmability of metamaterials, the recent development of reconfigurable metamaterial antennas provides a promising method to overcome the shortfalls of traditional antennas [208]. In particular, reconfigurable holographic surfaces (RHSs) composed of numerous metamaterial radiation elements can control beam patterns via software rather than costly hardware components [209], indicating their significant potential for serving as ultra-thin and lightweight antennas to support satellite communications. In a RHS, the radiation elements are laid on a waveguide attached by multiple feeds, such that the electromagnetic waves generated by the feeds are injected into the waveguide directly. The unique characteristic of RHS is that it constructs a *holographic pattern* on the metasurface according to the holographic interference principle [210], based on which each metamaterial element controls the *radiation amplitude* of the incident waves electrically to generate desired directional beams;[5] such a beamforming technique is also known as *holographic beamforming* [212].

Given the basics of the RHS, its advantages can be summarized as follows: First, based on the printed circuit board (PCB) technology, the size and weight of the RHS are small. Its compact design also enables the convenient implementation of the transceiver. Second, unlike the phased array, the RHS does not rely on active amplification internally and complex phase-shifting circuits, leading to low power consumption. Third, all components (such as diodes, PCBs, and DC control circuits) needed to fabricate the RHS are high-volume, commercial off-the-shelf parts, such that the manufacturing and hardware costs of the RHS are low [213]. Therefore, the RHS provides a promising alternative to traditional antennas for satellite communications.

Existing research on RHSs has focused primarily on the hardware structure design [214] and the software-defined beam controller to improve its performance in terms of directive gain and radiation efficiency [215]. In [214], a tunable metamaterial element with a radiation amplitude that can be controlled by a PIN diode's on/off state was designed for the fabrication of RHSs. In [215], an adaptive beam pattern controller was proposed for RHSs to cancel sidelobes and enhance the directive gain. Due to their unique working principles and ultra-thin structures, RHSs have drawn attention from industry and spawned various applications. For example, Pivotal Commwave has developed commercial customized RHS systems from 1 GHz to 70 GHz for the coverage extension of terrestrial communications and the construction of smart repeater ecosystems by leveraging the holographic beamforming technique [213]. In addition, RHSs can be applied to microwave imaging [216] and radar [217] due to their ability to achieve dynamic multibeam steering. However, most existing works only demonstrate the capability of the RHS to generate given object beam patterns in static terrestrial communication scenarios. Such methods cannot guarantee the QoS in LEO satellite communications because the high mobility of the satellites leads to time-varying beam patterns. Besides, novel holographic beamforming schemes are required for the LoS dominated channels in satellite communications, which are different from those in terrestrial communications.

To achieve this goal, in this section we consider RHS-aided LEO satellite communications by fully exploiting the potential of amplitude-controlled holographic beamforming. To support continuous communication with multiple LEO satellites, we propose a RHS-aided multisatellite communication scheme. Specifically, the satellites' positions are predicted according to the temporal variation law to avoid frequent satellite positioning, and the holographic beamforming is performed for data transmission. We then develop a holographic beamforming algorithm for sum-rate maximization and derive a closed-form optimal holographic beamformer. The robustness of the proposed algorithm can be guaranteed in that the tracking errors of the satellites' positions have little influence on the maximum sum rate. Simulation results verify the theoretical analysis regarding the number of radiation elements and the robustness of the holographic beamforming algorithm against the tracking errors of the satellites' positions.

8.3.1 System Model

In this subsection, we introduce a RHS-aided LEO satellite communication system where one UT equipped with a RHS transmits to multiple LEO satellites. Following that, the holographic beamforming model is presented.

Scenario Description

As shown in Fig. 8.9, we consider an uplink RHS-aided LEO satellite communication system, where a UT equipped with a RHS uploads data streams to Q LEO satellites denoted by $\mathcal{Q} = \{1, \ldots, q, \ldots, Q\}$ over the Ka-band spectrum for satellite-backhauled network access. Without loss of generality, it is assumed that each LEO satellite is equipped with K antennas. By leveraging the holographic principle, the RHS can realize dynamic beamforming to cope with the mobility of satellites without relying on complex phase-shifting circuits and bulky mechanics.

A RHS is a leaky-wave antenna mainly consisting of three parts: L feeds, a waveguide, and N subwavelength metamaterial radiation elements. The feeds are embedded in the bottom layer of the RHS and generate the incident electromagnetic waves, which are also called reference waves, carrying intended signals for satellites. The reference wave then propagates along the waveguide. During the propagation process,

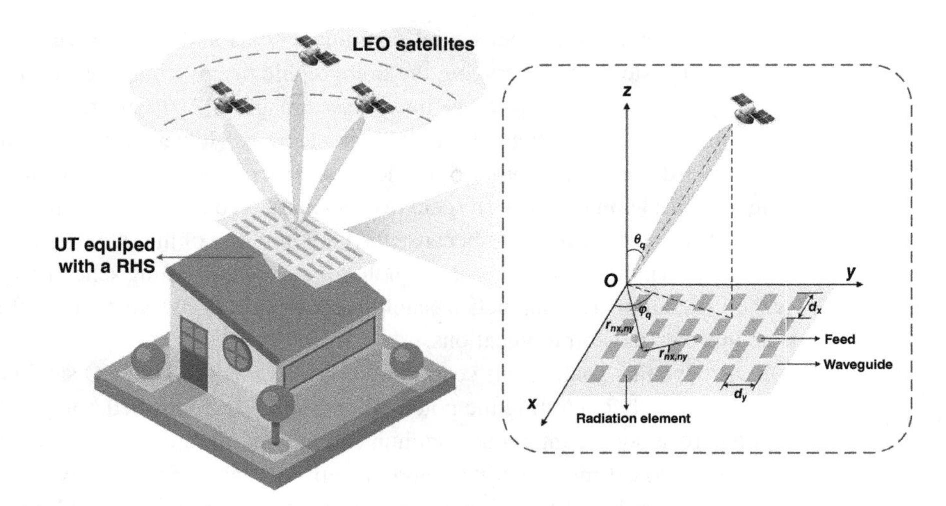

Figure 8.9 RHS-aided LEO satellite communication system

the reference wave radiates its energy through the radiation elements into free space, and the radiated wave is also known as the leaky wave [210], where the waveform of the leaky wave is the same as that of the transmit signal. The RHS can control the electromagnetic response (such as the radiation amplitude) of the elements to construct a holographic pattern, which records the interference between the reference wave and the desired object wave. Specifically, the radiation amplitude of the reference wave at each radiation element can be controlled to represent the amplitude of the interference via a simple diode-based amplitude controller [218]. Therefore, the leaked wave from each radiation element can be shaped independently according to the holographic pattern. The superposition of the leaked waves from different elements finally generates the object wave toward the satellite.

Holographic Beamforming Model

Without loss of generality, as shown in Fig. 8.9, we adopt the Cartesian coordinate where the $x-y$ plane coincides with the RHS, and the z axis is vertical to the RHS. We denote the number of radiation elements along the x axis and y axis as N_x and N_y, respectively, satisfying $N = N_x \times N_y$. We denote the interelement spacing of the RHS along the x axis and y axis as d_x and d_y, respectively. The position vector of the (n_x, n_y)th element can then be given by $\mathbf{r}_{n_x,n_y} = [n_x d_x, n_y d_y, 0]^T$, where $n_x = \{1,2,\ldots,N_x\}$ and $n_y = \{1,2,\ldots,N_y\}$.

At the (n_x, n_y)th radiation element, the reference wave generated by feed l and the wave propagating in free space with the desired direction of (θ, φ) can be given by [211]

$$\Psi_{ref}\left(\mathbf{r}^l_{n_x,n_y}\right) = \exp\left(-j\mathbf{k}_s \cdot \mathbf{r}^l_{n_x,n_y}\right), \tag{8.57}$$

$$\Psi_{obj}(\mathbf{r}_{n_x,n_y}, \theta, \varphi) = \exp(-j\mathbf{k}_f \cdot \mathbf{r}_{n_x,n_y}), \tag{8.58}$$

where $\mathbf{k}_s$ is the propagation vector of the reference wave in the waveguide, $\mathbf{r}^l_{n_x,n_y}$ is the distance vector from the feed l to the (n_x, n_y)th radiation element. The interference between the reference wave and the object wave is defined as

$$\Psi_{intf}\left(\mathbf{r}^l_{n_x,n_y}, \theta, \varphi\right) = \Psi_{obj}(\mathbf{r}_{n_x,n_y}, \theta, \varphi)\Psi^*_{ref}\left(\mathbf{r}^l_{n_x,n_y}\right). \tag{8.59}$$

When such interference is excited by the reference wave, the generated wave satisfies

$$\Psi_{intf}\Psi_{ref} \propto \Psi_{obj}|\Psi_{ref}|^2, \tag{8.60}$$

implying that it propagates in the desired direction of (θ, φ).

Therefore, the information contained in Ψ_{intf}, which is also called the holographic pattern, is supposed to be recorded by the radiation elements. For this purpose, each radiation element in the RHS controls the radiation amplitude of the reference wave instead of the traditional phase-controlled method. The main idea of the amplitude-controlled method is that the elements with reference waves that are in phase with the object wave are tuned to radiate strongly (large radiation amplitude), whereas the elements that are out of phase are detuned so as not to radiate (small radiation amplitude) [218].

Note that the real part of the interference ($\mathrm{Re}[\Psi_{intf}]$) (i.e., cosine value of the phase difference between the reference wave and the object wave), decreases as the phase difference grows, which exactly meets the amplitude control requirements. Therefore, $\mathrm{Re}[\Psi_{intf}]$ can represent the radiation amplitude of the reference wave at each element. To avoid a negative value, $\mathrm{Re}[\Psi_{intf}]$ is normalized to $[0, 1]$. The radiation amplitude of each element to generate the wave propagating in the direction (θ, φ) can then be parameterized mathematically by [215]

$$m\left(\mathbf{r}^l_{n_x,n_y}, \theta, \varphi\right) = \frac{\mathrm{Re}\left[\Psi_{intf}\left(\mathbf{r}^l_{n_x,n_y}, \theta, \varphi\right)\right] + 1}{2}. \tag{8.61}$$

According to (8.61), it can be seen that one holographic pattern corresponds to one desired wave direction. Considering the superposition of different holographic patterns corresponding to different desired directions, multiple beams directed toward the satellites can then be generated by the RHS. Specifically, such a superposed holographic pattern is calculated as a weighted summation of the radiation amplitude distribution corresponding to each object beam. Therefore, the normalized radiation amplitude of each element can be expressed as

$$m_{n_x,n_y} = \sum_{q=1}^{Q}\sum_{l=1}^{L} a_{q,l}\, m\left(\mathbf{r}^l_{n_x,n_y}, \theta_q, \varphi_q\right), \tag{8.62}$$

where (θ_q, φ_q) is the position of satellite q, and $a_{q,l}$ is the amplitude ratio for the beam toward satellite q from feed l satisfying

$$\sum_{q=1}^{Q}\sum_{l=1}^{L} a_{q,l} = 1. \tag{8.63}$$

This constraint is set to keep the radiation amplitude of each element m_{n_x,n_y} within [0,1] to guarantee the power radiated from the element is no larger than the power accepted by the element. Specifically, we denote the power accepted by each element as P_a and the efficiency of each element as η, which is defined as the ratio of the power accepted by each element to the total power P_t of the reference waves from all feeds[6] (i.e., $\eta = P_a/P_t$). When the radiation amplitude of each element is m_{n_x,n_y}, the radiated power of each element is $P_a \cdot m^2_{n_x,n_y}$.

Note that the sum of the radiated power from each RHS element should be no larger than P_t (i.e., $\sum_{n_x=1}^{N_x}\sum_{n_y=1}^{N_y} \eta P_t \cdot m^2_{n_y,n_z} \leq P_t$, where m_{n_x,n_y} varies with the desired beam directions). To guarantee the conservation of energy for each beam direction, we adopt the average value of $m^2_{n_x,n_y}$, and thus, in the RHS, the efficiency of the element η and the number of elements should satisfy[7]

$$\sum_{n_x=1}^{N_x}\sum_{n_y=1}^{N_y} \eta P_t \cdot \mathbb{E}\left(m^2_{n_x,n_y}\right) \leq P_t \Rightarrow \eta \cdot N_x N_y \leq \frac{8}{3}. \tag{8.64}$$

Because the RHS is a leaky-wave antenna, the leaky-wave effect makes the amplitude of the reference wave gradually decrease as it propagates along the waveguide [219]. We denote the leakage constant during the propagation process of the reference wave[8] as α, and the holographic beamforming matrix $\mathbf{M} \in \mathbb{C}^{N_x N_y \times L}$ is then formed by the following elements

$$M^l_{n_x,n_y} = \sqrt{\eta} \cdot m_{n_x,n_y} \cdot e^{-\alpha|\mathbf{r}^l_{nx,ny}|} \cdot e^{-j\mathbf{k}_s \cdot \mathbf{r}^l_{nx,ny}}, \tag{8.65}$$

where $e^{-\alpha|\mathbf{r}^l_{nx,ny}|}$ and $e^{-j\mathbf{k}_s \cdot \mathbf{r}^l_{nx,ny}}$ denote the amplitude attenuation and the phase of the reference wave when it propagates to the (n_x, n_y)th element from feed l, respectively.

8.3.2 RHS-Aided LEO Satellite Communication Scheme

In this subsection, we present an overview of the RHS-aided multisatellite communication scheme containing the LEO satellite tracking and data transmission. The detailed data transmission scheme and LEO satellite tracking scheme that can obtain the satellites' positions without frequent satellite positioning are then introduced separately.

Overview of Multisatellite Communication Scheme

Due to the high mobility of LEO satellites, the positions of the satellites change with the orbital motion, which influences the channel between the RHS and the satellite as well as the holographic beamforming scheme. To support the continuous communication between the UT and multiple LEO satellites, we develop a RHS-aided multisatellite communication scheme.

We assume that the revolution of each LEO satellite around the Earth is a uniform circular motion; this is widely utilized in the literature to make the theoretical analysis of the satellite communications tractable [220, 221]. The position of satellite q is

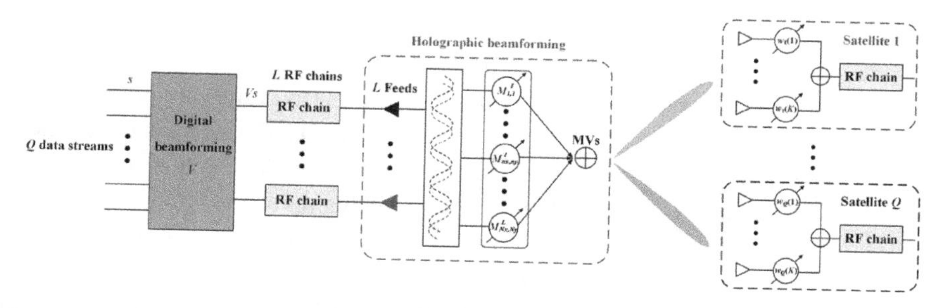

Figure 8.10 RHS-aided multisatellite transmission

Algorithm 8.1 RHS-aided multisatellite communication scheme

1 **Initialization**: The positions of the satellites in the first two time slots: each satellite's altitude H, zenith angle $\{\theta_q^{(1)}, \theta_q^{(2)}\}$, and azimuth angle $\{\varphi_q^{(1)}, \varphi_q^{(2)}\}$;
2 **repeat**
3 **Data transmission via holographic beamforming:**
4 The UT precodes the signals and then transmits the encoded signals to the satellites via the holographic beamformer **M**. Each LEO satellite processes the received signals via the receive combiner;
5 **LEO satellite tracking:**
6 The UT predicts the position of satellite q in time slot $t(t > 2)$ based on the temporal variation law;
7 **if** $mod(t, \tau) = 0$ **then**
8 **if** *RSS is smaller than the threshold* Υ. **then**
9 The UT reacquires the positions of the LEO satellites.
10 **end**
11 **end**
12 **until** *The satellites are outside the communication range of the UT.*

---

depicted by its zenith angle θ_q and azimuth angle ϕ_q. We divide the time of communication into T time slots denoted by $\{1, \ldots, t, \ldots, T\}$ with length Δ, during which the position of each satellite $(\theta_q^{(t)}, \varphi_q^{(t)})$ is considered unchanged. The overall RHS-aided multisatellite communication scheme is summarized in Algorithm 8.1. Specifically, in the first two time slots, the UT determines the data transmission scheme (which is illustrated in Fig. 8.10) according to the initial satellite positioning.[9] In each of the following time slots, the UT tracks the positions of the LEO satellites based on the temporal variation law [222] of the satellites' positions,[10] and updates the data transmission scheme accordingly.

Moreover, the LEO satellites are under the influence of various perturbation forces such as the atmospherical drag and the gravitational pull from other planets, which cause relative drifts and tracking errors to accumulate. To solve this problem, each LEO satellite will feed received signal strength (RSS) information back to the UT every τ time slots. As shown in Algorithm 8.1, once the RSS is smaller than the threshold Υ, the position of the satellite in the next two time slots will be reacquired, based on which its positions in the following time slots are predicted according to the temporal variation law of the positions of the LEO satellites. Given the complexity of Algorithm 8.1, we observe the following facts: In each time slot, the UT predicts the position of each satellite q based on the temporal variation law given by Proposition 8.1, which has a complexity of $\mathcal{O}(Q)$. Therefore, because there are T time slots in total, the complexity of Algorithm 8.1 is $\mathcal{O}(QT)$. In the following, we will elaborate on the data transmission scheme and the LEO satellite tracking scheme separately.

Data Transmission Scheme

As shown in Fig. 8.10, to upload data streams to the satellites simultaneously,[11] the UT first encodes the data streams for different satellites via a digital beamformer $\mathbf{V} \in \mathbb{C}^{L\times Q}$ at the baseband satisfying $\mathrm{Tr}(\mathbf{V}\mathbf{V}^H) \leq P$, where P is the transmit power at the UT. The processed signals are then up-converted to the carrier frequency by passing through L RF chains. Note that each feed of the RHS is connected to one RF chain.[12] After up-converting the transmitted signals to the carrier frequency, each RF chain sends the up-converted signals to its connected feed. The feed can transform the high-frequency current into the electromagnetic wave, which is also called a reference wave, propagating along the RHS. Because the feeds of the RHS are directly connected with the RF chains, there is no channel or attenuation between the UT and the RHS. The reference waves will be transformed into leaky waves through radiation elements and radiate energy into free space, where its radiation amplitude at each element is controlled via a holographic beamformer $\mathbf{M} \in \mathbb{C}^{N_x N_y \times L}$ to generate beams directed toward the satellites.

We denote the intended signal vector for Q satellites as $\mathbf{s} \in \mathbb{C}^{Q\times 1}$ satisfying $\mathbb{E}[\mathbf{s}\mathbf{s}^H] = \mathbf{I}_Q$. After the digital beamforming by the UT and the holographic beamforming by the RHS, the final transmitted signals are $\mathbf{MVs}$. Consider the case where the LoS link dominates the communication between the UT and satellites given frequency-flat fading due to the high-frequency bands.[13] Therefore, based on free-space propagation, the LoS channel between the (n_x, n_y)th element of the RHS and the kth antenna of satellite q can be given by

$$h_{n_x,n_y}^{q,k} = \frac{\lambda\sqrt{G_q}\sqrt{A_q}\cdot e^{-j\mathbf{k}_f\cdot \mathbf{d}_{n_x,n_y}^{q,k}}}{4\pi|\mathbf{d}_{n_x,n_y}^{q,k}|}, \tag{8.66}$$

where A_q is the effective aperture of each RHS element depending on the satellite's direction. The wavelength of the signal in free space is λ, G_q is the antenna gain of satellite q, and $\mathbf{d}_{n_x,n_y}^{q,k}$ is the distance vector[14] from the (n_x, n_y)th element to the kth antenna of satellite q. Therefore, the received signal at satellite q can be written as

$$\mathbf{y}_q = \mathbf{H}_q\mathbf{M}\mathbf{v}_q s_q + \mathbf{H}_q\mathbf{M}\sum_{q'\neq q}\mathbf{v}_{q'} s_{q'} + \mathbf{z}_q, \tag{8.67}$$

where $\mathbf{H}_q \in \mathbb{C}^{K\times N_x N_y}$ is the channel matrix between the RHS and satellite q composed of element $h_{n_x,n_y}^{q,k}$ given in (8.66), $\mathbf{M}$ is the holographic beamforming matrix given in (8.65), $\mathbf{v}_q$ is the qth column of the digital beamformer $\mathbf{V}$, and $\mathbf{z}_q \sim \mathscr{CN}(\mathbf{0}, \sigma^2)$ is the AWGN. After receiving the signals, satellite q first processes the signals using a receive combiner $\mathbf{w}_q \in \mathbb{C}^{K\times 1}$, implemented by relying on phase shifters such that $|\mathbf{w}_q(k)|^2 = 1$. Following that, the satellite q down-converts the signals to the baseband through the RF chain. The final recovered signal can then be given by

$$\widetilde{y}_q = \mathbf{w}_q^H \mathbf{H}_q \mathbf{M} \mathbf{v}_q s_q + \mathbf{w}_q^H \mathbf{H}_q \mathbf{M} \sum_{q' \neq q} \mathbf{v}_{q'} s_{q'} + \mathbf{w}_q^H \mathbf{z}_q. \tag{8.68}$$

LEO Satellite Tracking Scheme

We aim to track the satellites based on the temporal variation law of the satellites' positions such that frequent satellite positioning can be avoided. We denote the angular velocity of each satellite as ω_q. Because each satellite moves in a uniform circular motion, the angle between the position vectors of satellite q in two adjacent time slots is $\omega_q \Delta$, and the following lemma about ω_q can then be derived.

LEMMA 8.12 *The angular velocity of each LEO satellite can be derived from the positions of each LEO satellite in the first two time slots, which can be given by*

$$\begin{aligned}\omega_q = \frac{1}{\Delta} \arccos \Big\{ \Big[&\sin\theta_q^{(1)} \sin\theta_q^{(2)} \cos(\varphi_q^{(1)} - \varphi_q^{(2)}) |\boldsymbol{d}_q^{(1)}||\boldsymbol{d}_q^{(2)}| \\ &+ \left(R_e + |\boldsymbol{d}_q^{(1)}| \cos\theta_q^{(1)}\right)\left(R_e + |\boldsymbol{d}_q^{(2)}| \cos\theta_q^{(2)}\right)\Big] \Big/ (R_e + H)^2 \Big\},\end{aligned} \tag{8.69}$$

where R_e is the radius of the Earth, H is the altitude of the satellites, and $|\boldsymbol{d}_q^{(t)}|$ is the distance between the UT and satellite q in time slot t; that is,

$$|\boldsymbol{d}_q^{(t)}| = \sqrt{R_e^2 \cos^2\theta_q^{(t)} + 2R_e H + H^2} - R_e \cos\theta_q^{(t)}. \tag{8.70}$$

Proof According to the definition of $\theta_q^{(t)}$ and $\varphi_q^{(t)}$ shown in Fig. 8.9, the position vector of satellite q with respect to the center of the Earth in time slot t can be given by

$$\mathbf{r}_q^{(t)} = \left(|\mathbf{d}_q^{(t)}| \sin\theta_q^{(t)} \cos\varphi_q^{(t)}, |\mathbf{d}_q^{(t)}| \sin\theta_q^{(t)} \sin\varphi_q^{(t)}, |\mathbf{d}_q^{(t)}| \cos\theta_q^{(t)} + R_e\right). \tag{8.71}$$

We denote the altitude of the satellites as H. Because $|\mathbf{r}_q^{(t)}| = R_e + H$, the distance between the UT and satellite q in time slot t can be given by

$$|\mathbf{d}_q^{(t)}| = \sqrt{R_e^2 \cos^2\theta_q^{(t)} + 2R_e H + H^2} - R_e \cos\theta_q^{(t)}. \tag{8.72}$$

Therefore, ω_q can be given by

$$\begin{aligned}\omega_q &= \frac{1}{\Delta} \arccos \frac{\mathbf{r}_q^{(1)} \cdot \mathbf{r}_q^{(2)}}{|\mathbf{r}_q^{(1)}| \cdot |\mathbf{r}_q^{(2)}|} \\ &= \frac{1}{\Delta} \arccos \Big\{ \Big[\sin\theta_q^{(1)} \sin\theta_q^{(2)} \cos\left(\varphi_q^{(1)} - \varphi_q^{(2)}\right) d_q^{(1)} d_q^{(2)} \\ &\quad + \left(R_e + d_q^{(1)} \cos\theta_q^{(1)}\right)\left(R_e + d_q^{(2)} \cos\theta_q^{(2)}\right)\Big] \Big/ (R_e + H)^2 \Big\}.\end{aligned} \tag{8.73}$$

□

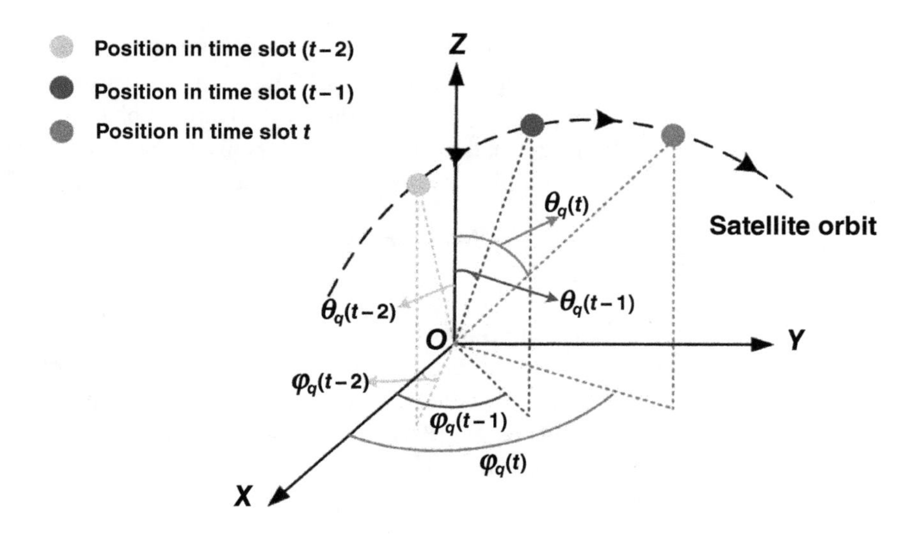

Figure 8.11 Motion trajectory of a LEO satellite

The following proposition about the temporal variation law of the satellites' positions can then be derived from Lemma 8.12.

PROPOSITION 8.10 *As shown in Fig. 8.11, considering the propagation delay in the UT-satellite link, the temporal variation law of the physical direction between the UT and each LEO satellite can be given by*

$$\theta_q^{(t)} = \arccos \frac{\frac{\sin \omega_q\left(2\Delta+|\boldsymbol{d}_q^{(t)}|/c\right)}{\sin \omega_q \Delta} D_q^{(t-1)} - \frac{\sin \omega_q\left(\Delta+|\boldsymbol{d}_q^{(t)}|/c\right)}{\sin \omega_q \Delta} D_q^{(t-2)} - R_e}{\sqrt{H^2+2R_e\left[H-\frac{\sin \omega_q\left(2\Delta+|\boldsymbol{d}_q^{(t)}|/c\right)}{\sin \omega_q \Delta} D_q^{(t-1)}+\frac{\sin \omega_q\left(\Delta+|\boldsymbol{d}_q^{(t)}|/c\right)}{\sin \omega_q \Delta} D_q^{(t-2)}+R_e\right]}}, \tag{8.74}$$

$$\varphi_q^{(t)} = \begin{cases} \arctan \dfrac{\sin\left[\omega_q\left(2\Delta+\left|\boldsymbol{d}_q^{(t)}\right|/c\right)\right]\left|\boldsymbol{d}_q^{(t-1)}\right|\Theta_q^{(t-1)} - \sin\left[\omega_q\left(\Delta+\left|\boldsymbol{d}_q^{(t)}\right|/c\right)\right]\left|\boldsymbol{d}_q^{(t-2)}\right|\Theta_q^{(t-2)}}{\sin\left[\omega_q\left(2\Delta+\left|\boldsymbol{d}_q^{(t)}\right|/c\right)\right]\left|\boldsymbol{d}_q^{(t-1)}\right|\Phi_q^{(t-1)} - \sin\left[\omega_q\left(\Delta+\left|\boldsymbol{d}_q^{(t)}\right|/c\right)\right]\left|\boldsymbol{d}_q^{(t-2)}\right|\Phi_q^{(t-2)}}, \\ \quad \text{if } \sin\left[\omega_q\left(2\Delta+\left|\boldsymbol{d}_q^{(t)}\right|/c\right)\right]\left|\boldsymbol{d}_q^{(t-1)}\right|\Phi_q^{(t-1)} - \sin\left[\omega_q\left(\Delta+\left|\boldsymbol{d}_q^{(t)}\right|/c\right)\right]\left|\boldsymbol{d}_q^{(t-2)}\right|\Phi_q^{(t-2)} > 0; \\ \arctan \dfrac{\sin\left[\omega_q\left(2\Delta+\left|\boldsymbol{d}_q^{(t)}\right|/c\right)\right] d_q^{(t-1)}\Theta_q^{(t-1)} - \sin\left[\omega_q\left(\Delta+\left|\boldsymbol{d}_q^{(t)}\right|/c\right)\right]\left|\boldsymbol{d}_q^{(t-2)}\right|\Theta_q^{(t-2)}}{\sin\left[\omega_q\left(2\Delta+\left|\boldsymbol{d}_q^{(t)}\right|/c\right)\right]\left|\boldsymbol{d}_q^{(t-1)}\right|\Phi_q^{(t-1)} - \sin\left[\omega_q\left(\Delta+\left|\boldsymbol{d}_q^{(t)}\right|/c\right)\right]\left|\boldsymbol{d}_q^{(t-2)}\right|\Phi_q^{(t-2)}} + \pi, \\ \quad \text{if } \sin\left[\omega_q\left(2\Delta+\left|\boldsymbol{d}_q^{(t)}\right|/c\right)\right]\left|\boldsymbol{d}_q^{(t-1)}\right|\Phi_q^{(t-1)} - \sin\left[\omega_q\left(\Delta+\left|\boldsymbol{d}_q^{(t)}\right|/c\right)\right]\left|\boldsymbol{d}_q^{(t-2)}\right|\Phi_q^{(t-2)} < 0, \end{cases} \tag{8.75}$$

where

$$\begin{aligned} \left|\boldsymbol{d}_q^{(t)}\right| &= \sqrt{(R_e+H)^2+R_e^2+2R_e D_q^{(t-2)} - 4R_e \cos \omega_q \Delta \cdot D_q^{(t-1)}}, \\ D_q^{(t)} &= \left|\boldsymbol{d}_q^{(t)}\right| \cos \theta_q^{(t)} + R_e, \quad \Theta_q^{(t)} = \sin \theta_q^{(t)} \sin \varphi_q^{(t)}, \quad \Phi_q^{(t)} = \sin \theta_q^{(t)} \cos \varphi_q^{(t)}. \end{aligned} \tag{8.76}$$

Proof We first derive $|\mathbf{d}_q^{(t)}|$ based on the positions of satellite q in time slots $(t-1)$ and $(t-2)$. Because the angle between $\mathbf{r}_q^{(t-2)}$ and $\mathbf{r}_q^{(t-1)}$ and the angle between $\mathbf{r}_q^{(t-1)}$ and $\mathbf{r}_q^{(t)}$ are both $\omega_q \Delta$, we have $\mathbf{r}_q^{(t)} = 2\mathbf{r}_q^{(t-1)} \cos \omega_q \Delta - \mathbf{r}_q^{(t-2)}$. Therefore, $|\mathbf{d}_q^{(t)}|$ can be obtained by

$$\begin{aligned}\left|\mathbf{d}_q^{(t)}\right| &= \left|2\mathbf{r}_q^{(t-1)} \cos \omega_q \Delta - \mathbf{r}_q^{(t-2)} - (0,0,R_e)\right| \\ &= \left[4(R_e+H)^2 \cos^2 \omega_q \Delta + (R_e+H)^2 + R_e^2 - 4(R_e+H)^2 \right. \\ &\quad \left. \cdot \cos^2 \omega_q \Delta - 4R_e \cos \omega_q \Delta \cdot \mathbf{r}_q^{(t-2)} \cdot (0,0,R) + 2\mathbf{r}_q^{(t-1)} \cdot (0,0,R_e)\right]^{\frac{1}{2}} \\ &= \sqrt{(R_e+H)^2 + R_e^2 + 2R_e D_q^{(t-2)} - 4R_e \cos \omega_q \Delta \cdot D_q^{(t-1)}},\end{aligned} \tag{8.77}$$

where $D_q^{(t)} = |\mathbf{d}_q^{(t)}| \cos \theta_q^{(t)} + R_e$.

We then derive the position of satellite q in time slot t. Considering the propagation delay in the UT-satellite link, the angle between $\mathbf{r}_q^{(t-1)}$ and $\mathbf{r}_q^{(t)}$ is refined as $\omega \Delta + \left|\mathbf{d}_q^{(t)}\right|/c$. Therefore, we have $\mathbf{r}_q^{(t)} = \frac{\sin(2\omega_q \Delta + |\mathbf{d}_q^{(t)}|/c)}{\sin \omega_q \Delta} \mathbf{r}_q^{(t-1)} - \frac{\sin(\omega_q \Delta + |\mathbf{d}_q^{(t)}|/c)}{\sin \omega_q \Delta} \mathbf{r}_q^{(t-2)}$, and $\mathbf{d}_q^{(t)}$ can be obtained by

$$\begin{aligned}\mathbf{d}_q^{(t)} &= \frac{\sin\left(2\omega_q \Delta + |\mathbf{d}_q^{(t)}|/c\right)}{\sin \omega_q \Delta} \mathbf{r}_q^{(t-1)} - \frac{\sin\left(\omega_q \Delta + |\mathbf{d}_q^{(t)}|/c\right)}{\sin \omega_q \Delta} \mathbf{r}_q^{(t-2)} - (0,0,R_e) \\ &= \begin{pmatrix} \frac{\sin\left(2\omega_q \Delta + \left|\mathbf{d}_q^{(t)}\right|/c\right)}{\sin \omega_q \Delta} \left|\mathbf{d}_q^{(t-1)}\right| \Phi_q^{(t-1)} - \frac{\sin\left(\omega_q \Delta + \left|\mathbf{d}_q^{(t-1)}\right|/c\right)}{\sin \omega_q \Delta} \left|\mathbf{d}_q^{(t-2)}\right| \Phi_q^{(t-2)} \\ \frac{\sin\left(2\omega_q \Delta + \left|\mathbf{d}_q^{(t)}\right|/c\right)}{\sin \omega_q \Delta} \left|\mathbf{d}_q^{(t-1)}\right| \Theta_q^{(t-1)} - \frac{\sin\left(\omega_q \Delta + \left|\mathbf{d}_q^{(t-1)}\right|/c\right)}{\sin \omega_q \Delta} \left|\mathbf{d}_q^{(t-2)}\right| \Theta_q^{(t-2)} \\ \frac{\sin\left(2\omega_q \Delta + \left|\mathbf{d}_q^{(t)}\right|/c\right)}{\sin \omega_q \Delta} D_q^{(t-1)} - \frac{\sin\left(\omega_q \Delta + \left|\mathbf{d}_q^{(t-1)}\right|/c\right)}{\sin \omega_q \Delta} D_q^{(t-2)} - R_e \end{pmatrix},\end{aligned} \tag{8.78}$$

where $\Theta_q^{(t)} = \sin \theta_q^{(t)} \sin \varphi_q^{(t)}$ and $\Phi_q^{(t)} = \sin \theta_q^{(t)} \cos \varphi_q^{(t)}$. The position of satellite q at time slot t can then be derived by

$$\begin{aligned}\theta_q^{(t)} &= \arccos \frac{\mathbf{d}_q^{(t)}(3)}{|\mathbf{d}_q^{(t)}|}, \\ \varphi_q^{(t)} &= \begin{cases} \arctan \mathbf{d}_q^{(t)}(2)/\mathbf{d}_q^{(t)}(1), & \mathbf{d}_q^{(t)}(1) > 0, \\ \arctan \mathbf{d}_q^{(t)}(2)/\mathbf{d}_q^{(t)}(1) + \pi, & \mathbf{d}_q^{(t)}(1) < 0. \end{cases}\end{aligned} \tag{8.79}$$

□

Therefore, once the positions of the satellites are obtained in the first two time slots, the position of each LEO satellite with respect to the origin in the following time slots can be predicted without satellite positioning.

8.3.3 Holographic Beamforming-Based Sum-Rate Maximization Problem Formulation

In this subsection, we present the sum rate with zero-forcing (ZF) digital beamforming of the RHS-aided LEO satellite system. A holographic beamforming problem is then formulated for sum-rate maximization.

ZF Digital Beamforming

Bases on (8.68), the sum rate of the RHS-aided LEO satellite communication system in time slot t can be given by

$$R\left(\theta_q^{(t)},\varphi_q^{(t)}\right)=\sum_{q=1}^{Q}\log_2\left(1+\frac{|\mathbf{w}_q^H\mathbf{H}_q\left(\theta_q^{(t)},\varphi_q^{(t)}\right)\mathbf{M}\mathbf{v}_q|^2}{K\sigma^2+\sum_{q'\neq q}|\mathbf{w}_q^H\mathbf{H}_q\left(\theta_q^{(t)},\varphi_q^{(t)}\right)\mathbf{M}\mathbf{v}_{q'}|^2}\right). \tag{8.80}$$

Because the position of the satellite in time slot t, $(\theta_q^{(t)},\varphi_q^{(t)})$, becomes a known quantity according to the temporal variation law given in Proposition 8.10, in the following, we omit the item $(\theta_q^{(t)},\varphi_q^{(t)})$ for notational convenience.

We adopt a ZF digital beamformer together with power allocation at the UT to alleviate the intersatellite interference [225]. The reason is that the ZF digital beamformer has low complexity, which is why it is widely utilized in the literature and can provide convenience for the follow-up optimization of the holographic beamformer [226, 227]. The ZF digital beamformer $\mathbf{V}\in\mathbb{C}^{L\times Q}$ can then be given by

$$\mathbf{V}=\mathbf{F}^H(\mathbf{F}\mathbf{F}^H)^{-1}\mathbf{P}^{\frac{1}{2}}=\widetilde{\mathbf{F}}\mathbf{P}^{\frac{1}{2}}, \tag{8.81}$$

where $\mathbf{F}=(\mathbf{M}^H\mathbf{H}_1^H\mathbf{w}_1,\ldots,\mathbf{M}^H\mathbf{H}_Q^H\mathbf{w}_Q)^H\in\mathbb{C}^{Q\times L}$ represents the equivalent channel matrix between the RHS and the satellites, and $\mathbf{P}=\mathrm{diag}(P_1,\ldots,P_q,\ldots,P_Q)$ is a diagonal matrix, where P_q denotes the received power at satellite q. Based on the properties of ZF beamforming (i.e., $\mathbf{w}_q^H\mathbf{H}_q\mathbf{M}\mathbf{v}_q=\sqrt{P_q}$ and $\mathbf{w}_q^H\mathbf{H}_q\mathbf{M}\mathbf{v}_{q'}=0,\forall q'\neq q$), the power allocation problem can be simplified as

$$\max_{\{P_q\}}\sum_{q=1}^{Q}\log_2\left(1+\frac{P_q}{K\sigma^2}\right),\quad s.t.\ \mathrm{Tr}(\mathbf{P}^{\frac{1}{2}}\widetilde{\mathbf{F}}^H\widetilde{\mathbf{F}}\mathbf{P}^{\frac{1}{2}})=P. \tag{8.82}$$

The optimal solution of (8.82) can be obtained by water-filling [223] as

$$P_q^*=\frac{1}{\mu_q}\max\left\{\frac{1}{\nu}-K\mu_q\sigma^2,0\right\}, \tag{8.83}$$

where μ_q is the qth diagonal element of $\widetilde{\mathbf{F}}^H\widetilde{\mathbf{F}}$, and ν is a normalized factor satisfying $\sum_{q=1}^{Q}\max\{\frac{1}{\nu}-K\mu_q\sigma^2,0\}=P$.

Because the power constraint given in (8.82) can be rewritten as $\sum_{q=1}^{Q}P_q[(\mathbf{F}\mathbf{F}^H)^{-1}]_{q,q}=P$, based on (8.81) and (8.83), the sum rate of the RHS-aided LEO satellite communication system can be given by

$$R=\sum_{q=1}^{Q}R_q=\sum_{q=1}^{Q}\log_2\left(1+\frac{\Lambda_q}{K\sigma^2[(\mathbf{F}\mathbf{F}^H)^{-1}]_{q,q}}\right), \tag{8.84}$$

where $\Lambda_q=P_q^*[(\mathbf{F}\mathbf{F}^H)^{-1}]_{q,q}$.

Holographic Beamforming Problem Formulation

We aim to maximize the sum rate of the RHS-aided LEO satellite communication system in each time slot by optimizing the holographic beamforming scheme $\mathbf{M}$, as follows:

$$\max_{\{\mathbf{M}\}} \sum_{q=1}^{Q} \log_2\left(1+\frac{\Lambda_q}{K\sigma^2[(\mathbf{FF}^H)^{-1}]_{q,q}}\right), \quad s.t. \sum_{q=1}^{Q}\sum_{l=1}^{L} a_{q,l}=1. \tag{8.85}$$

Because $[(\mathbf{FF}^H)^{-1}]_{q,q}$ does not have a closed-form expression, making the derivation of the optimal holographic beamformer difficult to conduct, we consider eliminating the item $[(\mathbf{FF}^H)^{-1}]_{q,q}$ by transforming R into its upper bound $\widehat{R}$. We then reformulate the sum-rate maximization problem to maximize $\widehat{R}$ and derive the optimal holographic beamformer. Finally, we will demonstrate that the derived optimal holographic beamformer is exactly the optimal solution of the original sum-rate maximization problem (8.85).

To eliminate of the item $[(\mathbf{FF}^H)^{-1}]_{q,q}$, we give the following proposition about $\widehat{R}$.

PROPOSITION 8.11 *The sum rate of the RHS-aided LEO satellite communication system satisfies*

$$R \le \sum_{q=1}^{Q} \log_2\left(1+\frac{\Lambda_q}{K\sigma^2}[\boldsymbol{F}\boldsymbol{F}^H]_{q,q}\right)=\widehat{R}, \tag{8.86}$$

where the equality holds when $\mathbf{FF}^H$ *is a diagonal matrix.*

Proof Because $\mathbf{FF}^H$ is a positive definite matrix, there exists another positive definite Hermitian matrix $\mathbf{G}$ such that $\mathbf{G}^2=\mathbf{FF}^H$. We can rewrite $[\mathbf{FF}^H]_{q,q}$ as $[\mathbf{FF}^H]_{q,q}=\mathbf{e}_q^H\mathbf{FF}^H\mathbf{e}_q=\mathbf{e}_q^H\mathbf{G}^H\mathbf{G}\mathbf{e}_q=|\mathbf{G}\mathbf{e}_q|^2$ and $[(\mathbf{FF}^H)^{-1}]_{q,q}=|\mathbf{G}^{-1}\mathbf{e}_q|^2$, where $\mathbf{e}_q\in\mathbb{R}^{Q\times 1}$ is a vector such that only its qth element is one and the other elements are zero. Based on the Cauchy–Schwartz inequality, we have

$$\begin{aligned}
[\mathbf{FF}^H]_{q,q}\cdot[(\mathbf{FF}^H)^{-1}]_{q,q} &= |\mathbf{G}\mathbf{e}_q|^2\cdot|\mathbf{G}^{-1}\mathbf{e}_q|^2\\
&\ge ((\mathbf{G}\mathbf{e}_q)^H\mathbf{G}^{-1}\mathbf{e}_q)^2\\
&= (\mathbf{e}_q^H\mathbf{e}_q)^2=1.
\end{aligned} \tag{8.87}$$

Therefore, we have $R=\sum_{q=1}^{Q}\log_2(1+\frac{\Lambda_q}{K\sigma^2[(\mathbf{FF}^H)^{-1}]_{q,q}})\le\sum_{q=1}^{Q}\log_2(1+\frac{\Lambda_q}{K\sigma^2}[(\mathbf{FF}^H)]_{q,q})$. The equality holds when $\mathbf{G}\mathbf{e}_q$ and $\mathbf{G}^{-1}\mathbf{e}_q$ are collinear. That is, there exists a constant λ_q such that $\mathbf{G}\mathbf{e}_q=\lambda_q\mathbf{G}^{-1}\mathbf{e}_q\Rightarrow\mathbf{FF}^H\mathbf{e}_q=\lambda_q\mathbf{e}_q$. This implies $[\mathbf{FF}^H]_{q,q}=\lambda_q$ and $[\mathbf{FF}^H]_{q',q}=0$ for $q'\neq q$ (i.e., $\mathbf{FF}^H$ is a diagonal matrix). □

Note that $\widehat{R}$ is exactly the capacity of the RHS-aided LEO satellite communication system that can be achieved by dirty paper coding (DPC) at the BS [228].

Because $[\mathbf{FF}^H]_{q,q}$ influences the upper bound of the sum rate, we present the following two lemmas about the properties of matrix $\mathbf{F}$ to derive the expression of $[\mathbf{FF}^H]_{q,q}$.

LEMMA 8.13 *The equivalent channel $[\boldsymbol{F}]_{q,l}$ can be expressed as*[15]

$$[\boldsymbol{F}]_{q,l} = \frac{\lambda\sqrt{G_q}\sqrt{A_q}\sqrt{\eta}\cdot e^{-j\boldsymbol{k}_f\cdot\boldsymbol{d}_q}}{4\pi|\boldsymbol{d}_q|d_x d_y}\cdot\left[\sum_{k=1}^{K}\boldsymbol{w}_q(k)\cdot e^{-j\boldsymbol{k}_f\cdot\Psi_{q,k}}\right]\cdot$$
$$\sum_{q'=1}^{Q}\sum_{l'=1}^{L}a_{q',l'}\left[\iint_{\mathscr{A}} e^{-\alpha\sqrt{(x-x_l)^2+(y-y_l)^2}-j\omega_{n_x,n_y}^{q,l}}\frac{1+\cos\omega_{n_x,n_y}^{q',l'}}{2}dxdy\right], \tag{8.88}$$

where $\mathscr{A} = \{(x,y) : 0 \le x \le N_x d_x, 0 \le y \le N_y d_y\}$ is the integral domain, and $\{e^{-j\boldsymbol{k}_f\cdot\Psi_{q,k}}\}$ is the antenna array response vector[16] *at satellite q. The position of feed l is (x_l, y_l), and $\omega_{n_x,n_y}^{q,l}$ is the phase difference between the reference wave generated by feed l and the object wave toward satellite q at the (n_x, n_y)th radiation element. That is,*

$$\begin{aligned}\omega_{n_x,n_y}^{q,l} &= \boldsymbol{k}_f\cdot\boldsymbol{r}_{n_x,n_y} - \boldsymbol{k}_s\cdot\boldsymbol{r}_{n_x,n_y}^{l} \\ &= |\boldsymbol{k}_f|\cdot(x\Phi_q + y\Theta_q) - |\boldsymbol{k}_s|\cdot\sqrt{(x-x_l)^2+(y-y_l)^2}.\end{aligned} \tag{8.89}$$

Proof Because $\mathbf{F} = (\mathbf{M}^H\mathbf{H}_1^H\mathbf{w}_1,\ldots,\mathbf{M}^H\mathbf{H}_Q^H\mathbf{w}_Q)^H$, based on the expressions of $\mathbf{M}$ and $\mathbf{H}$ given in (8.65) and (8.66), respectively, $[\mathbf{F}]_{q,l}$ can be given by

$$\begin{aligned}[\mathbf{F}]_{q,l} &= \sum_{n_x=1}^{N_x}\sum_{n_y=1}^{N_y}\frac{\lambda\sqrt{G_q A_q\eta}\cdot e^{-j\mathbf{k}_f\cdot\mathbf{d}_q}\cdot\sum_{k=1}^{K}\mathbf{w}_q(k)\cdot e^{-j\mathbf{k}_f\cdot\Psi_{q,k}}\cdot m_{n_y,n_z}\cdot e^{-\alpha|\mathbf{r}_{n_x,n_y}^{l}|}\cdot e^{-j\mathbf{k}_s\cdot\mathbf{r}_{n_x,n_y}^{l}}}{4\pi|\mathbf{d}_q|} \\ &= \frac{\lambda\sqrt{G_q A_q\eta}e^{-j\mathbf{k}_f\mathbf{d}_q}}{4\pi|\mathbf{d}_q|}\cdot\left[\sum_{k=1}^{K}\mathbf{w}_q(k)\cdot e^{-j\mathbf{k}_f\cdot\Psi_{q,k}}\right]\sum_{n_x=1}^{N_x}\sum_{n_y=1}^{N_y}e^{-\alpha\sqrt{(n_x d_x-x_l)^2+(n_y d_y-y_l)^2}-j\omega_{n_x,n_y}^{q,l}}\cdot m_{n_x,n_y},\end{aligned} \tag{8.90}$$

where $m_{n_x,n_y} = \sum_{q'=1}^{Q}\sum_{l'=1}^{L}\frac{a_{q',l'}(1+\cos\omega_{n_x,n_y}^{q',l'})}{2}$. We define the function $u(x,y)$ as

$$u(x,y) = e^{-\alpha\sqrt{(x-x_l)^2+(y-y_l)^2}-j\omega_{n_x,n_y}^{q,l}}\cdot\left[\sum_{q'=1}^{Q}\sum_{l'=1}^{L}\frac{a_{q',l'}\left(1+\cos\omega_{n_x,n_y}^{q',l'}\right)}{2}\right]. \tag{8.91}$$

Based on the definition of the double integral, we have

$$\sum_{n_x,n_y}u(n_x d_x, n_y d_y) \approx \frac{1}{d_x d_y}\iint_{0\le x\le N_x d_x, 0\le y\le N_y d_y}u(x,y)dxdy, \tag{8.92}$$

where $x = n_x d_x$, $y = n_y d_y$. Then, $[\mathbf{F}]_{q,l}$ can be expressed as

$$[\mathbf{F}]_{q,l} = \frac{\lambda\sqrt{G_q}\sqrt{A_q}\sqrt{\eta}\cdot e^{-j\mathbf{k}_f\cdot\mathbf{d}_q}}{4\pi|\mathbf{d}_q|d_x d_y}\cdot\left[\sum_{k=1}^{K}\mathbf{w}_q(k)\cdot e^{-j\mathbf{k}_f\cdot\Psi_{q,k}}\right]\cdot$$
$$\sum_{q'=1}^{Q}\sum_{l'=1}^{L}a_{q',l'}\left[\iint_{\mathscr{A}} e^{-\alpha\sqrt{(x-x_l)^2+(y-y_l)^2}-j\omega_{n_x,n_y}^{q,l}}\frac{1+\cos\omega_{n_x,n_y}^{q',l'}}{2}dxdy\right], \tag{8.93}$$

where $\mathscr{A} = \{(x,y) : 0 \le x \le N_x d_x, 0 \le y \le N_y d_y\}$ is the integral domain. □

The following lemma can derived from Lemma 8.13.

LEMMA 8.14 *The equivalent channel $[\mathbf{F}]_{q,l}$ can be approximated by*

$$[\mathbf{F}]_{q,l} = \frac{\lambda\sqrt{G_q}\sqrt{A_q}\sqrt{\eta}\cdot e^{-j\mathbf{k}_f\cdot\mathbf{d}_q}}{16\pi|\mathbf{d}_q|d_x d_y}\cdot\left[\sum_{k=1}^{K}\mathbf{w}_q(k)\cdot e^{-j\mathbf{k}_f\cdot\Psi_{q,k}}\right]$$
$$\cdot a_{q,l}\cdot\iint_{\mathscr{A}} e^{-\alpha\sqrt{(x-x_l)^2+(y-y_l)^2}}dxdy. \tag{8.94}$$

Proof Note that $e^{-j\omega^{q,l}_{n_x,n_y}}\cdot\frac{(1+\cos\omega^{q',l'}_{n_x,n_y})}{2}$ can be expressed as

$$\frac{\cos\omega^{q,l}_{n_x,n_y}}{2}+\frac{\cos\left(\omega^{q,l}_{n_x,n_y}+\omega^{q',l'}_{n_x,n_y}\right)}{4}+\frac{\cos\left(\omega^{q,l}_{n_x,n_y}-\omega^{q',l'}_{n_x,n_y}\right)}{4}$$
$$+j\cdot\left[\frac{\sin\omega^{q,l}_{n_x,n_y}}{2}+\frac{\sin\left(\omega^{q,l}_{n_x,n_y}+\omega^{q',l'}_{n_x,n_y}\right)}{4}+\frac{\sin\left(\omega^{q,l}_{n_x,n_y}-\omega^{q',l'}_{n_x,n_y}\right)}{4}\right]. \tag{8.95}$$

We need to prove that the integral of the items containing sine and cosine functions can be neglected unless $q=q', l=l'$, in which case only the item $\frac{\cos(\omega^{q,l}_{n_x,n_y}-\omega^{q',l'}_{n_x,n_y})}{4}=\frac{1}{4}$ remains. For brevity, we give a proof for the item $\frac{\cos\omega^{q,l}_{n_x,n_y}}{2}$. The proof for the other items is obtained similarly.

For a large-scale RHS, utilizing polar coordinate transformation $x-x_l=\rho\cos\vartheta$, $y-y_l=\rho\sin\vartheta$, the integral in (8.88) corresponding to the item $\frac{\cos\omega^{q,l}_{n_x,n_y}}{2}$ can be approximated by

$$\frac{1}{2}\iint_{\mathbb{R}^2} e^{-\alpha\sqrt{(x-x_l)^2+(y-y_l)^2}}\cos\omega^{q,l}_{n_x,n_y}dxdy \tag{8.96}$$
$$=\frac{1}{2}\int_0^{2\pi}\int_0^{+\infty}\rho e^{-\alpha\rho}\cos(U_q\rho+\eta_{q,l})d\rho d\vartheta, \tag{8.97}$$

where $U_q=|\mathbf{k}_s|-|\mathbf{k}_f|\cdot(\Phi_q\cos\vartheta+\Theta_q\sin\vartheta)$, and $\eta_{l,k}=|\mathbf{k}_f|(x_l\Phi_q+y_l\Theta_q)$. For a fixed ϑ, by utilizing integration by parts, we have

$$\int_0^{+\infty}\rho e^{-\alpha\rho}\cos(U_q\rho+\eta_{q,l})d\rho \tag{8.98}$$
$$=-\frac{1}{U_q}\int_0^{+\infty}(1-\alpha\rho)e^{-\alpha\rho}\sin(U_q\rho+\eta_{q,l})d\rho \tag{8.99}$$
$$=\frac{1}{U_q^2}\cos\eta_{q,l}-\frac{1}{U_q^2}\int_0^{+\infty}(\alpha^2\rho-2\alpha)e^{-\alpha\rho}\cos(U_q\rho+\eta_{q,l})d\rho \tag{8.100}$$
$$=\mathscr{O}(U_q^{-2}). \tag{8.101}$$

Because $U_q\geq|\mathbf{k}_s|-|\mathbf{k}_f|$ has a magnitude of 10^2, the integral in (8.88) corresponding to the item $\frac{\cos\omega^{q,l}_{n_x,n_y}}{2}$ has a magnitude of 10^{-4} and can be neglected. In addition, the correctness of this lemma for a moderate-sized RHS is also verified later on. □

Remark 8.5 Based on Lemma 8.14, the equivalent channel $\mathbf{F}$ can be rewritten as $\mathbf{F} = \mathbf{D}_q \cdot \mathbf{A} \cdot \mathbf{D}_l$, where $\mathbf{D}_q$ is a Q-by-Q diagonal matrix with $[\mathbf{D}_q]_{q,q} = \frac{\lambda\sqrt{G_q}\sqrt{A_q}\sqrt{\eta}\cdot e^{-j\mathbf{k}_f\cdot\mathbf{d}_q}}{16\pi|\mathbf{d}_q|d_x d_y}$. In addition, $[\sum_{k=1}^{K}\mathbf{w}_q(k)\cdot e^{-j\mathbf{k}_f\cdot\Psi_{q,k}}]$, $\mathbf{A}$ is a Q-by-L matrix with $[\boldsymbol{a}]_{q,l} = a_{q,l}$, and $\mathbf{D}_l$ is an L-by-L diagonal matrix with $[\mathbf{D}_l]_{l,l} = \iint_{\mathscr{A}} e^{-\alpha\sqrt{(x-x_l)^2+(y-y_l)^2}}dxdy$. Because the diagonal matrices $\mathbf{D}_q$ and $\mathbf{D}_l$ are invertible, the rank of $\mathbf{F}$ is equal to the rank of $\mathbf{A}$. Note that $\mathbf{A}$ is composed of elements $a_{q,l}$ satisfying

$$\sum_{q=1}^{Q}\sum_{l=1}^{L} a_{q,l} = 1. \tag{8.102}$$

It is shown in [229] that a random matrix satisfying (8.102) has full row rank with a probability of one if $L \geq Q$. Therefore, $\mathbf{F}\mathbf{F}^H$ has full rank and the ZF digital beamformer $\mathbf{V}$ exists.

The following proposition about the expression of $[\mathbf{F}\mathbf{F}^H]_{q,q}$ can then be derived from Lemma 8.14.

PROPOSITION 8.12 *The expression of* $[\mathbf{F}\mathbf{F}^H]_{q,q}$ *can be written as*

$$[\mathbf{F}\mathbf{F}^H]_{q,q} = \frac{\lambda^2 G_q A_q \eta \cdot \left|\sum_{k=1}^{K}\mathbf{w}_q(k)\cdot e^{-j\mathbf{k}_f\cdot\Psi_{q,k}}\right|^2}{256\pi^2|\mathbf{d}_q|^2 d_x^2 d_y^2}\cdot\sum_{l=1}^{L} a_{q,l}^2 B_l^2, \tag{8.103}$$

where

$$B_l = \iint_{\mathscr{A}} e^{-\alpha\sqrt{(x-x_l)^2+(y-y_l)^2}}dxdy. \tag{8.104}$$

According to Proposition 8.12, $\widehat{R}$ can be further expressed as

$$\widehat{R} = \sum_{q=1}^{Q}\log_2\left(1 + \frac{\Lambda_q\lambda^2 G_q A_q\eta\cdot\left|\sum_{k=1}^{K}\mathbf{W}_q(k)\cdot e^{-j\mathbf{k}_f\cdot\Psi_{q,k}}\right|^2}{256K\sigma^2\pi^2|\mathbf{d}_q|^2 d_x^2 d_y^2}\cdot\sum_{l=1}^{L} a_{q,l}^2 B_l^2\right) \tag{8.105}$$

$$\leq \sum_{q=1}^{Q}\log_2\left(1 + \frac{\Lambda_q\lambda^2 G_q A_q\eta K}{256\sigma^2\pi^2|\mathbf{d}_q|^2 d_x^2 d_y^2}\cdot\sum_{l=1}^{L} a_{q,l}^2 B_l^2\right), \tag{8.106}$$

where the equality holds when the receive combiner of satellite q, $\mathbf{w}_q$, is composed of the following elements:

$$\mathbf{w}_q^*(k) = e^{j\mathbf{k}_f\cdot\Psi_{q,k}}. \tag{8.107}$$

According to (8.62) and (8.65), the optimization of the holographic beamformer $\mathbf{M}$ is equivalent to the optimization of $\{a_{q,l}\}$ (i.e., the amplitude ratio for the beam toward satellite q from feed l). Based on the expressions of $\mathbf{M}$ and $\widehat{R}$ given in (8.65) and (8.105), respectively, we reformulate the sum-rate maximization problem as

$$\max_{\{a_{q,l}\}}\sum_{q=1}^{Q}\log_2\left(1 + \frac{\Lambda_q\lambda^2 G_q A_q\eta K}{256\sigma^2\pi^2|\mathbf{d}_q|^2 d_x^2 d_y^2}\cdot\sum_{l=1}^{L} a_{q,l}^2 B_l^2\right), \tag{8.108}$$

$$s.t. \sum_{q=1}^{Q}\sum_{l=1}^{L} a_{q,l} = 1. \tag{8.109}$$

In the next section, we will develop a holographic beamforming algorithm to solve the reformulated problem (8.108). We then demonstrate that the derived optimal holographic beamforming scheme together with the receive combining scheme given in (8.107) is exactly the optimal solution of the original sum-rate maximization problem (8.85).

8.3.4 Holographic Beamforming-Based Sum-Rate Maximization Algorithm

In this subsection, we decompose the sum-rate maximization problem into Q subproblems and derive closed-form optimal solutions to the subproblems. An iterative holographic beamforming algorithm based on dynamic programming is then developed for sum-rate maximization.

Problem Decomposition

To derive the optimal holographic beamformer, we first present the following proposition about the upper bound of $\widehat{R}$.

PROPOSITION 8.13 *$\widehat{R}$ can be upper bounded by*

$$\widehat{R} \le \sum_{q=1}^{Q} \log_2\left(1 + \frac{\Lambda_q \lambda^2 G_q A_q \eta K B^2}{256\sigma^2\pi^2|\mathbf{d}_q|^2 d_x^2 d_y^2}\cdot\left(\sum_{l=1}^{L} a_{q,l}\right)^2\right), \tag{8.110}$$

where $B = \iint_{\mathscr{A}} e^{-\alpha\sqrt{x^2+y^2}}dxdy$, and the equality holds when for each satellite q there exists a unique $l_q \in \{1,2,\ldots,L\}$ satisfying

$$\begin{cases} a_{q,l} \neq 0, & l = l_q, \\ a_{q,l} = 0, & l \neq l_q. \end{cases} \tag{8.111}$$

Proof B_l can be expressed as

$$\iint_{\mathscr{A}} e^{-\alpha\sqrt{(x-x_l)^2+(y-y_l)^2}}dxdy = d_q^2 \iint_{\mathscr{A}} e^{-\alpha d_q\sqrt{(u-x_l/d_q)^2+(v-y_l/d_1)^2}}dudv, \tag{8.112}$$

where $\mathscr{A} = \{(u,v) : 0 \le u \le \frac{N_x d_x}{d_q}, 0 \le v \le \frac{N_y d_y}{d_q}\}$. Because $x_l \ll d_q$ and $y_l \ll d_q$, B_l can be approximated by $B_l \approx d_q^2 \iint_{\mathscr{A}} e^{-\alpha d_q\sqrt{u^2+v^2}}dudv = \iint_{\mathscr{A}} e^{-\alpha\sqrt{x^2+y^2}}dxdy$, and $\widehat{R}$ can then be approximated by

$$\widehat{R} \approx \sum_{q=1}^{Q} \log_2\left(1 + \frac{\Lambda_q \lambda^2 G_q A_q \eta K B^2}{256\sigma^2\pi^2|\mathbf{d}_q|^2 d_y^2 d_z^2}\cdot\sum_{l=1}^{L} a_{q,l}^2\right), \tag{8.113}$$

where $B = \iint_{\mathscr{A}} e^{-\alpha\sqrt{x^2+y^2}} dxdy$. Because $\left(\sum_{l=1}^{L} a_{q,l}\right)^2 - \sum_{l=1}^{L} a_{q,l}^2 = \sum_{l \neq l'} a_{q,l} q_{q,l'} \geq 0$, the upper bound of $\widehat{R}$ can then be given by

$$\widehat{R} \leq \sum_{q=1}^{Q} \log_2 \left[1 + \frac{\Lambda_q \lambda^2 G_q A_q \eta K B^2}{256\sigma^2 \pi^2 |\mathbf{d}_q|^2 d_y^2 d_z^2} \cdot \left(\sum_{l=1}^{L} a_{q,l} \right)^2 \right]. \tag{8.114}$$

The equality holds when $a_{q,l} q_{q,l'} = 0$, $\forall l,\ l'$, which indicates that there exists a unique $l_q \in \{1, 2, \ldots, L\}$ satisfying $\begin{cases} a_{q,l} \neq 0, & l = l_q, \\ a_{q,l} = 0, & l \neq l_q. \end{cases}$ □

For notational convenience, $C_q = \frac{\Lambda_q \lambda^2 G_q A_q \eta K B^2}{256\sigma^2 \pi^2 |\mathbf{d}_q|^2 d_x^2 d_y^2}$, $l_q = q$, and $a_{q,l_q} = a_q$. Based on (8.108), the sum-rate maximization problem can then be simplified as

$$\max_{\{a_q\}} \sum_{q=1}^{Q} \log_2 \left(1 + C_q a_q^2\right), \quad s.t. \sum_{q=1}^{Q} a_q = 1. \tag{8.115}$$

Note that the maxima of (8.115) may be achieved on the boundary of the feasible region (i.e., in the optimal solution $\{a_q^*\}$, $|\{a_q : a_q = 0, q = 1, \ldots, Q\}| = i > 0$). Therefore, to obtain the global optimal solution, we develop a dynamic programming-based algorithm by traversing the number of nonzero items in $\{a_q\}_{q=1}^{Q}$ from one to Q, where the sum-rate maximization problem can then be divided into Q subproblems. Specifically, in the ith ($1 \leq i \leq Q$) iteration, it is assumed that the number of nonzero items in $\{a_q\}_{q=1}^{Q}$ is i, and thus, the subproblem i can be written as

$$\max_{\{a_q\}} \sum_{q=1}^{Q} \log_2 \left(1 + C_q a_q^2\right), \tag{8.116a}$$

$$s.t. \quad \sum_{q=1}^{Q} a_q = 1, \quad |\{a_q : a_q = 0, q = 1, \ldots, Q\}| = i. \tag{8.116b}$$

After solving these Q subproblems, the global optimal sum rate can be obtained by comparing the maximum sum rate corresponding to each subproblem. In the following, based on the derivation of the optimal solution to the subproblem given in (8.116), we present the whole holographic beamforming algorithm based on dynamic programming.

Optimal Solution to the Subproblems

To derive the optimal solution to subproblem i denoted as $\{a_{q,i}^*\}$, we first give the following lemma about the property of $\{a_{q,i}^*\}$.

LEMMA 8.15 *For $q_1 \neq q_2$, if $C_{q_1} \geq C_{q_2}$, then in the optimal solution $\{a_{q,i}^*\}$, $a_{q_1,i}^* \geq a_{q_2,i}^*$.*

Proof We prove by contradiction. Assume that for $q_1 \neq q_2$, when $C_{q_1} \geq C_{q_2}$, the optimal solution $\{a^*_{q,i}\}$ satisfies $a^*_{q_1,i} < a^*_{q_2,i}$. Note that

$$\begin{aligned} &\log_2\left[1+C_{q_1}\left(a^*_{q_1,i}\right)^2\right]+\log_2\left[1+C_{q_2}\left(a^*_{q_2,i}\right)^2\right] \\ &\quad -\log_2\left[1+C_{q_1}\left(a^*_{q_2,i}\right)^2\right]-\log_2\left[1+C_{q_2}\left(a^*_{q_1,i}\right)^2\right] \end{aligned} \tag{8.117}$$

$$= \log_2\left\{1+\frac{\left(C_{q_1}-C_{q_2}\right)\left[\left(a^*_{q_1,i}\right)^2-\left(a^*_{q_2,i}\right)^2\right]}{1+C_{q_1}\left(a^*_{q_2,i}\right)^2+C_{q_2}\left(a^*_{q_1,i}\right)^2+C_{q_1}C_{q_2}\left(a^*_{q_1,i}\right)^2\left(a^*_{q_2,i}\right)^2}\right\} < 0. \tag{8.118}$$

Therefore, exchanging the values of $a^*_{q_1,i}$ and $a^*_{q_2,i}$ will lead to an increase in the sum rate, which contradicts the optimality of $\{a^*_{q,i}\}$. □

Without loss of generality, we assume that $C_1 \geq C_2 \geq \cdots \geq C_Q$. Based on Lemma 8.15, we have $a^*_{i+1,i} = a^*_{i+2,i} = \cdots = a^*_{Q,i} = 0$, and the subproblem i can be simplified as

$$\max_{\{a_q\}_{q=1}^{i}} \sum_{q=1}^{i} \log_2\left(1+C_q a^2_{q,i}\right), \tag{8.119a}$$

$$s.t. \quad \sum_{q=1}^{i} a_{q,i} = 1, \quad a_{q,i} > 0. \tag{8.119b}$$

We adopt the Lagrangian dual form to relax the constraint (8.119b) with a multiplier. We denote β as the Lagrangian multiplier associated with the constraint (8.119b). The Lagrangian associated with subproblem i can then be given by

$$\mathscr{L}(a_{q,i},\beta) = \sum_{q=1}^{i} \log_2\left(1+C_q a^2_{q,i}\right) - \beta\left(\sum_{q=1}^{i} a_{q,i} - 1\right). \tag{8.120}$$

By setting $\partial\mathscr{L}/\partial a_{q,i} = 0$, the optimal $\{a^*_{q,i}\}$ and β^* can be obtained by solving the following system of equations:

$$\begin{cases} a^*_{q,i} = \frac{1}{\beta \ln 2} + \sqrt{\frac{1}{(\beta\ln 2)^2} - \frac{1}{C_q}}, \\ \sum_{q=1}^{i} a^*_{q,i} = 1. \end{cases}$$

Therefore, the maximum sum rate corresponding to subproblem i can be given by

$$R_{si} = \sum_{q=1}^{i} \log_2\left[1+C_q(a^*_{q,i})^2\right]. \tag{8.121}$$

Overall Holographic Beamforming Algorithm

The overall holographic beamforming algorithm for sum-rate maximization composed of Q iterations is summarized in Algorithm 8.2. In each iteration, the subproblem i is solved based on (8.11.4). After traversing i from one to Q, the global optimal sum rate $R_{\max}$ can be finally obtained by comparing the maximum R_{si} corresponding to

Algorithm 8.2 Holographic beamforming for sum-rate maximization

Input: The position of each satellite in time slot t $\{(\theta_q^{(t)}, \varphi_q^{(t)})\}$.

1 **Initialization**: Maximum sum rate $R_{\max} = 0$;
2 **for** $i = 1 : Q$ **do**
3 Compute the maximum sum rate R_{si} corresponding to subproblem i based on the optimal solution $\{a_{q,i}^*\}$ given in (8.11.4);
4 **if** $R_{si} \geq R_{\max}$ **then**
5 Update the maximum sum rate $R_{\max} \leftarrow R_{si}$;
6 Update the optimal holographic beamformer $\mathbf{M}^*$ by (8.122);
7 **end**
8 **end**

Output: The optimal holographic beamformer $\mathbf{M}^*$ and the maximum sum rate $R_{\max}$.

each subproblem (i.e., the global optimal solution $\{a_q^*\} = \{a_{q, \arg\max_i R_{si}}\}$). Therefore, the optimal holographic beamformer $\mathbf{M}^*$ can be obtained by substituting $\{a_q^*\}$ into (8.65); that is, $\mathbf{M}^*$ is formed by the following elements:

$$M_{n_x,n_y}^l = \sum_{q=1}^{Q} a_q^* \sqrt{\eta} \cdot m\left(\mathbf{r}_{n_x,n_y}^q, \theta_q, \varphi_q\right) \cdot e^{-\alpha |\mathbf{r}_{nx,ny}^l|} \cdot e^{-j\mathbf{k}_s \cdot \mathbf{r}_{nx,ny}^l}, \tag{8.122}$$

where $m(\mathbf{r}_{n_x,n_y}^q, \theta_q, \varphi_q)$ is given in (8.61).

Remark 8.6 The optimal $\{a_{q,l}\}$ given by (8.111) shows that $\{a_{q,l}\}$ is a diagonal matrix. Based on Lemma 8.14, we have

$$\begin{cases} a_{q,l}^* \neq 0, & l = q, \\ a_{q,l}^* = 0, & l \neq q. \end{cases} \Rightarrow \begin{cases} [\mathbf{F}]_{q,l} \neq 0, & l = q, \\ [\mathbf{F}]_{q,l} = 0, & l \neq q, \end{cases} \tag{8.123}$$

indicating that $\mathbf{F}$ is a diagonal matrix, and thus, $\mathbf{F}\mathbf{F}^H$ is also a diagonal matrix. Therefore, the derived optimal solution $\{a_q^*\}$ to (8.108) makes the equality hold in Proposition 8.11 (i.e., $R = \widehat{R}$). In other words, the optimal holographic beamforming scheme given in (8.122) together with the receive combining scheme given in (8.107) is exactly the optimal solution to the original sum-rate maximization problem (8.85). This indicates that, with simple ZF digital beamforming, the sum rate of the RHS-aided LEO satellite communication system achieves capacity.

Remark 8.7 The proposed holographic beamforming algorithm is also applicable to other communications with LoS-dominated channels such as UAV communications [230, 231]. Although tracking schemes for satellites and UAVs are different because the temporal variation law of satellites' positions and that of UAVs' positions may be different, the holographic beamforming algorithm is developed on the premise that the receivers' positions are known. Hence, it is still applicable to UAV communications.

8.3.5 Performance Analysis of RHS-Aided LEO Satellite Communications

Complexity Analysis of the Algorithm

Given the complexity of Algorithm 8.2, we observe the following facts. In each iteration, the maximum sum rate R_{si} and the optimal holographic beamformer $\mathbf{M}^*$ are computed. The computation of R_{si} has a complexity of $\mathcal{O}(Q)$ according to (8.121). Also, based on (8.122), the computation of each element in the optimal holographic beamformer $\mathbf{M}^*$ is of complexity $\mathcal{O}(Q)$, and thus, the computation of $\mathbf{M}^*$ has a complexity of $\mathcal{O}(NLQ)$. Because Q iterations are performed in Algorithm 8.2, its complexity is $\mathcal{O}(NLQ^2)$.

Robustness Analysis of the Algorithm

We define the robustness of the holographic beamforming algorithm as its ability to tolerate tracking errors of the satellites' positions that might affect the optimal holographic beamformer and the maximum sum rate.

Based on the optimal $\{a_q^*\}$ given in (8.3.4) (i.e., $a_{q,i}^* = \frac{1}{\beta \ln 2} + \sqrt{\frac{1}{(\beta \ln 2)^2} - \frac{1}{C_q}}$, $\sum_{q=1}^{i} a_{q,i}^* = 1$), the magnitude of $C_q = \frac{\Lambda_q \lambda^2 G_q A_q \eta K B^2}{256\sigma^2\pi^2|\mathbf{d}_q|^2 d_x^2 d_y^2}$ is about 10^3 under the parameter settings in satellite communications such that item $\frac{1}{C_q}$ in (8.3.4) can be neglected. Therefore, the following proposition stands.

PROPOSITION 8.14 *In the RHS-aided LEO satellite communication system, $\{a_q^*\}$ can be well approximated by $\{a_1 = a_2 = \cdots = a_{i^*} = \frac{1}{i^*}\}$, where i^* is the number of connected satellites. The maximum sum rate of the system can then be well approximated by*

$$R_{\max} \approx \sum_{q=1}^{i^*} \log_2 \left(1 + \frac{\Lambda_q \lambda^2 G_q A_q \eta K B^2}{256\sigma^2\pi^2|\mathbf{d}_q|^2 d_x^2 d_y^2 (i^*)^2} \right). \tag{8.124}$$

Proposition 8.14 reveals that the tracking errors of the satellites' positions have almost no influence on $\{a_q^*\}$, and thus, the maximum sum rate of the RHS-aided LEO satellite communication system is not significantly affected. This indicates that the proposed holographic beamforming algorithm is robust against the tracking errors of the satellites' positions. The correctness of Proposition 8.14 will also be verified in Fig. 8.12.

8.3.6 Simulation Results

In this section, we evaluate the performance of the proposed holographic beamforming algorithm for the RHS-aided LEO satellite system. We also compare the performance of the RHS-aided LEO satellite system and the traditional phased array-aided LEO satellite system in terms of the sum rate and cost efficiency. Specifically, in the phase-array-aided system, the physical dimensions of the RHS and the phased array are the same. The digital beamforming is also based on the ZF method, and the receive combiner is the same as that of the RHS-aided system. The traditional analog beamformer is optimized by the coordinate ascent algorithm [223]. Major simulation parameters

Table 8.4 Simulation parameters

Parameters	Values
Transmit power of the UT P (W)	10
Carrier frequency f (GHz)	30
Element spacing of the RHS d_x and d_y (cm)	0.25
Propagation vector in the free space k_f	200π
Propagation vector on the RHS k_s	$200\sqrt{3}\pi$
Number of LEO satellites Q	3
Altitude of LEO satellites H (Km)	600
Antenna gain of each satellite G_q (dBi)	60
Noise figure over Ka-band (dB)	1.2

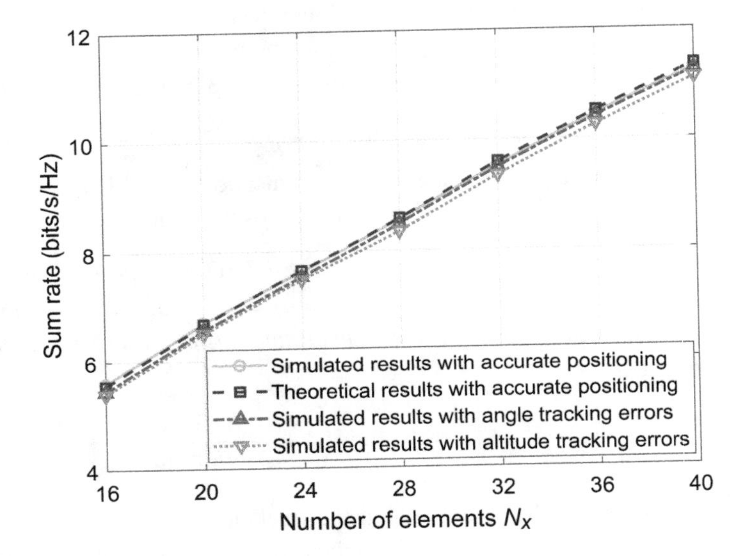

Figure 8.12 Sum rate versus number of RHS elements

are set up based on existing works [232, 233] and 3GPP specifications [234] as given in Table 8.4.

Figure 8.12 shows the sum rate versus the number of RHS elements N. For convenience, we assume that $N_x = N_y$ and utilize N_x to represent the number of RHS elements. It can be seen that the sum rate grows with the number of RHS elements. The accuracy of the approximation of the equivalent channel $[\mathbf{F}]_{q,l}$ given in Lemma 8.14 is also verified. Moreover, Fig. 8.12 evaluates the robustness of the proposed holographic beamforming algorithm against the satellite tracking errors where the satellites' angles and altitudes are both considered. We observe that there is only a slight drop in the sum rate even if the tracking errors of the satellites' angles are as large as 10° (i.e., $(\theta_q^{Tracking}, \varphi_q^{Tracking}) = (\theta_q^{Accurate} \pm 10°, \varphi_q^{Accurate} \pm 10°)$). Similarly, the tracking errors of the satellites' altitudes also have almost no influence on the

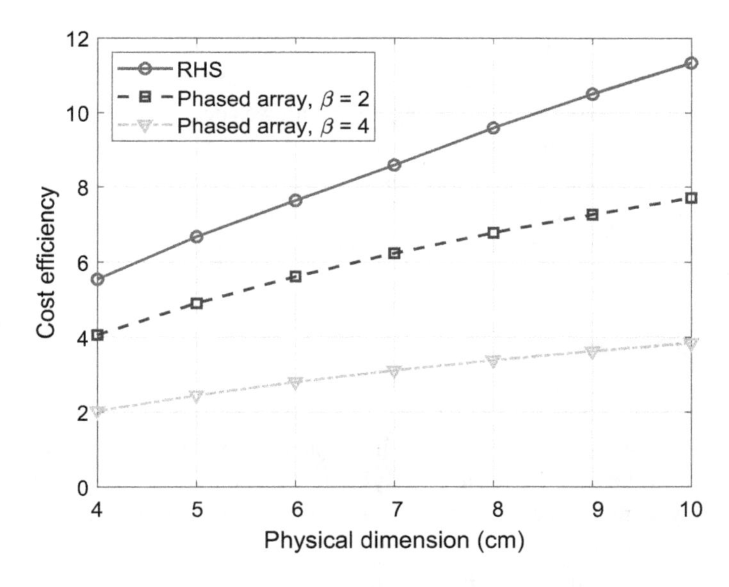

Figure 8.13 Cost efficiency versus the physical dimension $N_x d_x$ with different cost ratio β

sum rate even if the tracking errors are as large as 10 km. This indicates the robustness of the holographic beamforming algorithm against the tracking errors of the satellites' positions. The reason is given in Proposition 8.14: The tracking errors of the satellites' positions have almost no influence on $\{a_q^*\}$, and thus, the maximum sum rate is not significantly affected.

Figure 8.13 evaluates the cost efficiency of the RHS-aided LEO satellite communication system and the phased-array-aided LEO satellite communication system, where the cost-efficiency metric is defined as the ratio of the sum rate to the hardware cost. In general, because a phased array requires high-priced electronic components such as phase shifters at each antenna element, the hardware cost ratio β of a phased array to the RHS of the same number of elements is about 2–10 [213]. Specifically, in Fig. 8.13, we set the element spacing and phased array as equal (i.e., $d_x = d_x^p, d_y = d_y^p$) such that the number of elements in the RHS and the phased array is also the same. We observe that, with the same element spacing, the cost efficiency of the RHS-aided system is greater than that of the phased-array-aided system due to the low hardware cost of the RHS element. Moreover, the advantages of the RHS in cost savings become overwhelming when the number of elements and the cost ratio β is large.

8.3.7 Conclusion

In this section, we studied an RHS-aided LEO satellite communication system, where a UT uploads data streams to multiple LEO satellites via holographic beamforming. To support continuous connections with the satellites, we proposed an RHS-aided multisatellite communication scheme by jointly considering the mobility of satellites and holographic beamforming design. A temporal variation law of the satellites' positions is derived for satellite tracking such that frequent satellite positioning can be avoided. A holographic beamforming algorithm was developed for sum-rate maximization and a closed-form optimal holographic beamformer was derived. The robustness of the algorithm against the tracking errors of the satellites' positions was proved. Simulation results also showed that the RHS provides a powerful solution to reduce the hardware cost while guaranteeing a high sum rate in practice.

8.4 Summary

In this chapter, we first presented the general network architecture where the LEO satellite networks can work in conjunction with terrestrial networks. Then, we introduced a satellite formation-flying technique that can enable ultra-dense LEO satellite systems to accomplish various aerospace and earth science missions. The capacity bounds of a LEO satellite formation system were derived, and the influence of satellite distribution and satellite formation size on the channel capacity were analyzed. Next, we proposed a novel metamaterial-based antenna (i.e, RHS) to support satellite communications. A RHS-aided multisatellite communication scheme by jointly considering the mobility of satellites and holographic beamforming design was developed.

9 Ultra-dense LEO Satellite Constellation Design

To further improve the performance of ultra-dense LEO-based terrestrial-satellite networks, one of the most important issues to be considered in the construction of such networks is the constellation design [235]. The LEO satellite constellation contains a group of LEO satellites launched into preplanned orbits by the ground control center to form a network [236]. A well-designed LEO satellite constellation, with a proper constellation size, can improve the QoS for terrestrial cellular networks without wasting resources. Although several typical satellite constellations have been proposed to achieve global coverage such as the polar orbit constellation [237], Walker Delta constellation [238], and Flower constellation [239, 240], they can only satisfy basic communication needs without considering the QoS of terrestrial users and the cost of LEO satellite deployment. Therefore, rather than satisfy basic communication needs, the main target of ultra-dense LEO satellite constellation design is to balance the terrestrial-satellite network performance (such as the backhaul capacity of the user terminal and the coverage ratio) and the cost of LEO satellite deployment [241].

The rest of this chapter is organized as follows. We first design the single-layer LEO constellation with the minimum number of satellites while satisfying the backhaul requirement in Section 9.1. The megaconstellation design for a multilayer satellite network is then investigated in Section 9.2, and Section 9.3 provides a summary.

9.1 Single-Layer Ultra-dense LEO Satellite Constellation Design

Various aspects have been considered for LEO satellite constellation deployment in existing works such as satellite number minimization [242, 243], coverage maximization [244, 245], communication delay minimization [246], and heterogeneous networks construction [247]. Several intelligent algorithms have been utilized for satellite constellation optimization such as the genetic algorithm (GA), differential evolution (DE), immune algorithm, and particle swarm optimization (PSO) [248]. In [242], a nondominated sorting genetic algorithm for regional LEO satellite constellation design was proposed to meet user requirements with lower satellite costs. In [243], a satellite constellation based on the evolutionary optimization method was proposed for continuous mutual regional coverage. The relation between the coverage ratio and the number of satellites has also been discussed: Both [244] and [245] adopted the genetic algorithm for regional satellite constellation design to maximize the coverage of target areas. In [246], a progressive satellite constellation network construction scheme was developed to minimize the end-to-end delay. In [247], an LEO satellite constellation-based Internet-of-Things system was investigated. The LEO satellite constellation, spectrum allocation scheme, and routing protocols have been discussed for such a heterogeneous network. In [248], the performance of different intelligent algorithms for satellite constellation design (i.e., GA, DE, immune algorithm, and PSO) were compared for satellite coverage capability enhancement.

However, several critical issues in LEO satellite constellation design have not been considered in most existing works. First, they mainly consider the case in which satellites are uniformly distributed on orbital planes [242]. The coupling of neighboring orbits has been ignored, resulting in redundant satellite coverage. Second, some works mainly focus on the satellite constellation design to achieve regional coverage with a static or quasistatic topology of LEO satellite constellation without considering LEO satellites' mobility [243, 244]. Third, seamless global coverage (i.e., the coverage on the Earth is continuous) and backhaul requirements of the UT[1] have not been jointly considered due to the limited number of satellites [245]. Against this background, we mainly investigate two open issues in the literature:

- Considering the cost of LEO satellite deployment and LEO satellites' mobility, how many LEO satellites are enough to achieve seamless global coverage while satisfying the backhaul requirement of each UT?
- For any given coverage ratio requirement,[2] how do we design an ultra-dense LEO satellite constellation with the minimum number of LEO satellites?

To address these issues, in this section, we map demands to multiple criteria in the ultra-dense LEO satellite constellation design (i.e., satisfying UT backhaul requirements, seamless global coverage, and minimizing satellite numbers). We propose a 3D constellation optimization algorithm simultaneously handling these criteria to achieve an optimal trade-off between the cost of LEO satellite deployment and the LEO-based backhaul services provided to UTs.

9.1.1 Single-Layer Ultra-dense LEO Satellite Constellation Topology

As shown in Fig. 9.1, we consider an ultra-dense LEO-based terrestrial-satellite network [249] consisting of multiple UTs as transmitters and a set of satellites as

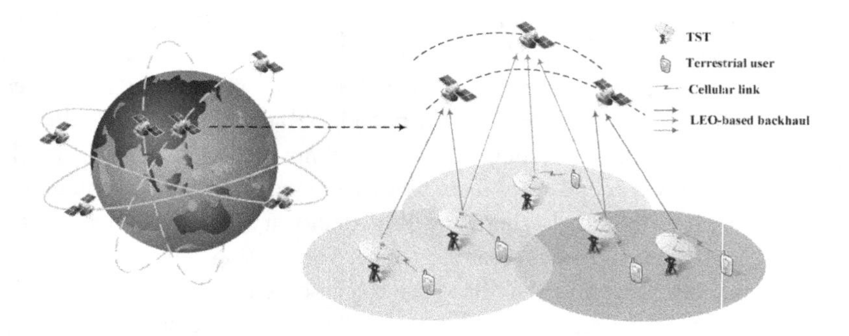

Figure 9.1 Ultra-dense LEO-based terrestrial-satellite network

receivers. To support seamless coverage for the UT, an ultra-dense LEO satellite constellation is configured to ensure that at least one satellite flies over the area of interest at each time slot. We assume that these satellites are operated in N orbital planes of altitude h, denoted by $\mathcal{N} = \{1, 2, \ldots, N\}$, and orbital plane n consists of M_n LEO satellites, denoted by $\mathcal{M}_n = \{1, 2, \ldots, M_n\}$. Each satellite associates with all the UTs in its coverage area, and the UTs share the available bandwidth B over the Ka-band equally for each LEO satellite.

LEO Satellite Orbit

We assume that the orbit of a LEO satellite is circular[1]. In order to describe the orientation of the satellite with respect to the equatorial coordinate system, as illustrated in Fig. 9.2, the traditional six orbital parameters are introduced as follows:

- Inclination angle i is the angle of intersection between the orbital plane and the equator. An inclination angle of more than 90° implies that the direction of the satellite's motion is the opposite of the Earth's rotation.
- The right ascension of ascending node Ω is the angle between the vernal equinox and the interactions of the orbital and the equatorial planes.
- Argument of the perigee ω is the angle between the ascending node and the perigee (i.e., the point where the satellite is the closest to the Earth, measured along the orbital plane).
- Eccentricity e is the eccentricity of the orbital ellipse. Because the LEO satellite orbit is assumed to be circular, the eccentricity e is 0.
- Semimajor axis a is half of the length of the orbit's major axis. In the circular orbit case, the semimajor axis a is equal to the radius of the orbit.
- True anomaly v is the geocentric angle between the perigee direction and satellite direction.

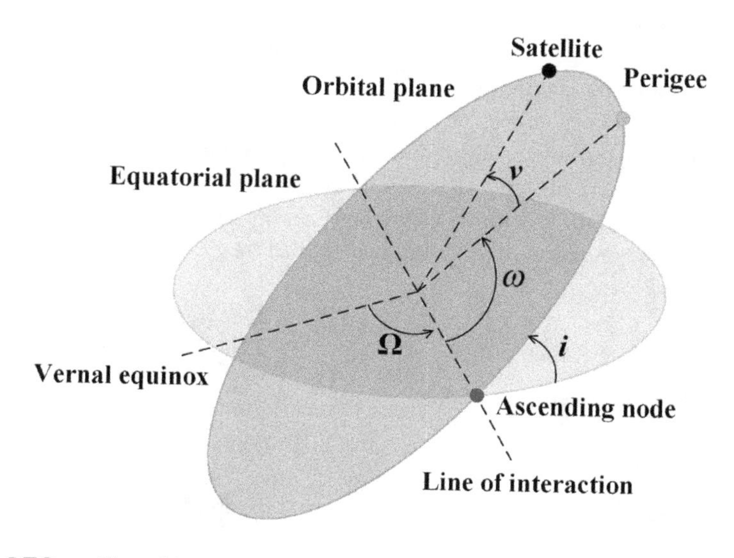

Figure 9.2 LEO satellite orbit

The position of each satellite can be expressed by a function of these angles. According to the results in [250], the Cartesian coordinate of each satellite can be given by

$$\begin{pmatrix} x^s \\ y^s \\ z^s \end{pmatrix} = (h + R_e) \begin{pmatrix} \cos(\omega+\nu)\cos\Omega - \sin(\omega+\nu)\cos i \sin\Omega \\ \cos(\omega+\nu)\sin\Omega + \sin(\omega+\nu)\cos i \cos\Omega \\ \sin(\omega+\nu)\sin i \end{pmatrix}, \tag{9.1}$$

where R_e is the radius of the Earth.

Orbital Period

For convenience, we divide the timeline into multiple time slots with length Δ_T, during which the position of a satellite is regarded to be unchanged. According to the results in [250], the orbital period (i.e., the time during which the mean anomaly changes by 2π) is $\frac{2\pi(h+R_e)^{\frac{3}{2}}}{\sqrt{GM_e}}$, where G is the gravitational constant and M_e is the value of the Earth's mass. Therefore, the number of time slots required for each revolution can be given by

$$T = \frac{2\pi(h + R_e)^{\frac{3}{2}}}{\Delta_T \sqrt{GM_e}}. \tag{9.2}$$

Because T indicates how many time slots are required for each revolution, T is an integer.[3]

Note that all the satellites will be in the same positions after T time slots. We only need to consider the time window containing T time slots, denoted by $\mathscr{T} = \{0, \ldots, T-1\}$. In time slot t, we can derive the position of satellite m in orbit n from (9.1) by setting $\Omega = \Omega_n$, $i = i_n$, and $\omega = \omega_n^m + 2\pi t/T, t \in \mathscr{T}$.

Satellite Coverage

The coverage area served by a satellite is dependent on the line-of-sight (LoS) propagation and the minimum elevation angle $\theta_{\min}$ at the UT.

As shown in Fig. 9.3, the satellite is operating on the orbit with altitude h. The coverage of the satellite decreases with the minimum elevation angle. Assuming that the surface of the Earth is an ideal sphere, the included angle φ from the point with the minimum elevation to the projection point of the satellite (i.e., the angular radius of the coverage circle) can be calculated by

$$\varphi = \arccos\left(\frac{R_e}{R_e + h}\cos\theta_{\min}\right) - \theta_{\min}, \tag{9.3}$$

and the coverage area S is

$$S = 2\pi R_e^2 (1 - \cos\varphi). \tag{9.4}$$

Therefore, a higher altitude can provide larger coverage. Besides, a larger value of $\theta_{\min}$ implies the decrease in the coverage.

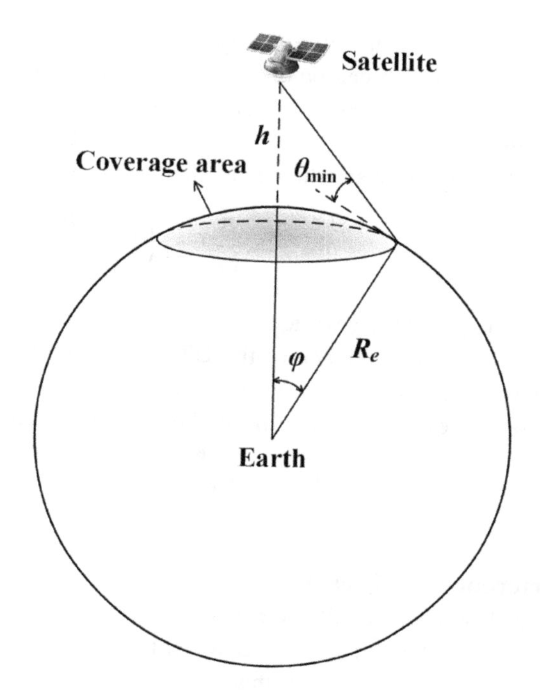

Figure 9.3 Satellite coverage area

9.1.2 LEO-Based Backhaul Capacity Analysis

Communication Model for LEO-Based Backhaul

In the ultra-dense LEO-based terrestrial-satellite network, each terrestrial user can access the UT with a large backhaul capacity supported by the satellite transmission over the Ka-band. Each UT uploads the users' data to the LEO satellite, and then the satellite forwards it to a terrestrial base station equipped with a communication interface over the Ka-band (connected to the core network). The backhaul capacity of each UT is related to the transmission model of the LEO satellite over the Ka-band.

In the following, we give the backhaul capacity over time slot t. For notation brevity, we omit index t. We define P as the transmit power for a UT associating with one LEO satellite, and $g_{m,q}$ as the channel gains of the UT q and satellite m link with only the large-scale fading taken into consideration.[4] Therefore, the channel gain can be expressed as

$$g_{m,q} = G d_{m,q}^{-\alpha}, \tag{9.5}$$

where $d_{m,q}$ is the distance between UT q and satellite m, α is the pathloss exponent, and G is the constant power gains factor introduced by amplifier and antenna. According to the results in [253], the distance $d_{m,q}$ can be given by

$$d_{m,q} = -R_e \sin\theta_{m,q} + \sqrt{(R_e \sin\theta_{m,q})^2 + h^2 + 2hR_e}, \tag{9.6}$$

where $\theta_{m,q}$ is the elevation angle of UT q with respect to satellite m.

Without loss of generality, we assume that all satellites share the same frequency resource pool, which can be divided into J orthogonal channels. Each satellite allocates orthogonal frequency resources to UTs within its coverage for backhaul. We denote the number of UTs in the coverage of satellite m as x_m, and the data rate for the link from UT q to satellite m can then be given by

$$R_{m,q} = \frac{B}{x_m} \log_2 \left(1 + \frac{P g_{m,q}}{\sigma^2 + I_{m,q}}\right), \tag{9.7}$$

where σ^2 is the additive white Gaussian noise (AWGN) variance at each LEO satellite and $I_{m,q}$ is the interference of the UT q and satellite m link. We define the set of satellites that can provide service to UT q as $\mathscr{S}_q = \{m | \theta_m \geq \theta_{\min}\}$, where θ_m is the elevation angle of satellite m. The total backhaul capacity of UT q can then be given by

$$C_q = \sum_{m \in \mathscr{S}_q} R_{m,q}. \tag{9.8}$$

Interference Analysis

Without loss of generality, we assume that the UTs are distributed according to a homogeneous Poisson point process (HPPP) Φ with density λ UTs/km^2. Because the LEO satellites allocate subchannels to UTs randomly, the possibility that other UTs occupy the same subchannel of UT q is $\frac{1}{J}$. Therefore, the interference of the UT q and satellite m link can be given by

$$I_{m,q} = P \sum_{q' \in \Phi'} g_{m,q'} = PG \sum_{q' \in \Phi'} d_{m,q'}^{-\alpha}, \tag{9.9}$$

where Φ' is a thinning HPPP of Φ with density $\frac{\lambda}{J}$.

We denote d_m as the maximum communication distance of satellite m. According to (9.6) and the definition of $\mathscr{S}_q$, d_m can be given by

$$d_m = -R_e \sin\theta_{\min} + \sqrt{(R_e \sin\theta_{\min})^2 + h^2 + 2hR_e}. \tag{9.10}$$

Based on this, we can have the following proposition.

PROPOSITION 9.1 *The average interference of the UT q and satellite m link can be given by*

$$\mathbb{E}(I_{m,q}) = \frac{2\pi R_e \lambda PK}{J(\alpha-2)(R_e+h)} \left[d_m^{2-\alpha} - (2R_e h + h^2)^{\frac{2-\alpha}{2}}\right]. \tag{9.11}$$

Proof Without loss of generality, we assume that UT q is located in the north pole. According to Campell's theorem (Chapter 4 in [254]), the interference of the UT q and satellite m link is $\mathbb{E}(I_{m,q}) = \lambda' PK \int_{q' \in D_{m,q}} d_{m,q'}^{-\alpha} dq'$, where $D_{m,q} = \{q' | d_m \leq d_{m,q'} \leq \sqrt{h^2 + 2R_e h}\}$ is the region with latitude between $(\arcsin \frac{R_e}{R_e+h})$ and $(\frac{\pi}{2} - \varphi)$. We then use the spherical segments between latitude lines as the partition of the integral region, and thus, $\mathbb{E}(I_{m,q})$ can be obtained by

$$\begin{aligned}\mathbb{E}(I_{m,q}) &= \lambda' PK \int_{\arcsin\frac{R_e}{R_e+h}}^{\frac{\pi}{2}-\varphi} 2\pi R_e^2 \cos\beta \cdot \left[R_e^2 + (R_e+h)^2 - 2R_e(R_e+h)\sin\beta\right]^{-\frac{\alpha}{2}} d\beta \\ &\overset{(a)}{=} \lambda' PK \int_{\frac{R_e}{R_e+h}}^{\cos\varphi} 2\pi R_e^2 \cdot \left[R_e^2 + (R_e+h)^2 - 2R_e(R_e+h)u\right]^{-\frac{\alpha}{2}} du \\ &\overset{(b)}{=} \frac{\lambda' PK}{2R_e(R_e+h)} \int_{d_m^2}^{h^2+2R_eh} 2\pi R_e^2 v^{-\frac{\alpha}{2}} dv \\ &= \frac{2\pi R_e \lambda PK}{J(\alpha-2)(R_e+h)} \left[d_m^{2-\alpha} - (2R_eh+h^2)^{\frac{2-\alpha}{2}}\right], \end{aligned} \tag{9.12}$$

where in (a) and (b) the substitutions $u = \sin\beta$ and $v = R_e^2 + (R_e+h)^2 - 2R_e(R_e+h)u$ are applied respectively.

□

Average Data Rate and Backhaul Capacity

According to the HPPP property the number of UTs in the coverage of a satellite is a random variable following the Poisson distribution. The probability of existing k UTs in the coverage can then be expressed as [255]

$$f(x_m = k) = (\lambda S)^k \frac{e^{-\lambda S}}{k!}, \tag{9.13}$$

where S is given in (9.4). Therefore, based on the Poisson probability, the average data rate for the UT q and satellite m link can be given by

$$\mathbb{E}(R_{m,q}) = \sum_{k=1}^{\infty} (\lambda S)^k \cdot \frac{e^{-\lambda S}}{k!} \cdot \frac{B}{k} \cdot \mathbb{E}\left[\log_2\left(1 + \frac{Pg_{m,q}}{\sigma^2 + \mathbb{E}(I_{m,q})}\right)\right]. \tag{9.14}$$

LEMMA 9.1 *The expectation of* $\log_2(1 + \frac{Pg_{m,q}}{\sigma^2+\mathbb{E}(I_{m,q})})$ *can be given by*

$$\begin{aligned}\Upsilon &= \mathbb{E}\left[\log_2\left(1 + \frac{Pg_{m,q}}{\sigma^2 + \mathbb{E}(I_{m,q})}\right)\right] \\ &= \frac{1}{2\ln 2 \cdot R_e(R_e+h)(1-\cos\varphi)} \cdot \left\{ d_m^2 \left[-\frac{\alpha}{2}\, {}_2F_1\left(1, -\frac{2}{\alpha}; 1-\frac{2}{\alpha}; -\frac{PGd_m^{-\alpha}}{\sigma^2 + \mathbb{E}(I_{m,q})}\right) \right.\right. \\ &\quad \left. + \ln\left(1 + \frac{PGd_m^{-\alpha}}{\sigma^2 + \mathbb{E}(I_{m,q})}\right) + \frac{\alpha}{2}\right] - h^2 \left[-\frac{\alpha}{2}\, {}_2F_1\left(1, -\frac{2}{\alpha}; 1-\frac{2}{\alpha}; -\frac{PGh^{-\alpha}}{\sigma^2 + \mathbb{E}(I_{m,q})}\right) \right. \\ &\quad \left.\left. + \ln\left(1 + \frac{PGh^{-\alpha}}{\sigma^2 + \mathbb{E}(I_{m,q})}\right) + \frac{\alpha}{2}\right]\right\}, \end{aligned} \tag{9.15}$$

where $\mathbb{E}(I_{m,q})$ *can be calculated by* (9.11) *and* ${}_2F_1(\cdot)$ *is the generalized hypergeometric function [256].*

Proof Without loss of generality, we assume that UT q is located in the north pole and the satellites are uniformly distributed on the sphere with radius $R_e + h$. For simplicity, we denote the region where satellites are visible to the UT in one snapshot as S_q. The area of S_q can then be given by $S_{S_q} = 2\pi(R_e+h)^2(1-\cos\varphi)$. Therefore, the expectation of $\log_2(1 + \frac{Pg_{m,q}}{\sigma^2+\mathbb{E}(I_{m,q})})$ can be given by

$$\Upsilon = \mathbb{E}\left[\log_2\left(1+\frac{Pg_{m,q}}{\sigma^2+\mathbb{E}(I_{m,q})}\right)\right] = \frac{1}{S_{S_q}\ln 2}\int_{m\in S_q}\ln\left(1+\frac{Pg_{m,q}}{\sigma^2+\mathbb{E}(I_{m,q})}\right)dm. \tag{9.16}$$

Similar to the partition of the integral region utilized in the proof of Proposition 9.1, Υ can be calculated by

$$\begin{aligned}\Upsilon = &\frac{1}{S_{S_q}\ln 2}\int_{\frac{\pi}{2}-\varphi}^{\frac{\pi}{2}} 2\pi(R_e+h)^2\cos\psi \ln\left\{1+\frac{PK}{\sigma^2+\mathbb{E}(I_{m,q})}\cdot\right.\\ &\left.\left[R_e^2+(R_e+h)^2-2R_e(R_e+h)\sin\psi\right]^{-\frac{\alpha}{2}}\right\}d\psi \qquad (9.17)\\ \overset{(a)}{=}&\frac{\pi(1+\frac{h}{R_e})}{S_{S_q}\ln 2}\int_{h^2}^{d_m^2}\ln\left[1+\frac{PKv^{-\frac{\alpha}{2}}}{\sigma^2+\mathbb{E}(I_{m,q})}\right]dv,\end{aligned}$$

where in (a) the substitution $v = R_e^2+(R_e+h)^2-2R_e(R_e+h)u$ is applied. By utilizing the integral formula

$$\int \ln\left(1+Av^B\right)dv = v\left[B\ {}_2F_1\left(1,\frac{1}{B};1+\frac{1}{B};-Av^B\right)+\log\left(Av^B+1\right)-B\right], \tag{9.18}$$

and Υ can then be obtained by setting $A = \frac{PK}{\sigma^2+\mathbb{E}(I_{m,q})}$ and $B = -\frac{\alpha}{2}$ in the integral formula. □

The following proposition can be derived from Lemma 9.1.

PROPOSITION 9.2 *The average data rate for the UT q and satellite m link can be given by*

$$\mathbb{E}(R_{m,q}) = \left(Ei(\lambda S)-\ln(\lambda S)-\gamma\right)e^{-\lambda S}B\Upsilon, \tag{9.19}$$

where $Ei(x) = \int_{-\infty}^{x}\frac{e^t}{t}dt$ is the exponential integral function (Chapter 5 in [257]), γ is the Euler constant, and Υ can be obtained by (9.15).

Proof Proposition 9.2 can be derived from Lemma 9.1 by utilizing the Puiseux series of the exponential integral (i.e., $Ei(x) = \gamma+\ln(x)+\sum_{k=1}^{\infty}\frac{x^k}{k\cdot k!}$), where γ is the Euler constant.

By substituting the expectation of $\log_2(1+\frac{Pg_{m,q}}{\sigma^2+\mathbb{E}(I_{m,q})})$ given in Lemma 9.1 into (9.14) (i.e., $\mathbb{E}(R_{m,q}) = \sum_{k=1}^{\infty}(\lambda S)^k\cdot\frac{e^{-\lambda S}}{k!}\cdot\frac{B}{k}\cdot\mathbb{E}[\log_2(1+\frac{Pg_{m,q}}{\sigma^2+\mathbb{E}(I_{m,q})})]$), we have $\mathbb{E}(R_{m,q}) = \sum_{k=1}^{\infty}(\lambda S)^k\cdot\frac{e^{-\lambda S}}{k!}\cdot\frac{B}{k}\cdot\Upsilon \overset{(a)}{=} \left(Ei(\lambda S)-\ln(\lambda S)-\gamma\right)e^{-\lambda S}B\Upsilon$, where in (a) the Puiseux series of the exponential integral is applied. □

We assume that the average number of satellites in $\mathscr{S}_q$ is k, and the average total backhaul capacity of each UT can be given by

$$\mathbb{E}(\tilde{C}_q) = \sum_{m\in\mathscr{S}_q}\mathbb{E}(R_{m,q}) = k\left(Ei(\lambda S)-\ln(\lambda S)-\gamma\right)e^{-\lambda S}B\Upsilon. \tag{9.20}$$

Remark 9.1 The average total backhaul capacity of each UT decreases as the density of UTs grows.

Proof On the one hand, because the average interference of the UT q and satellite m link given in (9.11) increases with the density of UTs λ, $\Upsilon = \mathbb{E}[\log_2(1+\frac{Pg_{m,q}}{\sigma^2+\mathbb{E}(I_{m,q})})]$ decreases as λ grows.

On the other hand, we have

$$\frac{d\left[\left(Ei(\lambda S)-\ln(\lambda S)-\gamma\right)e^{-\lambda S}\right]}{d(\lambda S)} = e^{-\lambda S}\left[1-\sum_{k=1}^{+\infty}\frac{(\lambda S)^k}{k!k(k+1)}\right]. \tag{9.21}$$

Note that when $\lambda S \geq 2$ (i.e., there are no less than two UTs within each LEO satellite's coverage, which is accordant with practical circumstances), we have

$$1-\sum_{k=1}^{+\infty}\frac{(\lambda S)^k}{k!k(k+1)} < 1-\frac{\lambda S}{2} \leq 0. \tag{9.22}$$

Therefore, $\left(Ei(\lambda S)-\ln(\lambda S)-\gamma\right)e^{-\lambda S}$ also decreases as λ grows, and the average total backhaul capacity of each UT given in (9.20) decreases as the density of UTs grows.

□

9.1.3 Ultra-dense LEO Satellite Constellation Design

In this section, we first formulate the global $k_{\min}$-coverage problem of LEO satellite constellation deployment to satisfy the backhaul requirement of Poisson-distributed UTs. We then propose three LEO satellite constellation design criteria, based on which we give a 3D constellation optimization algorithm to minimize the number of LEO satellites while guaranteeing seamless global coverage and satisfying the backhaul requirement of each UT.

Global $k_{\min}$-Coverage Problem Formulation

We assume that the backhaul requirement for a UT is C_{th}. Therefore, the expectation of the backhaul capacity for a UT should satisfy the backhaul requirement for any time slot $t \in \mathcal{T}$(i.e., $\mathbb{E}(\tilde{C}_q^t) \geq C_{th}, \forall t \in \mathcal{T}, \forall q$). According to (9.20), the minimum number of satellites providing service to each UT can be given by

$$k_{\min} = \frac{C_{th}}{\left(Ei(\lambda S)-\ln(\lambda S)-\gamma\right)e^{-\lambda S}B\Upsilon}. \tag{9.23}$$

Note that each satellite will associate with all the UTs in its coverage area, the first criterion for LEO satellite constellation design can then be given as follows:

For criterion 1 (backhaul requirement), each UT should be covered by at least $k_{\min}$ satellites in any time slot $t \in \mathcal{T}$.

Moreover, because the region on the orbit sphere where satellites are visible to the UT in one snapshot is $2\pi(R_e+h)^2(1-\cos\varphi)$, the minimum density of LEO satellites can be given by

$$\rho = \frac{k_{\min}}{2\pi(R_e+h)^2(1-\cos\varphi)} = \frac{C_{th}}{2\pi(R_e+h)^2(1-\cos\varphi)\left[Ei(\lambda S)-\ln(\lambda S)-\gamma\right]e^{-\lambda S}B\Upsilon}, \tag{9.24}$$

and the minimum number of LEO satellites guaranteeing seamless global coverage while satisfying the backhaul requirement of each UT can then be given by

$$\left(\sum_{n=1}^{N} M_n\right)_{\min} = \rho \cdot 4\pi (R_e + h)^2 = \frac{2C_{th}}{(1-\cos\varphi)\left[Ei(\lambda S) - \ln(\lambda S) - \gamma\right] e^{-\lambda S} B\Upsilon}. \tag{9.25}$$

DEFINITION 9.1 *A point q in a region A is said to be k-covered if it is covered by at least k satellites, and k is said to be the coverage degree of point q. A region A is said to be k-covered if every point $q \in A$ is k-covered.*

Because the UTs are distributed according to a HPPP, their locations are random points. We aim to minimize the number of LEO satellites under the constraint that the LEO satellite constellation deployment must guarantee that the Earth is $k_{\min}$ covered in any time slot $t \in \mathcal{T}$. Such a *global $k_{\min}$-coverage problem* can be formulated as

$$\begin{aligned} &\min \sum_{n \in \mathcal{N}} M_n \\ &s.t. \quad \mathcal{S}_q \geq k_{\min}, \quad \forall t \in \mathcal{T}, \forall q. \end{aligned} \tag{9.26}$$

Note that the problem of selecting a minimum subset of $\mathcal{M}_{\min} \subset \mathcal{M}$ of LEO satellites to achieve global $k_{\min}$ coverage is NP-hard. To solve this global $k_{\min}$-coverage problem, we first give an initial constellation deployment that can guarantee global $k_{\min}$-coverage. Following that, we design a 3D constellation optimization algorithm to find extra satellites that can be removed from the initial constellation deployment without degrading the global $k_{\min}$ coverage.

Initial LEO Satellite Constellation Deployment

We adopt the typical LEO polar-orbit constellation for the initial constellation deployment[5] [237]. We first consider the global one-coverage with polar orbits, which has been investigated in the previous work. Specifically, as shown in Fig. 9.4a, each orbit

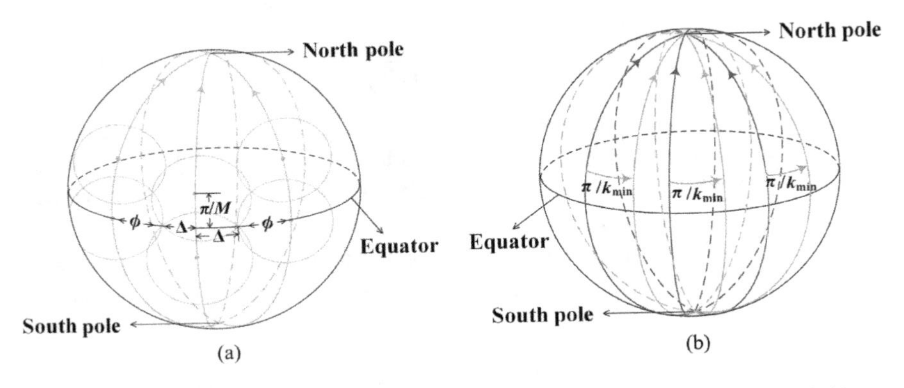

Figure 9.4 Initial LEO satellite constellation deployment: (a) global one-coverage and (b) global $k_{\min}$ coverage

consists of the same number of satellites (i.e., $M_1 = M_2 = \cdots = M_n = M$). Satellites in adjacent orbits move in the same direction with a phase difference of π/M. The number of orbital planes N_0 and the angular radius of the coverage circle φ satisfies

$$(N_0 - 1)\varphi + (N_0 + 1)\Delta = \pi, \tag{9.27}$$

where $\Delta = \cos^{-1}[\cos\varphi/\cos(\pi/M)]$, and φ can be given by (9.3). Therefore, the N orbital planes are separated in angle by ϕ, where $\phi = \varphi + \Delta$, and the relation between the number of orbital planes N, the number of satellites in each orbit M, the altitude of each satellite h, and the minimum elevation angle $\theta_{\min}$ at the UT satisfy the following equation:

$$\begin{aligned}(N_0+1)\cos^{-1}&\left\{\frac{\cos\left[\cos^{-1}\left(\frac{R_e\cos\theta_{\min}}{R_e+h}\right)-\theta_{\min}\right]}{\cos(\frac{\pi}{M})}\right\}\\ &+(N_0-1)\left[\cos^{-1}\left(\frac{R_e\cos\theta_{\min}}{R_e+h}\right)-\theta_{\min}\right]=\pi.\end{aligned} \tag{9.28}$$

Note that the satellites in adjacent orbits moving in the same direction can cover an area with a longitude span of $(\varphi + \Delta)$, whereas the satellites in adjacent orbits moving in the opposite direction can cover an area with a longitude span of 2Δ [237]. Therefore, the global one-coverage can be guaranteed by the polar-orbit LEO satellite constellation satisfying (9.27). After determining the global one-coverage polar-orbit LEO satellite constellation deployment, we can copy the polar-orbit LEO satellite constellation illustrated earlier $k_{\min} - 1$ times on the same location and then rotate the ith $(1 \le i \le k_{\min} - 1)$ duplicate in the north-south direction by an angle $i\pi/k_{\min}$. Therefore, as shown in Fig. 9.4b, we obtain the initial polar-orbit LEO satellite constellation consisting of $N = N_0 k_{\min}$ polar orbits and M satellites in each orbit, based on which the global $k_{\min}$ coverage can be ensured. The upper bound of the minimum number of LEO satellites guaranteeing global $k_{\min}$ coverage can then be given by $(\sum_{n=1}^{N} M_n)_{\max} = M N_0 k_{\min}$.

Three-Dimensional Constellation Optimization Algorithm

Eligibility of a Satellite

Note that for a given initial LEO satellite constellation, as mentioned earlier, there are several unnecessary LEO satellites resulting in redundant coverage. In other words, the backhaul requirement of each UT can also be satisfied even if these LEO satellites are removed from the initial constellation. Therefore, we first determine the eligibility of each satellite in the initial constellation. For convenience, we define the following notations.

DEFINITION 9.2 *Satellite m is* ineligible *and can be removed from the initial constellation if every location within satellite m's coverage is already $k_{\min}$ covered by other satellites in its neighborhood. Otherwise, satellite m is* eligible *and can be selected into the minimum subset of LEO satellites.*

DEFINITION 9.3 *A point p on Earth is called an* intersection point *between two satellites if p is an intersection point of the coverage area's boundaries of two satellites.*

Specifically, the coordinate of the intersection point between two satellites located at (x_1, y_1, z_1) and (x_2, y_2, z_2) can be obtained by solving the following equations with respect to (x, y, z):

$$\begin{cases} xx_1 + yy_1 + zz_1 = R_e(R_e + h)\cos\varphi, \\ xx_2 + yy_2 + zz_2 = R_e(R_e + h)\cos\varphi, \\ x^2 + y^2 + z^2 \quad = R_e^2. \end{cases} \tag{9.29}$$

Without loss of generality, we assume that $x_1 y_2 \neq x_2 y_1$. The closed-form solution of the coordinate of the intersection point can be given by

$$\begin{pmatrix} x \\ y \\ z \end{pmatrix} = \begin{pmatrix} \frac{(y_2 - y_1)R_e(R_e+h)\cos\varphi}{x_1 y_2 - x_2 y_1} \\ \frac{(x_1 - x_2)R_e(R_e+h)\cos\varphi}{x_1 y_2 - x_2 y_1} \\ 0 \end{pmatrix} + \kappa \begin{pmatrix} y_1 z_2 - y_2 z_1 \\ z_1 x_2 - z_2 x_1 \\ x_1 y_2 - x_2 y_1 \end{pmatrix}, \tag{9.30}$$

where κ can be obtained by solving the following quadratic equation:

$$\begin{aligned} &\kappa^2 \left[(y_1 z_2 - y_2 z_1)^2 + (z_1 x_2 - z_2 x_1)^2 + (x_1 y_2 - x_2 y_1)^2 \right] \\ &+ \kappa \frac{2R_e(R_e + h)\cos\varphi}{x_1 y_2 - x_2 y_1} [(y_2 - y_1)(y_1 z_2 - y_2 z_1) \\ &+ (x_1 - x_2)(z_1 x_2 - z_2 x_1)] + \left[\frac{R_e(R_e + h)\cos\varphi}{x_1 y_2 - x_2 y_1} \right]^2 \\ &\cdot \left[(y_2 - y_1)^2 + (x_1 - x_2)^2 \right] - R_e^2 = 0. \end{aligned} \tag{9.31}$$

Before presenting the 3D constellation optimization algorithm, we introduce the following proposition, which states a sufficient condition for global $k_{\min}$ coverage.

PROPOSITION 9.3 *The Earth's surface is k-covered by a set of LEO satellites if (a) there exist intersection points between two satellites and (b) all intersection points are at least k-covered [259].*

Proof We prove by contradiction. As shown in Fig. 9.5, assume that point q has the smallest coverage degree $k_q < k$ on the Earth's surface, and all intersection points between two satellites on the Earth's surface are at least k-covered. The set of LEO satellites' coverage area can partition the Earth's surface into a collection of coverage patches, which are bounded by arcs of the coverage area. All points in each coverage patch have the same coverage degree. We assume that point q is located in coverage patch S.

We first prove by contradiction that the interior arc of any coverage area cannot serve as the boundary of S. We assume that there exists an interior arc serving as the boundary of S. Note that crossing this arc can reach an area that has a smaller coverage degree than point q, which contradicts the assumption that point q has the smallest coverage degree on the Earth's surface. Therefore, the boundary of S consists of exterior arcs of the set of LEO satellites' coverage area as illustrated in Fig. 9.. Moreover, because the arcs of each LEO satellite's coverage area are outside the satellite's coverage range, the entire boundary of S, including the intersection poin

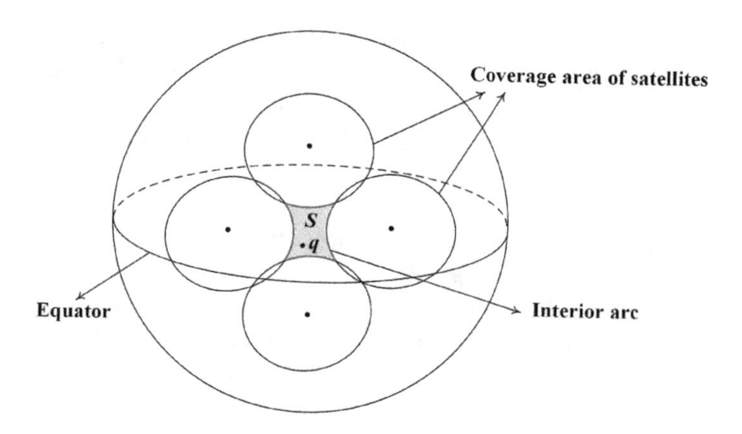

Figure 9.5 Coverage patch bounded by exterior arcs of satellites' coverage area

of between two satellites defining the boundary, has the same coverage degree as point q. This contradicts the assumption that q has a smaller coverage degree than k and all intersection points are at least k-covered. □

According to Proposition 9.3, the $k_{\min}$-*covered eligibility rule* determining whether a satellite in the initial polar-orbit LEO satellite constellation should be removed can be obtained. Specifically, a satellite is ineligible and can be removed from the initial constellation if all the intersection points inside its coverage area are at least $k_{\min}$-covered. Otherwise, the satellite cannot be removed. Therefore, we propose the second criterion for LEO satellite constellation design to determine whether a satellite should be removed from the initial constellation.

Criterion 2 involves the satellite's $k_{\min}$-covered eligibility: For each LEO satellite in the initial polar-orbit LEO satellite constellation, whether it should be removed from the initial constellation is determined by the coverage-based eligibility rule.

Regarding execution of Criterion 2, note that, if the eligibility of all satellites is determined based on this $k_{\min}$-covered eligibility rule in the first time slot, the number of ineligible satellites flying over the high latitude area will be larger than that of the low latitude area. When satellites in the high latitude area fly over the low latitude area in later time slots, it is likely that global $k_{\min}$ coverage cannot be achieved. Therefore, we consider reducing the number of ineligible satellites flying over the high latitude area to guarantee the global $k_{\min}$ coverage in any time slot $t \in \mathcal{T}$. Specifically, we define a larger required coverage degree over high latitude area $k_{\max}$ ($k_{\max} \geq k_{\min}$). The eligibility of satellites flying over the high latitude area is determined based on the $k_{\max}$-covered eligibility rule.

Because the number of eligible satellites that cannot be removed from the initial constellation increases with $k_{\max}$, the coverage ratio averaged over all time slots of the optimized LEO satellite constellation will increase correspondingly. Therefore, the value of $k_{\max}$ can be determined by the coverage ratio requirement, and the third criterion for LEO satellite constellation design can then be given as follows.

Criterion 3 involves latitude-based k-covered eligibility selection: The eligibility of satellites flying over the high latitude area is determined by the k_{max}-covered eligibility rule ($k_{max} \geq k_{min}$), and the eligibility of satellites flying over the low latitude area is determined by the k_{min}-covered eligibility rule. The value of k_{max} is determined by the coverage ratio requirement.

Three-Dimensional Constellation Optimization Algorithm Description

According to the three criteria (i.e., backhaul requirement, satellite's k_{min}-covered eligibility, and latitude-based k-covered eligibility selection) for LEO satellite constellation design, the 3D constellation optimization algorithm is summarized in Algorithm 9.1 where the bisection method [260] is adopted to find the proper value of k_{max}. In the initialization step (line 1), the LEO satellites are deployed according to the global k-coverage LEO polar-orbit constellation shown in Section 9.1.3. The search interval of k_{max} for the bisection method is set as $[k_{min}, k_{max}]$. The following LEO satellite constellation optimization process (lines 2–23) consists of multiple iterations. In each iteration, the selection of a minimum subset of LEO satellites following the coverage-based eligibility rule (lines 4–15) is performed based on which the proper value of k_{max} is selected using the bisection method (lines 16–22). The iterations will not stop until the search interval width of the bisection method is reduced to zero (i.e., the average coverage ratio of the optimized constellation η meets the coverage ratio requirement η_0). Therefore, for any given coverage ratio requirement, the proposed 3D constellation optimization algorithm can give a corresponding optimized LEO satellite constellation by choosing a proper value of k_{max}.[6]

Although the eligibility of all satellites is determined in the first time slot, the seamless global k_{min} coverage can be guaranteed, which will be verified in Section 9.1.4. Moreover, because the initial polar-orbit constellation is deterministic, the Cartesian coordinate of each satellite in the optimized LEO satellite constellation can be given explicitly after the 3D constellation optimization algorithm is performed.

Complexity Analysis

On the convergence of Algorithm 9.1, we observe the following facts. The complexity of Algorithm 9.1 is determined by the calculation of each intersection point's coverage degrees in line 5. We assume that the coverage area of each satellite is intersected with at most L coverage areas of other satellites. For each satellite, the complexity of finding all intersection points inside its coverage area is $\mathcal{O}(NML^2)$, where N is the number of orbits and M is the number of satellites operated in each orbit in the initial polar-orbit constellation. The calculation of each intersection point's coverage degree has a complexity of $\mathcal{O}(L)$. Therefore, the complexity of line 5 is $\mathcal{O}(NML^3)$, and thus, the whole algorithm has a complexity of $\mathcal{O}(N^2M^2L^3)$.

9.1.4 Performance Evaluation

In this section, we present the simulation results of the LEO satellite constellation with a minimum number of satellites based on the proposed 3D constellation optimization

Algorithm 9.1 3D Constellation Optimization

Input: Coverage degree for low latitude area $k_{\min}$, coverage degree for high latitude area $k_{\max}$, coverage ratio requirement η_0.

1 **Initialization**: Global k-coverage LEO polar-orbit constellation deployment shown in Section 9.1.3, $k_{\max}$'s lower bound $k_m = k_{\min}$, $k_{\max}$'s upper bound $k_M = k_{\max}$;
2 **while** $k_M > k_m$ **do**
3 Set $k_{\max} = [\frac{k_m+k_M-1}{2}] + 1$;
4 **for** *Satellite m in* $\mathcal{M}$ **do**
5 Find all intersection points inside the coverage area of m and calculate their coverage degrees;
6 **if** *m flies over high latitude area* **then**
7 Determine the eligibility of m based on the $k_{\max}$-covered eligibility rule;
8 **end**
9 **else if** *m flies over low latitude area* **then**
10 Determine the eligibility of m based on the $k_{\min}$-covered eligibility rule;
11 **end**
12 **if** *m is ineligible* **then**
13 Remove m from the satellite set $\mathcal{M}$;
14 **end**
15 **end**
16 Calculate the average coverage ratio η;
17 **if** $\eta < \eta_0$ *or* $k_{\max} = k_M$ **then**
18 Set $k_m = k_{\max}$;
19 **end**
20 **else**
21 Set $k_M = k_{\max}$;
22 **end**
23 **end**

Output: The minimum subset $\mathcal{M}_{\min} \subset \mathcal{M}$ of LEO satellites guaranteeing the given coverage ratio requirement.

algorithm. Simulation parameters are set up based on the 3GPP specifications [204] as given in Table 9.1.

9.1.5 Average Total Backhaul Capacity of Each UT

Figure 9.6 illustrates the average total backhaul capacity of each UT versus the density of UTs λ. It shows that the average total backhaul capacity of each UT decreases as λ grows, which verifies the statement in Remark 9.1. The main reason is that the

Table 9.1 Simulation parameters

Parameters	Values
Pathloss exponent α	2
Transmit power of each UT P (W)	2
Constant power gains factor G (dBi)	43.3
Minimum elevation angle $\theta_{\min}$ at the UT	$10^\circ - 18^\circ$
Number of subchannels J	1,000
Density of UTs λ (km^{-2})	4×10^{-6}
Backhaul requirement for each UT C_{th} (Mbps)	60–160
Altitude of LEO satellites (km)	900
Bandwidth for Ka-band communications B (MHz)	800
Noise density for Ka-band communications σ^2 (dBm/Hz)	−203

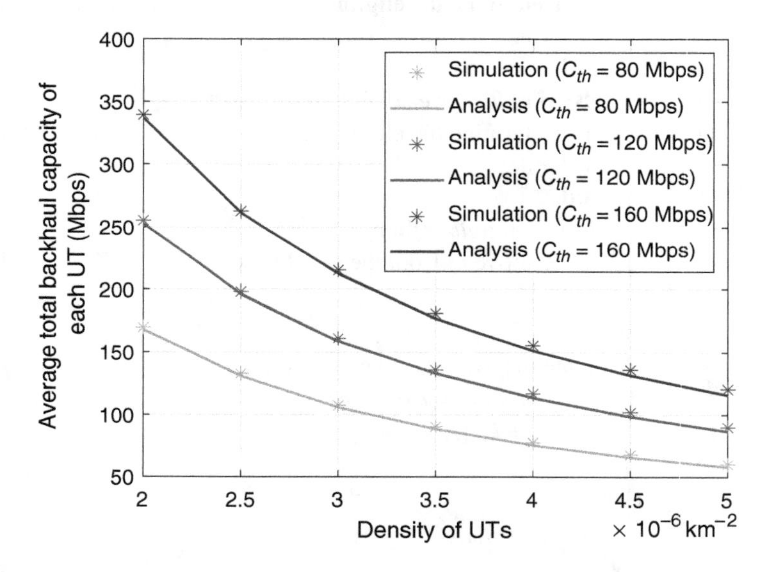

Figure 9.6 Average total backhaul capacity of each UT versus the density of UTs

available bandwidth for each UT decreases, and more UTs result in more interference links. We also observe that the average total backhaul capacity of each UT increases with the number of satellites providing data services. It can also be seen that our theoretical result in (9.20) perfectly matches the simulation results.

Number of Deployed LEO Satellites

Note that there are various factors influencing the number of deployed (i.e., eligible) LEO satellites such as the backhaul requirement of each UT, the average coverage ratio requirement, the density of UTs, the initial constellation deployment, the altitude of satellites, and the minimum elevation angle at the UT. We first evaluate the influence of the required backhaul capacity of each UT C_{th} and the coverage ratio averaged

Table 9.2 Minimum number of LEO satellites guaranteeing coverage ratio over 90% and 95%

Required coverage degree $k_{\min}$		3	4	5	6	7	8
Number of LEO satellites	Coverage ratio over 90%	193	259	311	389	440	503
	Coverage ratio over 95%	202	270	328	408	468	528

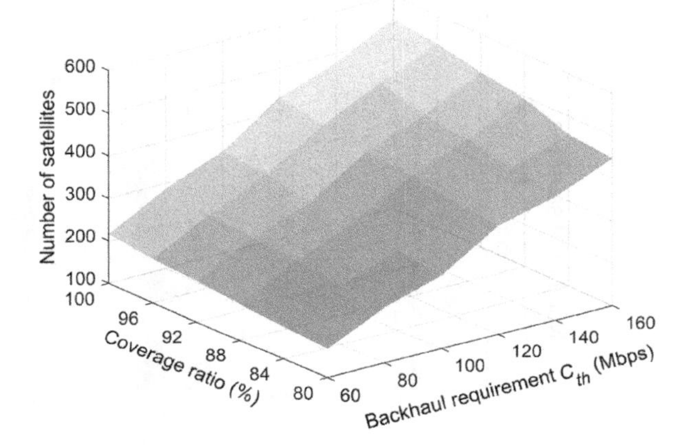

Figure 9.7 Number of deployed LEO satellite versus coverage degree

over all time slots and then investigate how the rest factors influence the number of deployed LEO satellites.

Figure 9.7 shows a 3D surface plot of the deployed LEO satellite number versus the required backhaul capacity of each UT C_{th} and the coverage ratio averaged over all time slots. The specific value of the minimum LEO satellite number guaranteeing a coverage ratio over 90% and 95% is also given in Table 9.2. It can be seen that the number of deployed LEO satellites increases with C_{th}. The coverage ratio averaged over all time slots also increases with the number of deployed LEO satellites when $k_{\min}$ is fixed. Moreover, Fig. 9.7 shows that, for any given coverage ratio requirement, the corresponding optimized LEO satellite constellation can be obtained by the proposed 3D constellation optimization algorithm.

Note that there is a close relation between the number of orbits in the initial global one-coverage polar-orbit constellation (i.e., N_0), the number of satellites in each orbit M, the satellites' altitude h, and the minimum elevation angle $\theta_{\min}$ at the UT as given in (9.28). Because the satellites' altitude h and the minimum elevation angle $\theta_{\min}$ at the UT also influences the average total backhaul capacity of each UT, we consider evaluating the influence of (N_0, M) and $(h, \theta_{\min})$ separately. Figure 9.8 illustrates the number of deployed LEO satellites versus N_0 when h and $\theta_{\min}$ are fixed.[7] We observe that the number of deployed LEO satellites first decreases and then increases as N_0 grows. The main reason is that when h and $\theta_{\min}$ are fixed, based on the relation between N_0 and M given in (9.28), the total number of satellites in the initial constellation MN (i.e., $MN_0k_{\min}$) first decreases and then increases as N_0 grows, as

Table 9.3 Relation between N_0 and initial number of LEO satellites MN

Number of orbits for global 1-coverage N_0		5	6	7	8	9
Initial number of LEO satellites MN	$C_{th} = 100$ Mbps	456	392	384	396	440
	$C_{th} = 140$ Mbps	570	490	480	495	550

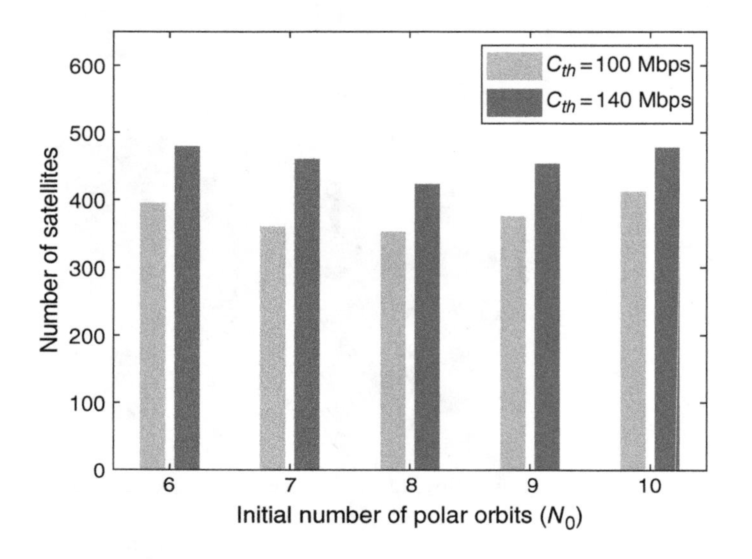

Figure 9.8 Number of deployed LEO satellite versus N_0

shown in Table 9.3. This implies that the number of deployed LEO satellites in the optimized constellation has a close relation to the initial LEO satellite constellation. The initialization of LEO satellite constellation can be optimized to minimize the number of LEO satellites.

Figure 9.9 shows a 3D surface plot of the deployed LEO satellite number versus the altitude of satellites h and the minimum elevation angle $\theta_{\min}$ at the UT when the backhaul requirement for each UT C_{th} is 100 Mbps.[8] It can be seen that the number of deployed LEO satellites first increases and then decreases as h grows when $\theta_{\min}$ is fixed. The main reason is that each satellite can provide services to more UTs as h grows. However, when h continues to grow, severe pathloss results in the sharp decline of the backhaul capacity, and thus, more satellites are needed.

We can also observe from Fig. 9.9 that the number of deployed LEO satellites first increases and then decreases as $\theta_{\min}$ grows when h is fixed. This is because the average data rate of the UT-satellite link increases with $\theta_{\min}$. However, the decline in the number of UTs within each satellite's coverage becomes the main influencing factor when $\theta_{\min}$ continues to grow. This implies there exists an optimal combination of the satellites' altitude and minimum elevation angle at the UT minimizing the number of LEO satellites. Under the parameter settings in the simulation, the optimal combination of h and $\theta_{\min}$ is (800 km, 14°).

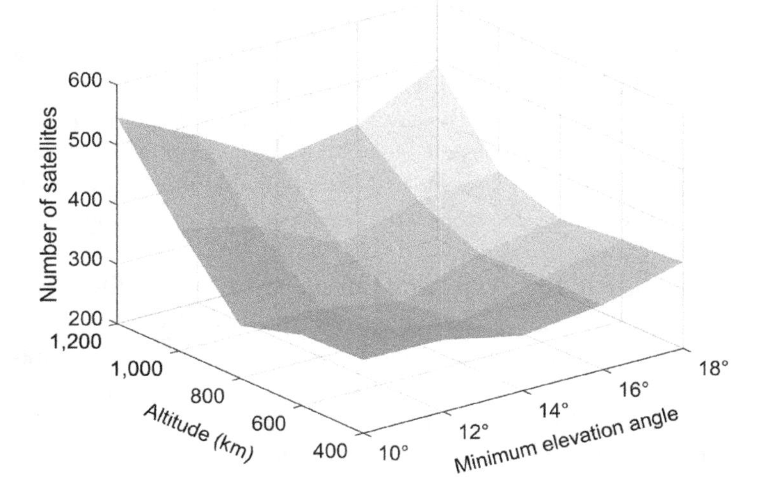

Figure 9.9 Number of deployed LEO satellite versus satellites' altitude and minimum elevation angle

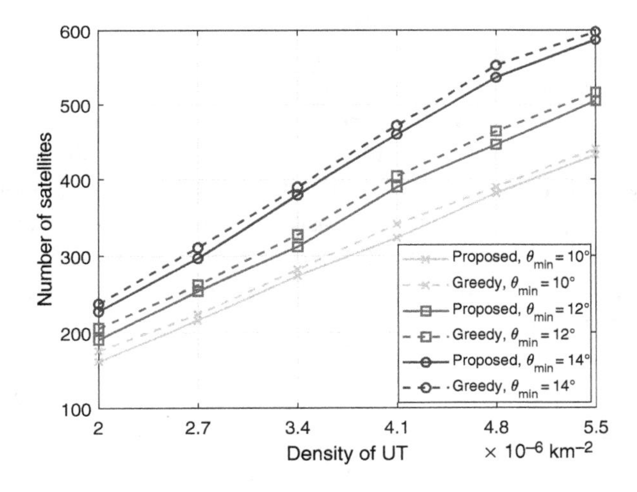

Figure 9.10 Number of deployed LEO satellite versus the density of UTs

Figure 9.10 depicts the number of deployed LEO satellites versus the density of UTs λ when the average data rate of the UT-satellite link and the backhaul requirement of each UT are fixed. We observe that the number of deployed LEO satellites increases with the density of UTs. The main reason is that the total backhaul capacity requirement of all UTs is higher. It can also be seen that the number of deployed LEO satellites increases with the minimum elevation angle $\theta_{\min}$ at the UT. This is because the number of satellites that can provide service to each UT decreases with the minimum elevation angle $\theta_{\min}$.

Another important observation in Fig. 9.10 is that the proposed 3D constellation optimization algorithm outperforms the greedy constellation algorithm, where the satellites in the initial constellation are removed based on the greedy method. Specifically, compared with the LEO satellite constellation optimized by the greedy algorithm, the satellite constellation optimized by the proposed 3D constellation optimization algorithm can meet the backhaul requirement with a smaller number of satellites. This shows the effectiveness of our proposed 3D constellation optimization algorithm.

Stability of the Proposed Optimized LEO Satellite Constellation

Because LEO satellites are under the influence of various perturbative forces such as the gravitational pull from other planets and atmospherical drag, which affect their orbits and cause relative drifts [261], Fig. 9.11 evaluates the stability of the proposed optimized LEO satellite constellation in different coverage ratios. It can be seen that there is only a slight drop in the coverage ratio even as the satellites' maximum radial offset grows to 10 km,[9] indicating the stability of the proposed optimized LEO satellite constellation.

Comparison with Other LEO Satellite Constellations

To further evaluate the performance of the proposed 3D constellation optimization algorithm, we consider comparing the LEO satellite constellation optimized by the algorithm with other representative global LEO satellite constellations.

Figure 9.12 compares the LEO satellite constellation optimized by the proposed algorithm with a modified Telesat constellation.[10] Specifically, the modified Telesat constellation consists of no less than 117 LEO satellites distributed on two sets of orbital planes. The first set consists of 6 polar orbits with at least 12 satellites on each orbital plane. The orbital inclination is 99.5° and the altitude is 1,000 km. The second set consists of no less than 5 inclined orbits with a minimum of 10 satellites on each orbital plane. The orbital inclination is 45° and the altitude is 1,200 km. It can be seen that for a given coverage degree and the total number of deployed LEO satellites, the

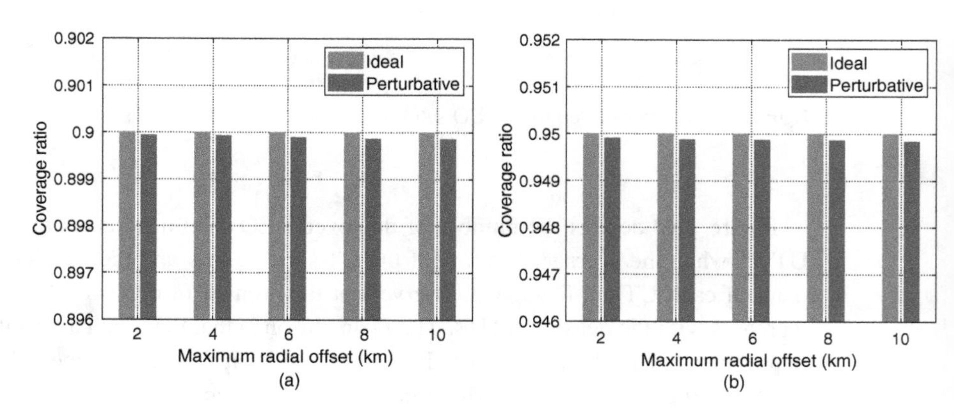

Figure 9.11 Coverage ratio versus the satellites' maximum radial offset: Coverage ratio of (a) 90% and (b) 95%

Table 9.4 Average coverage ratio of the OneWeb constellation

Backhual capacity requirement C_{th} (Mbps)	160	180	200	220	240
Average coverage ratio (%)	100	96	93	90	84

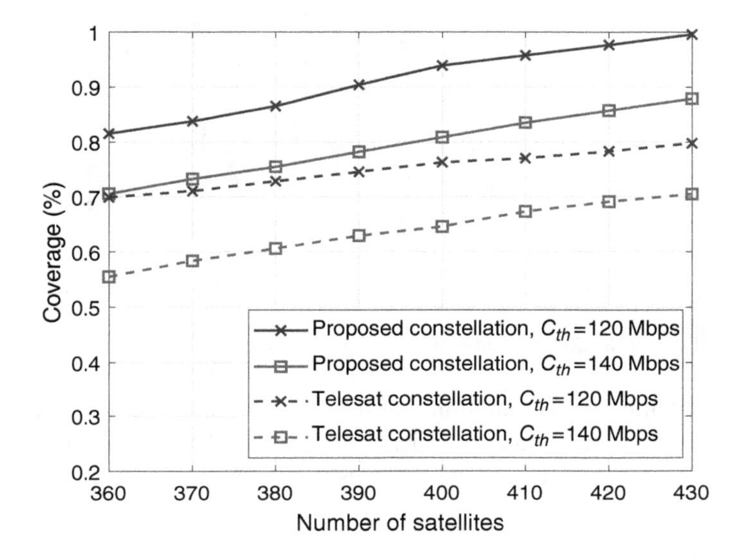

Figure 9.12 Comparison between the proposed constellation and the modified Telesat constellation

average coverage ratio of the proposed optimized LEO satellite constellation is at least 10 percentage points higher than that of the modified Telesat constellation, indicating the effectiveness of the proposed 3D constellation optimization algorithm.

Figure 9.13 compares the LEO satellite constellation optimized by the proposed 3D constellation optimization algorithm with the OneWeb constellation. The OneWeb constellation consists of 720 LEO satellites distributed on 18 near polar orbits. Each near polar orbit consists of 40 uniformly distributed LEO satellites. The orbital inclination is 87° and the altitude is 1,200 km. Because the total number and the distribution of LEO satellites are fixed in the OneWeb constellation, the average coverage ratio of the OneWeb constellation varies with the backhaul capacity requirement of the UT. Specifically, the average coverage ratio corresponding to different backhaul capacity requirements of the UT is given in Table 9.4. To increase the fairness of the comparison, under the same backhaul capacity requirement C_{th}, we set the average coverage ratio of the proposed constellation the same as that of the OneWeb constellation. It can be seen that for a given backhaul capacity requirement and average coverage ratio, the number of deployed LEO satellites in the proposed optimized LEO satellite constellation is smaller than that of the OneWeb constellation, which also verifies the effectiveness of the proposed 3D constellation optimization algorithm. Both Figs. 9.12 and 9.13 imply that a well-designed asymmetrical satellite constellation (i.e., where

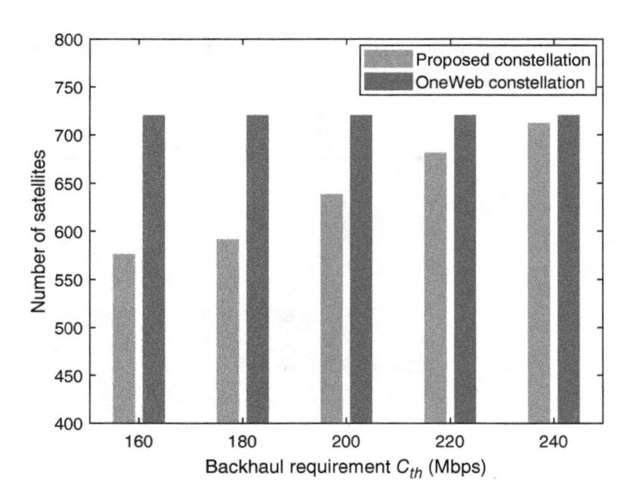

Figure 9.13 Comparison between the proposed constellation and the OneWeb constellation

satellites are not uniformly distributed on orbits) can meet the backhaul requirement with a smaller number of deployed satellites than the traditional symmetrical satellite constellation (i.e., where satellites are uniformly distributed on orbits).

9.1.6 Conclusion

In this section, we considered the ultra-dense LEO satellite constellation deployment to satisfy the backhaul capacity requirement of UTs. We have derived the average total backhaul capacity and the required coverage degree (i.e., $k_{\min}$) of each UT, based on an initial constellation deployment guaranteeing global $k_{\min}$ coverage. To minimize the number of LEO satellites, a 3D constellation optimization algorithm jointly considering the seamless global coverage and the backhaul requirement of UTs was designed based on which the corresponding optimized LEO satellite constellation can be obtained for any given coverage ratio requirement. Specifically, three design criteria about backhaul requirement satisfaction, satellite's eligibility, and latitude-based eligibility selection were applied to optimize the LEO satellite constellation. The Cartesian coordinate of each satellite in the optimized LEO satellite constellation can be given explicitly after the proposed algorithm is performed. Simulation results showed that the coverage ratio of the proposed LEO satellite constellation remains almost unchanged even when the satellites' maximum radial offset is large, thereby verifying the constellation's stability. It has also been proved that the proposed LEO satellite constellation can improve the average coverage ratio of Telesat constellation by at least 10 percentage points.

9.2 Multilayer Satellite Constellation Design

Given the limited orbit resources on a single satellite layer, multilayer satellite constellations such as Starlink and Telesat, are widely applied in practice, and thus, their design needs to be considered. Moreover, the influence of the terrestrial network on constellation design is ignored in current works. In practice, the satellite network is constructed on top of the terrestrial network, and thus, the satellite deployment relies heavily on the distribution and terrestrial transmission capacity of user terminals. Therefore, the satellite and terrestrial networks need to be considered jointly for the satellite constellation design.

To handle these issues, we discuss the megaconstellation design for the integrated satellite-terrestrial network given the terrestrial terminal deployment, aiming to support both global seamless connectivity and high-rate backhaul transmission. Against this background, two questions remain to be answered:

- How do we depict the uplink capacity of the integrated backhaul network, which is influenced by the number of satellites and the terrestrial transmission capacity of UTs?
- How do we design the megaconstellation for multilayer satellites to fulfill the rate-differentiated transmission demands with the minimum number of satellites?

To solve these problems, we propose an uplink capacity theoretical framework where the average uplink capacity is analyzed given the terrestrial distribution density of the UTs. Based on the data rate of satellites analyzed in the theoretical framework, we optimize the megaconstellation of multilayer satellites to minimize the total number of required satellites. Considering the transmission capacity of the terrestrial network, the proposed novel megaconstellation can satisfy the backhaul transmission rate requirements of all UTs even in cases with the maximum amount of interference, providing an upper bound on the satellite number for the actual deployed constellation.

9.2.1 System Model

Network Structure

As illustrated in Fig. 9.14, we consider an integrated satellite-terrestrial network consisting of the UTs as well as multiple layers of LEO satellites at different altitudes. The terrestrial users transmit the data to the UTs over the C-band spectrum. The UTs

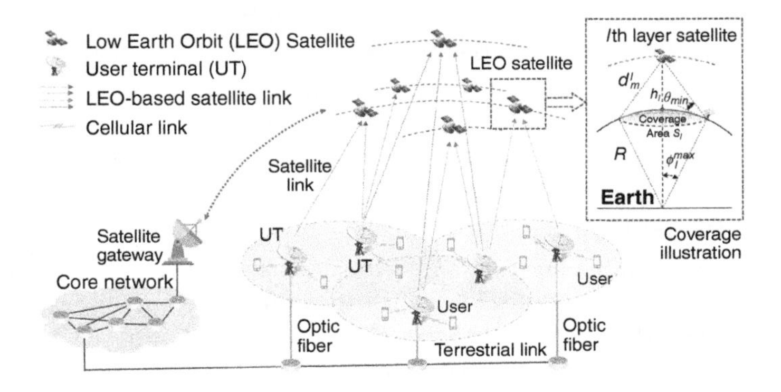

Figure 9.14 Integrated satellite-terrestrial network model

then access the core network via two types of backhaul links: through the *terrestrial link* once the UT is within the range of the optical network and through the *satellite link* via the layer of LEO satellite.[11] Specifically, the satellite links are established over the Ka-band spectrum, and all satellites deliver the data to the core network via the satellite gateways.

Satellite Coverage Model

We discuss the coverage of the satellite on layer l assuming that the Earth is a sphere with radius R. The coverage area of a satellite is determined by its altitude h_l and the elevation angle range of the LoS terrestrial-satellite link. Let $\theta_{\min}$ represent the minimum elevation angle of the UT, so the UT is considered within the coverage of a satellite when its elevation angle $\theta_l \geq \theta_{\min}$. Therefore, the coverage area S_l of a satellite is the area of a spherical crown with the angular radius $\phi_l^{\max} = \arccos(\frac{R}{R+h_l}\cos\theta_{\min}) - \theta_{\min}$, expressed by

$$S_l = 2\pi R^2(1 - \cos\phi_l^{\max}). \tag{9.32}$$

Satellite Transmission Model

We discuss the data transmission of the UT to the satellite within time slot[12] t. For the brevity of expression, the index t is omitted. The antenna arrays are assumed at both satellites and UTs for the directional beamforming. For the tractability of the analysis, an approximated sectored array model is widely adopted for formulating the antenna gain of each beam [267]. A single beam is discussed for the satellite, where all the UTs connected to that satellite are within its main lobe with the gain G_{sate}^M. A multibeam is assumed for each UT to connect with multiple satellites, where the antenna gain of UT's each beam is G_{ut}^M for the desired link in the main lobe and G_{ut}^S for the interference link in the side lobe. Considering the satellite link from UT q to satellite m on the lth layer, the signal to interference plus noise ratio (SINR) $\gamma_{m,q}^l$ at the satellite can be expressed by

$$\gamma_{m,q}^l = \frac{PG_{ut}^M G_{sate}^M h_{r,qm,p}(d_{m,q}^l)^{-\alpha}}{(\sum_{n=1}^L k_n)(\sigma^2 + I_{m,q}^l)}, \tag{9.33}$$

where P is the UT's total transmitting power and k_n denotes the average number of connected satellites on the nth layer. An equal power scheme is used by the UT to allocate the same transmitting power to each satellite backhaul link. The rain fading $h_{r,qm,p} = 10^{-\frac{A_{r,qm,p}}{10}}$ is considered over the Ka-band, where $A_{r,qm,p}$ (dB) denotes the rain attenuation exceeded for $p\%$ of an average year, which is determined by the geometrical and electrical characteristics of the antenna [268]. For convenience, we define $G_0 = G_{sate}^M G_{ut}^M h_r$, where h_r is the expectation of the rain fading. The distance between UT q and satellite m on the lth layer is $d_{m,q}^l$, and α is the pathloss exponent. The AWGN follows the $\mathcal{CN}(0, \sigma^2)$ distribution, and $I_{m,q}^l$ is the interference.

We assume that all satellites share the same spectrum resource pool with total bandwidth B for the data transmission. In total J orthogonal subchannels are divided and allocated by each satellite to the UTs within its coverage according to their satellite transmission rate requirements. Thus, the transmission rate $R^l_{m,q}$ of the satellite link from UT q to satellite m on the lth layer is influenced by the number $V^l_{m,q}$ of subchannels allocated to UT q, which is expressed by

$$R^l_{m,q} = \frac{BV^l_{m,q}}{J}\log(1+\gamma^l_{m,q}). \tag{9.34}$$

Note that any UT q can be served by multiple satellites simultaneously once the elevation angle of the satellite is not smaller than $\theta_{\min}$ for UT q. Let $\mathscr{A}^q_l$ denote the set of serving satellites on layer l for UT q, where $\mathscr{A}^q_l = \{m \mid \theta^l_m \geq \theta_{\min}\}$. Considering L satellite layers, the total data rate of UT q is

$$R_q = \sum_{l=1}^{L}\sum_{m\in\mathscr{A}^q_l} R^l_{m,q}. \tag{9.35}$$

9.2.2 Uplink Capacity Theoretical Framework

Interference Model

Given the on-demand deployment of the UTs, their locations are commonly assumed to follow the homogeneous Poisson point process (HPPP) Φ with density λ UTs/km^2. The rate requirement of the UT's backhaul service is represented by C_{th}. The average coverage ratio of the terrestrial network is denoted by ρ. Each UT within the coverage of the terrestrial network is provided by the transmission capacity of C_{terr}.

The orthogonal subchannels are allocated by the satellite to the UTs randomly within its coverage, and the allocated number is proportional to the satellite rate requirements. Thus, the distribution of the interfering UT on the same subchannel for UT q follows the HPPP with density $\lambda' = \frac{\lambda}{J}$ UTs/km^2. The interference of UT q and satellite m (layer l) link is

$$I^l_{m,q} = V^l_{m,q}\sum_{n=1}^{L}\sum_{q'\in D^n_{m,q}} \frac{PG^S_{ut}G^M_{sate}h_{r,q'm,p}}{K}(d^l_{m,q'})^{-\alpha}, \tag{9.36}$$

where $K = \sum_{n=1}^{L} k_n$ and $D^n_{m,q}$ is given in Proposition 9.1. Based on Campell's theorem, the average interference of that satellite backhaul link $I^{l,(in)}_{m,q}$ (within the terrestrial coverage) and $I^{l,(out)}_{m,q}$ (outside the terrestrial coverage) is expressed by

$$\mathbb{E}[I^{l,(s)}_{m,q}] = \frac{PG_I}{K}\sum_{n=1}^{L}\frac{C^{(s)}_{sate}J\lambda'}{(C_{th}-C_{terr}\rho)\lambda S_l}\int_{q'\in D^n_{m,q}}(d^l_{m,q'})^{-\alpha}dq', \tag{9.37}$$

where $s \in \{in, out\}$ and $G_I = G^S_{ut}G^M_{sate}h_r$. We define the rate requirement of the satellite backhaul link by $C^{(s)}_{sate}$, where $C^{(in)}_{sate} = C_{th} - C_{terr}$ when $s = in$ and $C^{(out)}_{sate} = C_{th}$ when $s = out$. Given the orthogonal subchannel among different UTs, the average allocated subchannel number is proportional to the ratio of the rate requirement of UT q and the total rate requirement in the coverage area S_l of the same satellite.

PROPOSITION 9.4 *The interference region $D^n_{m,q}$ of colayer and cross-layer interfering UTs is*

$$D^n_{m,q} = \begin{cases} \left\{q' \mid d^l_o \le d^l_{m,q'} \le \sqrt{h_l^2 + 2Rh_l}\right\}, & n = l, \\ \left\{q' \mid h_l \le d^n_{m,q'} \le \sqrt{h_l^2 + 2Rh_l}\right\}, & n \ne l. \end{cases}$$

The smallest distance d^l_o from the UT within coverage of the neighboring satellite to satellite m satisfies

$$(d^l_o)^2 = R^2 + (R + h_l)^2 - 2R(R + h_l)\cos\phi^o_l, \tag{9.38}$$

where $\phi^o_l = \max\{0, \sqrt{\frac{8\pi R^2(1-\cos\phi_l^{\max})}{2\sqrt{3}k_l}} - R \cdot \phi_l^{\max}\}/R$ and k_l is the connected satellite number on layer l for each UT. The longest distance of the LoS path between the interfering UT q' and satellite m on layer l is indicated by $\sqrt{h_l^2 + 2Rh_l}$.

Proof *Interference from colayer satellite links ($n = l$):* The interfering range to satellite m is the area where the UTs can transmit data to the neighboring colayer satellites while simultaneously maintaining a LoS path to satellite m. Let $d^l_{m,q}$ denote the distance between the UT q and satellite m on the lth layer. The region of co-layer interfering UTs is expressed by $D^l_{m,q} = \{q' \mid d^l_o \le d^l_{m,q'} \le \sqrt{h_l^2 + 2Rh_l}\}$, where d^l_o represents the smallest distance from the UT of the neighboring satellite cell to satellite m, and $\sqrt{h_l^2 + 2Rh_l}$ indicates the longest distance of the interference LoS path.

We then derive the expression of d^l_o by geometry. Given the uniform distribution, let K_l denote the total number of satellites on layer l expressed by $K_l = \frac{k_l}{2\pi(R+h_l)^2(1-\cos\phi_l^{\max})}$. Here, $4\pi(R + h_l)^2 = \frac{2k_l}{1-\cos\phi_l^{\max}}$ and the distance between the coverage center of any satellite on the Earth's surface is equidistant to its surrounding coverage centers, expressed by $b_l = \sqrt{\frac{8\pi R^2}{\sqrt{3}K_l}}$ according to the geometry relation. The radius of each satellite is coverage circle is $r_l = R \cdot \phi_l^{\max}$. Thus, the radius of the interference-free coverage of each satellite is $r^o_l = \max\{0, b_l - r_l\}$ and its corresponding angular radius $\phi^o_l = \frac{r^o_l}{R}$. According to the law of cosines, we can derive the distance $d^l_o = \sqrt{R^2 + (R + h_l)^2 - 2R(R + h_l)\cos\phi^o_l}$.

Interference from cross-layer satellite links ($n \ne l$): The serving range of satellites on different layers completely overlap to ensure the seamless coverage of the Earth. Therefore, the interfering range to satellite m on layer l is the whole LoS area. Let $D^n_{m,q}$ indicate the region of UTs inducing the cross-layer interference, which is expressed by $D^n_{m,q} = \{q' \mid h_l \le d^l_{m,q'} \le \sqrt{h_l^2 + 2Rh_l}\}$, $n \ne l$, where h_l indicates the altitude of the satellite m on layer l, and $\sqrt{h_l^2 + 2Rh_l}$ indicates the longest distance of the LoS path between the interfering UT q' and satellite m.

□

Therefore, we can obtain the average interference in (9.37):

$$\mathbb{E}[I_{m,q}^{l,(s)}] = \frac{2\pi R P G_I C_{sate}^{(s)}}{(\alpha-2)(C_{th}-C_{terr}\rho)K S_l}\left[\eta_l \frac{(d_o^l)^{2-\alpha}-(2Rh_l+h_l^2)^{\frac{2-\alpha}{2}}}{R+h_l} + \left(\sum_{n\neq l}\eta_n \frac{(h_l)^{2-\alpha}-(2Rh_l+h_l^2)^{\frac{2-\alpha}{2}}}{R+h_l}\right)\right], \tag{9.39}$$

where $\eta_l = k_l/(\sum_{n=1}^{L} k_n)$ represents the proportion of the satellite link number to the lth layer of satellite (denoted by k_l) in all the satellite links of the same UT.

Uplink Capacity Analysis

In this section, we first derive the average spectrum efficiency $\mu_l^{(s)}$ of the satellite link from the UT (where $s \in \{in, out\}$) to the satellite on the layer l and then derive the average data rate $C_{UT}^{(s)}$ of the UTs considering all the terrestrial fiber links. The uplink capacity of the integrated satellite-terrestrial network is calculated considering all UTs.

PROPOSITION 9.5 *The average spectrum efficiency $\mu_l^{(s)}$ is*

$$\mu_l^{(s)} = \frac{\mathscr{F}((d_m^l)^2, X, Y) - \mathscr{F}((h_l)^2, X, Y)}{2R(R+h_l)(1-\cos\phi_l^{\max})\ln 2}, \tag{9.40}$$

where the longest distance between satellite m (on layer l) and its serving UT is $d_m^l = \sqrt{(R\sin\theta_{\min})^2 + h_l^2 + 2Rh_l} - R\sin\theta_{\min}$. The function is defined by

$$\mathscr{F}(v, X, Y) = v\left[Y\,{}_2F_1\left(1, \frac{1}{Y}; 1+\frac{1}{Y}; -Xv^Y\right) + \ln\left(1+Xv^Y\right) - Y\right], \tag{9.41}$$

and we set that $X = \frac{PG_0}{K(\sigma^2+\mathbb{E}[I_{m,q}^l])}$, $Y = -\frac{\alpha}{2}$.

Proof According to the geometry relation, the area of serving satellites on layer l for UT q (in set $\mathscr{A}_q^l$) can be calculated by $S_q^l = 2\pi(R+h_l)^2(1-\cos\phi_l^{\max})$. Thus, the average spectrum efficiency $\mu_l^{(s)}$ for layer l can be calculated by

$$\begin{aligned}\mu_l^{(s)} &= \frac{1}{S_q^l \ln 2}\int_{m\in A_q^l} \ln\left(1+\frac{PG_0(d_{m,q}^l)^{-\alpha}}{K(\sigma^2+\mathbb{E}[I_{m,q}^l])}\right) dm \\ &\overset{(a)}{=} \frac{\pi(R+h_l)}{S_q^l R\cdot\ln 2}\int_{(h_l)^2}^{(d_m^l)^2} \ln\left(1+\frac{PG_0\cdot v^{-\frac{\alpha}{2}}}{K(\sigma^2+\mathbb{E}[I_{m,q}^l])}\right) dv,\end{aligned} \tag{9.42}$$

where in (a) we let $v = R^2 + (R+h_l)^2 - 2R(R+h_l)\sin\beta$. We denote the longest distance of between satellite m (on layer l) and its serving UT by $d_m^l = -R\sin\theta_{\min} + \sqrt{(R\sin\theta_{\min})^2 + h_l^2 + 2Rh_l}$.

By utilizing the integral formula, let $X = \frac{PG_0}{K(\sigma^2+\mathbb{E}[I^l_{m,q}])}$ and $Y = -\frac{\alpha}{2}$. We define that $\mathscr{F}(v,X,Y) = \int \ln(1+Xv^Y)dv = v\left[Y \cdot {}_2F_1(1,\frac{1}{Y};1+\frac{1}{Y};-Xv^Y)+\ln(1+Xv^Y)-Y\right]$. Thus, the average spectrum efficiency $\mu_l = \frac{\mathscr{F}((d^l_m)^2,X,Y)-\mathscr{F}((h_l)^2,X,Y)}{2R(R+h_l)(1-\cos\phi^{\max}_l)\ln 2}$. □

The average data rate of each satellite link $\mathbb{E}[R^l_{m,q}]$ from the UT to the satellite on the lth layer, denoted by $C^{(s)}_l$, is the product of the available bandwidth for the UT and the average spectrum efficiency $\mu^{(s)}_l$. According to the properties of HPPP, we express $C^{(s)}_l = \mathbb{E}[R^l_{m,q}]$ as

$$C^{(s)}_l = \sum_{u_1=1}^{\infty}\sum_{u_2=1}^{\infty}\left[(\rho\lambda S_l)^{u_1}\frac{e^{-\rho\lambda S_l}}{u_1!}\right]\left[((1-\rho)\lambda S_l)^{u_2}\frac{e^{(\rho-1)\lambda S_l}}{u_2!}\right] \cdot \frac{C^{(s)}_{sate}B\cdot\mu_l}{(C_{th}-C_{terr})\rho u_1 + C_{th}(1-\rho)u_2}. \tag{9.43}$$

Given that the UT connects with k_l satellites on the lth layer, the average data rate of the UT within and outside the terrestrial network coverage are be expressed by

$$C^{(in)}_{UT} = \sum_{l=1}^{L} k_l C^{(in)}_l + C_{terr}, \quad C^{(out)}_{UT} = \sum_{l=1}^{L} k_l C^{(out)}_l. \tag{9.44}$$

Thus, the uplink capacity C for the entire integrated satellite-terrestrial network can be modeled by the sum rate of both satellite and terrestrial fiber links of all UTs:

$$C = \rho\lambda S_E \cdot C^{(in)}_{UT} + (1-\rho)\lambda S_E \cdot C^{(out)}_{UT}, \tag{9.45}$$

where S_E denotes the area of the whole Earth. The uplink capacity is related to the average number of satellites k_l $(1 \le l \le L)$ serving the same UT as well as the coverage ratio ρ and the transmission capacity C_{terr} of the terrestrial network. Given the density λ and data rate requirement C_{th} of the UTs, the uplink capacity C can be calculated by solving (9.39) through (9.45).

9.2.3 Megaconstellation Design

In this section, we first formulate the satellite number minimization problem and then design the megaconstellation (including the layer number of deployed satellites, the height of satellites, the number of orbits, and the locations of satellites on each orbit) using the dynamic programming based method.

Satellite Number Minimization Problem Formulation

We aim to minimize the total number of deployed satellites in all layers given the terrestrial optical network while satisfying the rate requirements of all UTs for backhaul transmission. Let N_l represent the number of orbits on layer l (with altitude h_l) and let M^l_i denote the number of satellites in orbit i on layer l. Thus, the satellite number minimization problem is

$$\min_{\boldsymbol{h},\boldsymbol{N},\boldsymbol{M}} \sum_{1\le l\le L}\sum_{i\in\mathcal{N}_l} M_i^{(l)}, \tag{9.46}$$

$$s.t.\ C_{UT}^{(s)} \ge C_{th}, \quad \forall s\in\{in,out\}, \quad \forall\, UT, \forall t, \tag{9.46a}$$

where $C_{UT}^{(s)}$ is the average data rate of the UT inside and outside of the terrestrial network coverage.

Megaconstellations Design

In this part, we design the megaconstellation for the integrated satellite-terrestrial network via dynamic programming, aiming to minimize the number of total deployed satellites.

One-Coverage Satellite Constellation Deployment

We first consider a one-coverage satellite constellation deployment (i.e., $L = 1, k = 1$) with polar orbits [269] on the same satellite layer l. For the seamless coverage, the number of satellites deployed on each orbit is assumed to be same, expressed by $M^{(l)} = M_1^{(l)} = M_2^{(l)} = \cdots = M_i^{(l)}$. The satellites moves in the same direction with a phase difference $\pi/M^{(l)}$ between two neighboring orbits. Considering the hemisphere divided by one polar orbit, to satisfy the seamless coverage of the equator, the coverage angular radius $\phi_l^{\max}$ and the number of orbits N satisfy the relation that

$$(N-1)\phi_l^{\max} + (N+1)\Delta_l = \pi, \tag{9.47}$$

where $\Delta_l = \arccos(\cos\phi_l^{\max}/\cos(\pi/M^{(l)}))$ according to the geometry relation. We can express the relation among $\theta_{\min}$, h_l, N, and satellites number $M^{(l)}$ on each orbit as

$$(N+1)\arccos\left[\cos\left[\arccos\left(\frac{R\cos\theta_{\min}}{R+h_l}\right)-\theta_{\min}\right]\Big/\cos\left(\frac{\pi}{M^{(l)}}\right)\right]$$
$$+(N-1)\left[\arccos\left(\frac{R\cos\theta_{\min}}{R+h_l}\right)-\theta_{\min}\right]=\pi. \tag{9.48}$$

DEFINITION 9.4 *A UT q is $\boldsymbol{k}$-covered (where $\boldsymbol{k} = \{k_l\}, 1 \le l \le L$) indicates that UT q is covered by k_l satellites on layer l, where k_l represents the lth coverage degree of UT q for the satellites on layer l. The region is $\boldsymbol{k}$-covered if each UT q within this region is $\boldsymbol{k}$-covered.*

Based on the uplink capacity framework in Section 9.2.2, the average data rate $C_{UT}^{(s)}(\boldsymbol{k})$ of the UT can be reached when the whole discussed region is $\boldsymbol{k}$-covered. Therefore, to satisfy the rate requirement C_{th} for any UT in any position, we need to realize that the whole Earth is seamlessly $\boldsymbol{k}_{th}$-covered so that $C_{UT}^{(s)}(\boldsymbol{k}_{th}) \ge C_{th}, \forall s \in \{in,out\}, \forall UT$.

Global Coverage for the Required Data Rate

Considering L satellite layers and terrestrial capacity C_{terr}, for providing the average data rate C_{th}, the satellite minimization problem (9.46) can be decoupled from reformulated as

$$\min_{\boldsymbol{h},\boldsymbol{k}} \sum_{l=1}^{L} k_l N \cdot M^{(l)} \tag{9.49}$$

$$s.t.\ C_{UT}^{(s)} \geq C_{th}, 0 \leq k_l \leq \left\lceil \frac{C_{sate}^{(s)}}{C_l^{(s)}(h_l)} \right\rceil, \quad \forall s, \forall\, UT, \tag{9.49a}$$

where k_l is the coverage degree of the lth satellite layer. Equation (9.49) is a complete knapsack problem, where the volume of each object is $NM^{(l)}$ and the weight of each object is $C_l^{(s)}(h_l)$. We aim to minimize the total number of satellites (i.e., total weight of the knapsack) while providing C_{th} data rate to each UT in any position (i.e., filling the knapsack with size C_{th}). Given that the number of satellites $M^{(l)}$ on each orbit is a natural number, the possible value of satellite altitude h_l is discrete with each feasible $M^{(l)}$ according to (9.48), and the reachable rate $C_l^{(s)}(h_l)$ of each satellite link is also discrete.

To solve the complete knapsack problem (9.49), we then propose a dynamic programming based satellite number minimization (DP-SNM) algorithm. Let $\chi(l,c)$ denote the minimum satellite number when deploying the satellites from layer 1 to layer l (initialed by L), where t denotes the remaining data rate required by the UTs (initialed by $C_{sate}^{(s)}$). The recursive relation between the minimum satellite number can be expressed by the following equation where $0 <= k_l C_l^{(s)} <= c$:

$$\chi(l,c) = \min_{k_l} \left\{ \chi(l-1, c - k_l C_l^{(s)}) + k_l N M^{(l)}, \left\lceil \frac{C_{sate}^{(s)}}{C_l^{(s)}} \right\rceil N M^{(l)} \right\}. \tag{9.50}$$

The DP-SNM algorithm is described in Algorithm 9.2. First, given $M^{(l)}$ for layer l, we calculate the optimal altitude h_l for each layer l and derive the data rate $C_l^{(s)}$ provided by the lth layer of satellites to each satellite. Second, we initialize the minimum satellite number $\chi(1,c)$ when deploying the satellites on only one layer. Third, we calculate the total satellite number with increasing l according to the dynamic programming based recurrence relation (9.50). Using the DP-SNM algorithm, we can obtain the optimal $\boldsymbol{k}_{th}$ that provides a seamless C_{th} data rate for any UT.

The complexity of Algorithm 9.2 is then analyzed. We denote the maximum number of each UT's connected colayer satellites by $W = \max_{l,s} \left\{ \left\lceil C_{sate}^{(s)} / C_l \right\rceil \right\}$. Using the dynamic programming method, the optimal vector $\boldsymbol{k}$ can be obtained by calculating $\chi(L, C_{th})$ recursively according to relation (9.50). Thus, the complexity of the DP-SNM algorithm is $O(SLW^2)$ for the L deployable satellite layers.

9.2.4 Performance Evaluation

In this section, we present the optimal satellite deployment scheme for global connectivity and show the minimum number of required satellites satisfying the backhaul rate requirement of any UT. The influences of terrestrial capacity on the satellite constellation deployment are then evaluated.

Algorithm 9.2 Satellite number minimization

Input: Maximum satellite layer L; satellite orbit number N; satellite number per orbit $M^{(l)}$ for each layer; required average data rate $C_{th}^{(s)}, \forall s$.

Output: Optimal satellite altitude $\boldsymbol{h}$, optimal coverage degree $\boldsymbol{k}_f$.

1 Calculate the altitude h_l for each layer l given $M^{(l)}$ according to (9.48);

2 Calculate the average rate $C_l^{(s)}$ of the l-th layer satellite link given in (9.43);

3 Initialize satellite number: $\chi(1,t) = \left\lceil t/C_l^{(s)} \right\rceil NM^{(1)}, 1 \leq t \leq C_{th}^{(s)}$;

4 **for** *Layer number* $2 \leq l \leq L$ **do**

5 Calculate the minimum satellite number $\chi(l, C_{th}^{(s)})$ according to (9.50);

6 Update the optimal coverage degree $\boldsymbol{k}_{th} = \arg\min_{\boldsymbol{k}} \chi(l, C_{th}^{(s)})$;

7 $l \leftarrow l+1$;

8 **end**

Table 9.5 Parameter settings

Parameters	Value
Total transmitting power of each UT P	2 W
Antenna gain of the UT G_{ut}^{M}, G_{ut}^{S}	31.2 dBi, −10 dBi
Antenna gain of the satellite $G_{sate}^{M}, G_{sate}^{S}$	37.1 dBi, −6.75 dBi
Main lobe beamwidth of the satellite and UT	$60^\circ, 2.5^\circ$
Bandwidth for satellite data transmission B	800 MHz
Number of subchannels J for each satellite	1,000
Noise density for Ka-band transmission σ^2	−174 dBm/Hz
Density of UTs λ	4×10^{-6} km^{-2}
Minimum elevation angle $\theta_{\min}$	10°
Number of one-coverage satellite orbits N	6
Number of feasible satellites layers	40

The main parameters and their default values are set according to the 3GPP standard (release 15) in Table 9.5 [204]. To fulfill the rate requirements of the backhaul services, we consider that $C_{th} = 200$ Mbps for all the UTs. We apply the rain fading parameters given by the ITU-R P.618 standard [268]. We discuss the number of deployed satellites given different terrestrial capacity $C_{terr} \sim (100, 180)$ Mbps and different global average coverage ratio of the terrestrial network $\rho \sim (50\%, 60\%)$ with the pathloss factor $\alpha = 2.5$.

The optimal constellation deployment solutions are presented in Fig. 9.15 given different terrestrial capacity C_{terr}. Compared with the satellite-only network, the required number of satellites largely decreases when considering the terrestrial network. As the coverage ratio of the terrestrial network increases, the altitude of satellites rises and the number of satellites decreases. Figure 9.16 presents the optimal deployed altitude and number of satellites for each orbit given different terrestrial capacities with the coverage ratio $\rho = 50\%$. We can observe from the

Terrestrial coverage ratio		50%					60%					
Terrestrial capacity (Mbps)		100		120		140	100		120		140	
Satellite altitude (km)		663	677	663	694	714	677	714	694	740	714	867
Orbit number		12	6	12	6	18	12	6	6	12	12	6
Satellite number/orbit		18	17	18	16	15	17	15	16	14	15	11
Total number of satellites	Integrated	318		294		270	294		264		246	
	Satellite only	420										

Figure 9.15 Optimal constellation deployment solutions

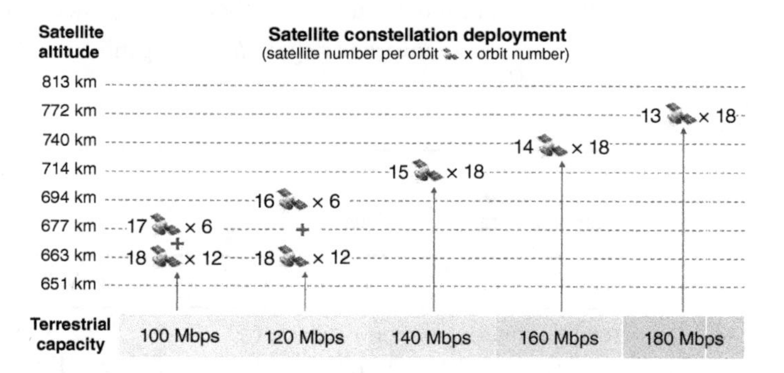

Figure 9.16 Satellite altitude of the optimized constellation versus terrestrial capacity

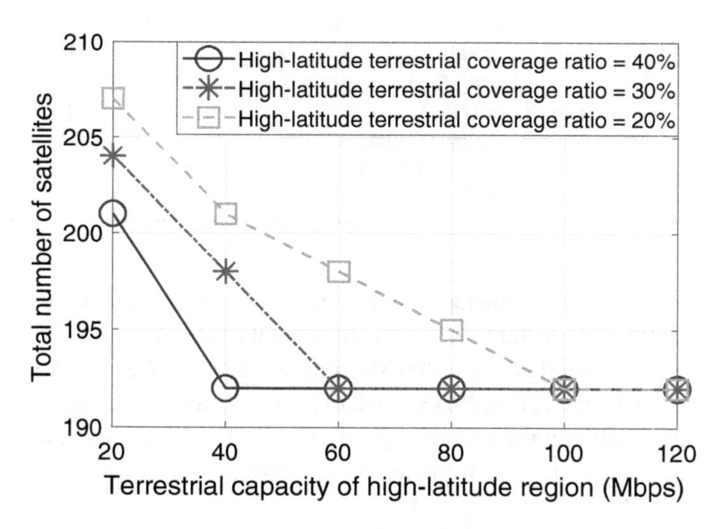

Figure 9.17 Total number of satellites versus high-latitude terrestrial capacity

optimal constellation that, as the terrestrial capacity increases, the deployed satellite altitude increases so that fewer satellites are required to deploy.

In Fig. 9.17, we evaluate the required satellite number given different terrestrial capacities in different regions. Assume that $C_{terr}^{low} = 180$ Mbps and $\rho = 60\%$ in the low-latitude and mid-latitude regions (i.e., between 60°S and 60°N). Figure 9.17 shows the total number of satellites versus the terrestrial capacity in high-latitude regions (i.e., larger than 60°) given different high-latitude terrestrial network coverage ratios. It is shown that the satellite number decreases as the high-latitude terrestrial capacity C_{terr}^{high} grows. A maximum high-latitude terrestrial capacity C_o^* exists where the satellite number cannot be further reduced and it remains at the minimum when $C_{terr}^{high} > C_o^*$, representing that a larger terrestrial capacity in the high-latitude regions cannot contribute to further reducing satellites. Moreover, with the larger coverage ratio of the terrestrial network, the satellite number reaches the minimum at a smaller high-latitude terrestrial capacity.

9.2.5 Conclusion

In this section, we studied the integrated satellite-terrestrial network. A theoretical framework for the uplink capacity analysis was proposed considering the terrestrial capacity. A megaconstellation with multilayer satellites was designed given any terrestrial capacity, where the differentiated data-rate requirements of all UTs can be satisfied even in the maximum interference case. The following conclusions were drawn. First, the optimal constellation deployment is obtained with the minimum number of satellites, where the deployed altitude increases with the terrestrial capacity. Second, as the terrestrial infrastructures are distributed more evenly in latitude, the megaconstellation first requires fewer satellites and then remains unchanged regardless of the terrestrial transmission capacity of high-latitude user terminals.

9.3 Summary

In this chapter, we first proposed a 3D constellation optimization algorithm for single-layer LEO constellation design with the minimum number of satellites while satisfying the backhaul requirement. We then designed a megaconstellation for a multilayer satellite network to realize the global connectivity with the minimum satellite number.

Part V

Integration of UAVs, HAPs, and Satellites

10 Integrating Terrestrial-Satellite Networks into 5G and Beyond

The LEO satellite access network (SAN) has shown potential as an expansion of terrestrial networks to provide reliable global services and massive connectivity. Several companies have been promoting their plans to launch thousands of LEO satellites [2, 270], including SpaceX, Kepler, and SPUTNIX. These projects aim to provide seamless and high-capacity global communication services by constructing an ultra-dense LEO constellation that cooperates with terrestrial operators. To utilize the high-frequency band in space, a terrestrial-satellite terminal (TST) equipped with phase antenna arrays acts as the access point (AP) for users. Each TST supports both the user, TST links over the C-band, and the TST-satellite links over the Ka-band, enabling terrestrial small-cell coverage for users. Benefiting from high altitudes, broad operating spectrum, and an ultra-dense topology, LEO satellite networks can support a massive number of users with their high-capacity backhaul, vast coverage, and more flexible access technique, which is less dependent on instantaneous radio environments.

To fully exploit the LEO constellation technique, in this chapter, we propose a terrestrial-satellite architecture to integrate the LEO satellite networks and the terrestrial networks for traffic offloading. Each user is scheduled to upload its data via the macro cell, the traditional small cell (TSC), or the LEO-based small cell (LSC). Each small-cell base station (SBS) and TST then upload the collected data to the core network via the traditional backhaul and LEO-based backhaul, respectively. As part of this ultra-dense LEO (UD-LEO) constellation, we assume that each TST is allowed to connect to multiple satellites simultaneously, thereby improving the resilience to the frequent handover of satellites [271]. Deployed by the same operator, all cells share the same C-band frequency resources for terrestrial communications. Multiple TST-satellite backhaul links over the Ka-band are scheduled for each LSC. The network aims to maximize the sum rate of all users and accommodate as many users as possible. Therefore, the user association and resource allocation of multiple cells should be optimized subject to the backhaul capacity constraint of each cell. At the same time, the backhaul capacity of each LSC also needs to be maximized via the satellite selection and resource allocation.

The rest of this chapter is organized as follows. In Section 10.1, we describe the model of integrated UD-LEO based terrestrial-satellite networks. In Section 10.2, we formulate a joint optimization problem for the terrestrial-satellite networks and propose a framework to iteratively solve two decoupled subproblems from Lagrangian dual decomposition. In Sections 10.3 and 10.4, we convert these two subproblems into

Table 10.1 Nonstandard abbreviations

Abbreviation	Meaning
LITS	LEO-based integrated terrestrial-satellite scheme
LBCO	LEO-based backhaul capacity optimization
LSC	LEO-based small cell
PC	Power control
NITS	Nonintegrated terrestrial-satellite networks
SBS	Small-cell base station
SMPC	Swap-matching algorithm with power control
TSC	Traditional small cell
TST	Terrestrial-satellite terminal
TTH	Traditional terrestrial heterogeneous networks
TTO	Terrestrial traffic offloading
TUASA	Terrestrial user association and subchannel allocation
UD	Ultra-dense

two different matching problems and solve them. Performance evaluation is presented in Section 10.5. The conclusion is drawn in Section 10.6.

10.1 System Model

In this section, we first introduce the UD-LEO based integrated terrestrial-satellite network in which the users can access the network via a macro BS, the TSCs, or the LSCs. We then provide the transmission models of both the terrestrial and satellite communications. For convenience, we summarize all nonstandard abbreviations in Table 10.1.

10.1.1 Scenario Description

Consider an UD-LEO based integrated terrestrial-satellite network as shown in Fig. 10.1, where one macro BS and a large number of small cells are deployed to serve the uplink ground users.[1] Each small cell assists the macro cell in offloading the traffic and is connected to the core networks via either wired or wireless backhauls. Therefore, each user, such as a mobile device or a sensor, can access the network via one of the following three cells: (1) the macro cell with large backhaul capacity supported by fiber links from the macro BS directly to the core network, (2) the TSC with limited backhaul capacity connected to the core network via multihop wired or wireless backhaul links, and (3) the LSC with large backhual capacity supported by the Ka-band transmission. For LSC backhaul, each TST uploads the user's data to the LEO satellites, and then each satellite forwards it to either an earth gateway station (connected to the core network) or a TST-equipped macro BS.

The ultra-dense LEO topology ensures that multiple satellites fly over the area of interest at each time slot, providing seamless coverage for the mobile users. Equipped

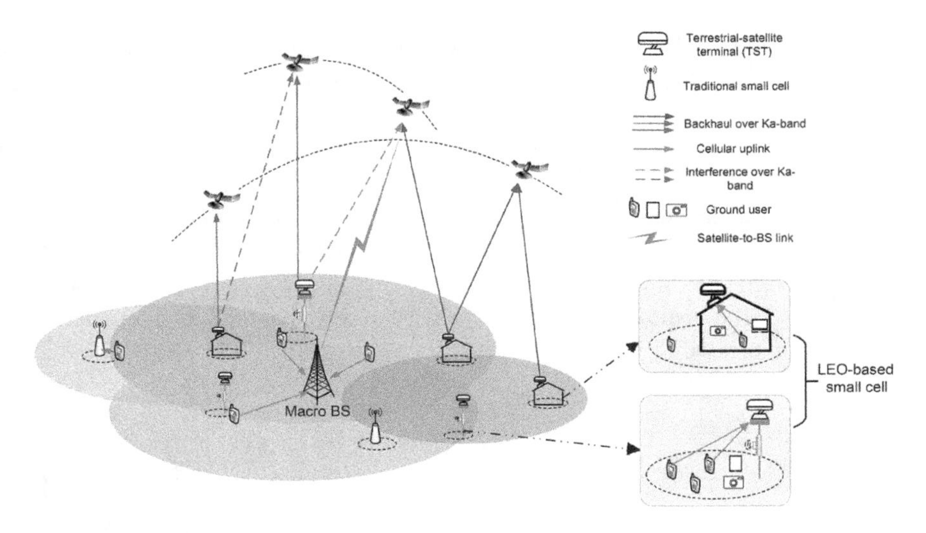

Figure 10.1 System model of the ultra-dense LEO-based integrated terrestrial-satellite network

with multiple independent antenna apertures [1, 2], each TST can connect to multiple satellites simultaneously, which further improves the backhaul capacity of the LSCs. Based on the current satellite technique [1], we assume that each satellite serves as a remote radio head of the BS with no onboard processing capacity. Therefore, the access control is located at the macro BS equipped with a TST [253]. For intercell interference management, we assume that the satellite operator and the traditional terrestrial operator cooperate to serve the users in a centralized manner.

10.1.2 Transmission Model for Terrestrial Communications

We denote the set of cells as $\mathcal{M} = \{0, 1, \ldots, M\}$ in which $m = 0$ represents the macro BS, $1 \leq m \leq M'$ represents the TSCs, and $M' + 1 \leq m \leq M$ indicates the LSCs. To describe the relationship between the positions of J users and the coverage of each cell, a binary coverage matrix $\boldsymbol{A}$ of size $(M + 1) \times J$ is introduced where $a_{m,j} = 1$ indicates that user j lies within the coverage of cell m, and $a_{m,j} = 0$ otherwise.

For frequency reuse, all cells share the same frequency resource pool, which can be divided into K subchannels. To better depict the user association and subchannel allocation, we introduce a binary matrix $\boldsymbol{X}$ of size $(M+1) \times J \times K$ in which $x_{m,j,k} = 1$ indicates that user j is served by BS m over subchannel k, and $x_{m,j,k} = 0$ otherwise. The received signal of BS m sent by user j over subchannel k is then given by

$$\begin{aligned} y_{m,j,k} = & \sqrt{p_u} h^B_{m,j,k} x_{m,j,k} s_j \\ & + \underbrace{\sum_{m' \neq m} \sum_{j' \neq j} \sqrt{p_u} h^B_{m,j',k} x_{m',j',k} s_{j'}}_{\text{intercell cochannel interference}} + N_m, \end{aligned} \tag{10.1}$$

where p_u is the transmit powers of each user, s_j (or $s_{j'}$) is the transmitted signal of user j (or user j') with unit energy, and the corresponding channel coefficient is $h^B_{m,j,k}$ (or $h^B_{m,j',k}$). Specifically, we denote $h^B_{m,j,k} = g_{m,j,k} \cdot \beta_{m,j} \cdot \left(d_{m,j}\right)^{-\alpha}$, where $g_{m,j,k} \sim \mathcal{CN}(0,1)$ is a complex Gaussian variable representing Rayleigh fading, $\beta_{m,j}$ follows log-normal distribution representing shadowing fading, $d_{m,j}$ is the distance between users j and m, and α represents the pathloss exponent. The additive white Gaussian noise (AWGN) at BS m is denoted by $N_m \sim \mathcal{CN}\left(0, \sigma^2\right)$, and σ^2 is the noise variance.

The achievable rate of each user j served by BS m over subchannel k can be expressed by

$$R^B_{m,j,k} = \log_2\left(1 + \frac{p_u\left|h^B_{m,j,k}\right|^2}{\sigma^2 + \sum\limits_{j' \neq j}\sum\limits_{m' \neq m} x_{m',j',k}\, p_u\left|h^B_{m,j',k}\right|^2}\right), \tag{10.2}$$

and thus, the data rate of each cell m can be obtained by

$$R^B_m = \sum_{j=1}^{J}\sum_{k=1}^{K} x_{m,j,k} R^B_{m,j,k}. \tag{10.3}$$

In practice, the data rate of each cell is required to be no larger than the backhaul capacity of this cell, which also influences the traffic offloading scheme of the system. For convenience, we denote the backhaul capacity of each cell m as C_m, which is fixed when $0 \leq m \leq M'$.

10.1.3 Transmission Model for LEO-Based Backhaul

Instead of a fixed value, the backhaul capacity of each LSC is related to the transmission model of the LEO-based backhaul over the Ka-band. Note that the satellite-to-ground links usually occupy a wider bandwidth than the TST-to-satellite links and the transmit power of satellites is usually larger than that of the TSTs [2]. The LEO backhaul capacity is thus constrained by the TST-to-satellite links, as illustrated next.

We assume that there are N satellites flying over the area of interest. Due to the preplanned orbit of each satellite, its altitude, speed, and position information are known to all cells in each time slot. For convenience, we adopt a qausistatic method to split a time period into multiple time slots[2] during each of which the position of a satellite is unchanged. We divide the available bandwidth over the Ka-band spectrum as a set of subchannels $\mathcal{Q} = \{1, ..., Q\}$. A binary matrix depicting the TST-LEO association and subchannel allocation is defined as $\boldsymbol{B}$, where the element $b_{m,n,q} = 1$ indicates that TST m is associated with satellite n over subchannel q, and $b_{m,n,q} = 0$ otherwise. The received signal of satellite n sent by TST m over subchannel q can be given by

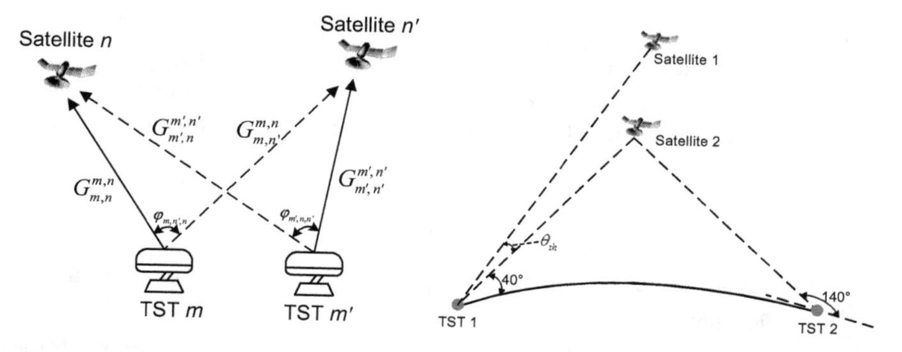

Figure 10.2 (a) Each TST's antenna gain. (b) Each satellite's service range and the angular separation

$$y_{m,n,q} = \sqrt{p_{m,n,q}^T G_{m,n}^{m,n}} h_{m,n,q}^T s_{m,n}^T b_{m,n,q} + N_n + \underbrace{\sum_{m=M'+1}^{M} \sum_{n' \neq n} \sqrt{p_{m',n',q}^T G_{m',n}^{m',n'}} h_{m',n,q}^T s_{m',n'}^T b_{m',n',q}}_{\text{intersatellite cochannel interference}}, \tag{10.4}$$

where $p_{m,n,q}^T$ (or $p_{m',n',q}^T$) is the transmit power of TST m to satellite n (or TST m' to satellite n') over subchannel q, $s_{m,n}^T$ (or $s_{m',n'}$) is the transmitted signal of TST m to satellite n (or TST m' to satellite n'). The channel gain of the TST m and satellite n link over subchannel q is denoted by $h_{m,n,q}^T$, with both the large-scale fading and the shadowed-Rician fading [273] taken into consideration.[3] As shown in Fig. 10.2a, $G_{m,n}^{m,n}$ is the antenna gain of TST m toward satellite n and $G_{m',n}^{m',n'}$ is the off-axis antenna gain of TST m' toward the direction of satellite n when the target direction of TST m' is toward satellite n'. We denote the angular separation between the TST m' and satellite n link and the TST m' and satellite n' link as $\varphi_{m',n,n'}$. The item $G_{m',n}^{m',n'}$ is a function of $\varphi_{m',n,n'}$, as shown in [Attachment III, Appendix 8 in 278]. At satellite n, AWGN is $N_n \sim \mathcal{CN}(0, \sigma^2)$.

The achievable rate of the TST m and satellite n link over subchannel k can be obtained by

$$R_{m,n,q}^T = \log_2 \left(1 + \frac{b_{m,n,q} p_{m,n,q}^T G_{m,n}^{m,n} \left| h_{m,n,q}^T \right|^2}{\sum_{n' \neq n} \sum_{m=M'+1}^{M} b_{m',n',q} p_{m',n',q}^T G_{m',n}^{m',n'} \left| h_{m',n,q}^T \right|^2 + \sigma^2} \right). \tag{10.5}$$

Note that each TST-satellite link suffers from propagation delay due to the long distance compared to the terrestrial communications. Therefore, given the traffic load of each TST, L_m, the equivalent backhaul capacity C_m can be given by

$$C_m = \sum_{n=1}^{N} \frac{1}{1/C_{m,n} + T_{trip,n}/L_{m,n}}, \tag{10.6}$$

where $C_{m,n} = \sum_{q=1}^{Q} R_{m,n,q}^{T}$ is the capacity of the TST m and satellite n link, $L_{m,n}$ is the traffic load over the TST m and satellite n link, and $T_{trip,n}$ is the propagation delay calculated by $2H_n/c$ with H_n being the altitude of satellite n and c being the speed of light. The values of $L_{m,n}$ can be obtained by solving the following equations:

$$L_{m,n}/C_{m,n} + T_{trip,n} = L_{m,n'}/C_{m,n'} + T_{trip,n'}, \quad \forall n, n' \in \mathcal{N}_m$$
$$\sum_{n \in \mathcal{N}_m} L_{m,n} = L_m, \tag{10.7}$$

where $\mathcal{N}_m = \{1 \leq n \leq N | b_{m,n,q} = 1, \forall q\}$ is the set of satellites associated with TST m. In practice, the traffic load L_m is set as $\min\{R_m^B, \sum_{j \in \mathcal{J}_m} D_j\}$, where $\mathcal{J}_m = \{1 \leq j \leq J | x_{m,j,k} = 1, \forall k\}$ is the set of users associating to cell m and D_j is the amount of data generated by each user j. For convenience, we denote the total backhaul capacity without considering the propagation delay as $\tilde{C}_m = \sum_{n=1}^{N} C_{m,n}$.

10.2 Problem Formulation and Decomposition

In this section, we aim to maximize the sum rate and number of accessed users by jointly optimizing the terrestrial data offloading, TST-satellite association, and resource allocation.

10.2.1 Angular Constraints for LEO Backhaul

To further depict the characteristics of UD-LEO networks such as angular sensitivity on satellite selection, the angular constraint on the terrestrial-satellite links are illustrated next.

As shown in Fig. 10.2b, due to different altitudes of the satellites, the elevation angles of two satellites serving the same TST may be the same. When a TST transmits to two such satellites over the same subchannel, they appear to be "in line," thereby leading to transmission failure. To cope with this issue, we define the angular separation as θ_{th} within which two satellites cannot provide service to the same TST over the same subchannel. Based on the preplanned satellite orbits, we introduce an angle matrix Θ, where the elevation angle of the TST m and satellite n link is denoted as $\theta_{m,n}$. The angle constraint can then be presented as

$$b_{m,n_1,q} b_{m,n_2,q} \theta_{th} \leq \left|\theta_{m,n_1} - \theta_{m,n_2}\right|,$$
$$\forall 1 \leq n_1, n_2 \leq N, \ M' + 1 \leq m \leq M, \ 1 \leq q \leq Q. \tag{10.8}$$

10.2.2 Problem Formulation

We consider two performance metrics in this work. First, we aim to maximize the sum rate of all cells over the C-band spectrum subject to the backhaul capacity constraints. Second, to extend the coverage of the traditional terrestrial cellular networks, it is necessary to maximize the number of active users in the network. The problem formulation can be shown as

$$\max_{\{x_{m,j,k},b_{m,n,q},p^T_{m,n}\}} \sum_{m=0}^{M} R_m^B + \mu \sum_{m=0}^{M}\sum_{j=1}^{J}\sum_{k=1}^{K} x_{m,j,k} \tag{10.9}$$

$$s.t.\ (10.8),$$

$$x_{m,j,k} \le a_{m,j}, \quad \forall 0 \le m \le M, 1 \le j \le J, 1 \le k \le K, \tag{10.9a}$$

$$\sum_{m=0}^{M}\sum_{k=1}^{K} x_{m,j,k} \le 1, \quad \forall 1 \le j \le J, \tag{10.9b}$$

$$\sum_{j=1}^{J} x_{m,j,k} \le 1, \quad \forall 0 \le m \le M, 1 \le k \le K, \tag{10.9c}$$

$$R_m^B \le C_m, \quad \forall 0 \le m \le M, \tag{10.9d}$$

$$\sum_{m=M'+1}^{M} b_{m,n,q} \le 1, \quad \forall 1 \le n \le N, 1 \le q \le Q, \tag{10.9e}$$

$$\sum_{n=1}^{N}\sum_{q=1}^{Q} b_{m,n,q} \le N_r, \quad \forall M'+1 \le m \le M, \tag{10.9f}$$

$$\sum_{n=1}^{N}\sum_{q=1}^{Q} p^T_{m,n,q} \le P^T, \quad \forall M'+1 \le m \le M, \tag{10.9g}$$

$$x_{m,j,k}, b_{m,n,q} \in \{0,1\}, \tag{10.9h}$$

where μ is the conversion parameter. Equation (10.9a) implies that each user j can only access a cell within its reach (i.e., user j is within the range of its subscribed cell). For fairness, we assume that each user can only access one cell and be assigned one subchannel, as shown in (10.9b). As is typical for the uplink in 3GPP, we consider orthogonal use of frequency resources within each cell, which is guaranteed by (10.9c). The relationship between the data rate of each cell and its backhaul capacity is shown in (10.9d), constructing the coupling between the terrestrial and satellite networks. Without loss of generality, we assume that each subchannel of satellite n can only be assigned to at most one TST, and each TST is associated with at most N_r satellite links simultaneously for backhaul, as presented in (10.9e) and (10.9f). The transmit power constraint is presented in (10.9g) where P^T is the maximum transmit power of each TST for backhaul.

10.2.3 Lagrangian Dual Decomposition

As shown in (10.9), the terrestrial data offloading and the satellite selection are coupled with each other via the backhaul capacity constraint. Therefore, we decompose the original problem into two subproblems connected by the Lagrangian multiplexers.

We denote $\Lambda = \{\lambda_m, 0 \le m \le M\}$ as a vector containing the Lagrangian multiplexers associated with constraint (10.9a). The Lagrangian associated with (10.9) is then defined as

$$\begin{aligned}
&\mathscr{L}(\boldsymbol{X},\boldsymbol{B},\boldsymbol{p}^T,\boldsymbol{\lambda}) \\
&= \sum_{m=0}^{M} R_m^B + \mu \sum_{m=0}^{M}\sum_{j=1}^{J}\sum_{k=1}^{K} x_{m,j,k} + \sum_{m=0}^{M} \lambda_m \left(C_m - R_m^B\right) \\
&= \sum_{m=0}^{M} (1-\lambda_m) R_m^B + \mu \sum_{m=0}^{M}\sum_{j=1}^{J}\sum_{k=1}^{K} x_{m,j,k} + \sum_{m=0}^{M} \lambda_m C_m .
\end{aligned} \tag{10.10}$$

Correspondingly, the dual optimum is given by

$$g^* = \min_{\boldsymbol{\lambda}\succeq 0} g(\boldsymbol{\lambda}) = \min_{\boldsymbol{\lambda}\succeq 0} \max_{\{\boldsymbol{X},\boldsymbol{B},\boldsymbol{p}^T\}} \mathscr{L}\left(\boldsymbol{X},\boldsymbol{B},\boldsymbol{p}^T,\boldsymbol{\lambda}\right). \tag{10.11}$$

For given $\boldsymbol{\lambda}$, the first item in (10.10) is only determined by $\boldsymbol{X}$, and the second item is only determined by $\boldsymbol{B}$ and $\boldsymbol{p}^T$. Therefore, we can divide the original problem into two subproblems: the terrestrial traffic offloading problem

$$\begin{aligned}
(TTO:)\ &\max_{\{x_{m,j,k}\}} \sum_{m=0}^{M}(1-\lambda_m)R_m^B + \mu \sum_{m=0}^{M}\sum_{j=1}^{J}\sum_{k=1}^{K} x_{m,j,k} \\
&s.t.\ (10.9a), (10.9b), (10.9c),
\end{aligned} \tag{10.12}$$

and the LEO-based backhual capacity optimization problem

$$\begin{aligned}
(LBCO:)\ &\max_{\left\{b_{m,n,q},\, p_{m,n,q}^T\right\}} \sum_{m=M'+1}^{M} \lambda_m C_m \\
&s.t.\ (10.8), (10.9e), (10.9f), (10.9g),
\end{aligned} \tag{10.13}$$

where the summation is counted from $M'+1$ because $\tilde{C}_m$ $(0 \le m \le M)$ is fixed.

The optimization process consists of multiple iterations, and in each iteration t, two steps are performed: (1) given $\boldsymbol{\lambda}^{(t)}$, the terrestrial traffic offloading (TTO) problem and the LEO-based backhaul capacity optimization problem are solved, respectively, and (2) update $\boldsymbol{\lambda}$ by $\lambda_m^{(t+1)} = \lambda_m^{(t)} - \delta^{(t)}(C_m^{(t)} - R_m^{B,(t)}), \forall 0 \le m \le M$ in which $\delta^{(t)}$ is a monotonically decreasing exponential function of t. The iterations will not stop until $|\delta^{(t+1)} - \delta^{(t)}| < \varepsilon$.

For those optimization problems with continuous variables, the outcome obtained from the Lagrangian dual method always satisfies the constraints [279]. However, this cannot be guaranteed for our formulated problem because there are only binary variables such that the data rate of each cell m does not vary continuously. Therefore, we adjust the user scheduling strategy after step one if the constraint (10.9d) is violated. For those cells violating (10.9d), the associating users are removed one by one following the order of increasing data rate of each user-BS link until (10.9d) is satisfied.

10.3 Algorithm Design for Terrestrial Data Offloading

In this section, we formulate the TTO problem in (10.12) as a one-to-one matching with externalities and propose a modified Gale–Shapley algorithm with redefined preference relations.

10.3.1 Matching Problem Formulation for Terrestrial Traffic Offloading

Note that the TTO problem is a 3D integer programming problem with a nonconvex objective function. Aiming to solve this problem with a low-complexity algorithm, we recognize that the user association and subchannel allocation can be regarded as a multivariate matching process. Specifically, the users, BSs, and the subchannels are three sets of players to be matched with each other to maximize the utility, while the interdependencies exist among the users due to the cochannel interference. This enables us to solve the problem by utilizing matching theory, as shown in the following sections.

10.3.2 Definitions

Consider the set of users, BSs, and the subchannels as $\mathscr{J}$, $\mathscr{M}$ and $\mathscr{K}$, which are disjointed from one another. By associating each BS with each subchannel, we construct a new set $\mathscr{S} = \mathscr{M} \times \mathscr{K}$ of size $(M+1) \times K$, where each BS-subchannel unit can be represented by (m,k). A *matching* Ψ is defined as a mapping from the set $\mathscr{J} \cup \mathscr{S} \cup \varnothing$ into itself such that for each user j and each BS-subchannel unit (m,k), we have $\Psi(j) = (m,k)$ if and only if $\Psi(m,k) = j$. In other words, if $\Psi(j) = (m,k)$, then user j is associated with BS m over subchannel k. Such a one-to-one matching naturally satisfies constraints (10.9b) and (10.9c).

To construct a matching Ψ, each BS-subchannel unit selects a user from $\mathscr{J} \cup \varnothing$ to match with such that the weighted sum rate in (10.12) can be maximized. For convenience, when $\Psi(j) = (m,k)$, we denote the utility of the BS-subchannel unit (m,k) and user j as $(1-\lambda_m) R^B_{m,j,k}$.

10.3.3 Preference Relation

Due to the intercell cochannel interference, there exist externalities [280] in this matching problem. That is, the utility of (m,k) is influenced by other users matched with (m',k). Therefore, each BS-subchannel unit actually has preference over the matching pairs due to different interferences levels brought by them.

Specifically, given a matched BS-subchannel unit (m,k), it prefers to be cohabitated with a new pair $(j',(m',k))$ with a high-quality user j' and BS m' link and a poor-quality user j' and BS m interference link over subchannel k. To depict such mutual effect, we construct a preference matrix $\mathbf{Z}_{m,k}$ of size $(M+1) \times J$ for each BS-subchannel unit (m,k), evaluating all possible matching pairs over subchannel k. Each element in $\mathbf{Z}_{m,k}$ is defined as

$$\zeta^{(m,k)}_{m',j',k} = \left|h^B_{m',j',k}\right|^{\rho_1 \lambda_{m'}} \Big/ \left|h^B_{m,j',k}\right|^{\rho_2}, \tag{10.14}$$

where ρ_1 and ρ_2 are preference parameters. We then say that a BS-subchannel unit (m,k) prefers $(j_1, (m_1,k))$ over $(j_2, (m_2,k))$ if $\zeta^{(m,k)}_{m_1,j_1,k} > \zeta^{(m,k)}_{m_2,j_2,k}$. That is,

$$\begin{aligned}&(j_1, (m_1,k)) \succ_{(m,k)} (j_2, (m_2,k)) \Leftrightarrow \zeta^{(m,k)}_{m_1,j_1,k} > \zeta^{(m,k)}_{m_2,j_2,k},\\ &\forall j_1 j_2 \in \mathscr{J}_{un}, \forall m_1, m_2 \neq m,\end{aligned} \tag{10.15}$$

where $\mathcal{J}_{un}$ is the set of unmatched users. Each BS-subchannel unit's attitude toward the potential matching pairs is affected by the preference parameters ρ_1 and ρ_2 in (10.14). When $\rho_1 = 0$, the unit has a *pessimistic attitude* (i.e., it only cares to minimize the interference brought by a potential matching pair). When $\rho_2 = 0$, the unit has an *optimistic attitude* (i.e., it only cares to maximize the benefit brought by the new matching pair). Accordingly, $\rho_1 = \rho_2$ reflects a *neutral attitude.*

The traditional preference of each subchannel over different matchings is also introduced. We define the utility of each subchannel k as the sum rate of all BSs over this subchannel (i.e., $R_k^B = \sum_{m=0}^{M}(1-\lambda_m)\sum_{j=1}^{J} R_{m,j,k}^B$). The preference of subchannel k is then given by

$$\Psi_1 \succ_k \Psi_2 \Leftrightarrow R_k^B(\Psi_1) > R_k^B(\Psi_2), \tag{10.16}$$

where $R_k^B(\Psi_1)$ is the utility obtained by subchannel k under Ψ_1.

Due to the logarithmic-form utility function, the externalities in our formulated problem do not share the additive characteristic with other well-defined one-to-one matchings with externalities [281].

10.3.4 Algorithm Design

We adopt a greedy algorithm to initialize the matching where each subchannel is matched with the most preferred combination of a BS and a user that can accept it. Specifically, each unmatched subchannel k proposes to an unmatched user j and BS m to match by satisfying

$$j^*, m^* = \arg \max_{j \in \mathcal{J}_{un}, 0 \le m \le M} \left| h_{m,j,k}^B \right|^2. \tag{10.17}$$

If user j is proposed to by more than one subchannel, then it selects a unit (m,k) with the largest channel gain from the candidates and rejects the others.

10.3.5 Propose-and-Reject Operation

Based on the defined preference relationship, we introduce the key operation of each matching pair, consisting of one proposing phase and two rejecting phases. Unlike traditional matchings, each matched BS-subchannel unit is allowed to propose to potential matching pairs instead of users.

- BS-subchannel unit proposing: Each matched BS-subchannel unit (m,k) in the current matching Ψ selects its most preferred matching pair $(j',(m',k))$ satisfying

$$j', (m',k) = \arg \max_{j' \in \mathcal{J}_{un}, m' \in \mathcal{M}_{un,k}} \left\{ \zeta_{m',j',k}^{(m,k)} | a_{m',j'} = 1 \right\}, \tag{10.18}$$

where $\mathcal{M}_{un,k}$ is the set of unmatched BSs over subchannel k. A set of candidate pairs $\mathcal{S}_k$ consisting of $(j',(m',k))$ is then constructed for each subchannel k.

- Subchannel rejecting: Each subchannel k selects one candidate matching pair $(j'', (m'', k))$ from $\mathscr{S}_k$ such that

$$\begin{aligned} &\{j'', (m'', k)\} \cup \Psi \succ_k \Psi, \\ &\{j'', (m'', k)\} \cup \Psi \succeq_k \{j', (m', k)\} \cup \Psi, \quad \forall (j', (m', k)) \in \mathscr{S}_k. \end{aligned} \tag{10.19}$$

  That is, the matched pair brings the highest positive utility to subchannel k. Other matching pairs in $\mathscr{S}_k$ are then rejected.
- User rejecting: If an unmatched user j'' is proposed more than once, it only accepts the BS-subchannel unit with the highest utility and rejects all the other proposals. Once a user is matched, it is removed from $\mathscr{J}_{un}$.

10.3.6 Algorithm Description

The whole matching algorithm for terrestrial user association and subchannel allocation (TUASA) is presented in detail in Algorithm 10.1. In the initialization step (lines 2–14), each subchannel is matched with a combination of one user and one BS with the best channel condition that it can achieve. The following matching process (lines 16–35) consists of multiple iterations in which the propose-and-reject operation (lines 19–35) is performed. The iterations will not stop until no matched BS-subchannel unit is left to propose to the users.

10.3.7 Algorithm Analysis

The propose-and-reject operation offers the optimal strategy to maximize the number of accessed users in the TTO problem because each user j is compulsively served by a BS as long as there exist an unmatched BS-subchannel unit (m, k) satisfying $a_{m,j} = 1$. In other words, the proposed TUASA algorithm guarantees the maximum number of accessed users while improving the sum rate.

As proved next, our proposed TUASA algorithm is guaranteed to converge to a final matching after a limited number of iterations.

Proof The convergence of the TUASA algorithm depends on the matching process. According to the coverage matrix $\boldsymbol{A}$, the number of potential users associated with each BS is limited, and thus, the preference list of each BS-subchannel unit is complete and transparent. In each iteration, the matched BS-subchannel unit proposes to an unmatched user and a BS in its preference list. As the number of iterations grows, the set of remaining choices for each BS-subchannel unit becomes smaller. Therefore, the iterations stop when there are no available choices in the preference list for each matched BS-subchannel unit. □

Due to the complicated interaction relationship between different pairs brought by the cochannel interference, the preference in our formulated problem does not satisfy the substitutability condition, which is usually a sufficient condition for the existence of a stable matching. Therefore, the traditional solution concepts of blocking pair and

Algorithm 10.1 Terrestrial user association and subchannel allocation (TUASA)

Input: Sets of users, BSs, and subchannels $\mathcal{J}$, $\mathcal{M}$, and $\mathcal{K}$; coverage matrix $\mathbf{A}$.
Output: A final matching Ψ^*.

1 **Initialization**
2 Record current matching as Ψ. Construct $\mathcal{J}_{un} = \mathcal{J}$ and $\mathcal{M}_{un,k} = \mathcal{M}, \forall 1 \leq k \leq K$;
3 **while** *at least one subchannel is unmatched* **do**
4 | Each unmatched subchannel k proposes to user j^* and BS m^* according to (10.17);
5 | **for** *each user receiving proposals* **do**
6 | | **if** *user j is proposed by more than one subchannel* **then**
7 | | | User j selects (m,k) with the highest channel gain $\left|h^B_{m,j,k}\right|^2$ from the candidates;
8 | | | User j rejects the other proposals;
9 | | **else**
10 | | | User j is matched with the proposing BS-subchannel unit;
11 | | **end**
12 | **end**
13 | User j is removed from $\mathcal{J}_{un}$;
14 **end**
15 **Matching Process**
16 **while** $\mathcal{J}_{un} \neq \varnothing$ *or some matched BS-subchannel unit still tries to propose* **do**
17 | Set $\mathcal{S}_k = \varnothing$ for all $1 \leq k \leq K$;
18 | **for** *each subchannel k* **do**
19 | | **for** *each matched BS-subchannel unit (m,k) in Ψ* **do**
20 | | | Propose to its most preferred pair $\left(j', \left(m',k\right)\right)$ according to (10.18);
21 | | | Add $\left(j', \left(m',k\right)\right)$ to $\mathcal{S}_k$;
22 | | **end**
23 | | Select one candidate pair $\left(j'', \left(m'',k\right)\right)$ from $\mathcal{S}_k$ according to (10.19);
24 | | Remove other candidate pairs from $\mathcal{S}_k$;
25 | **end**
26 | **for** *each user j included in $\mathcal{S}_k$ $(1 \leq k \leq K)$* **do**
27 | | **if** *it is proposed more than once* **then**
28 | | | User j accepts the best BS-subchannel unit and rejects others;
29 | | **else**
30 | | | User j accepts the received proposal;
31 | | **end**
32 | | User j is removed from $\mathcal{J}_{un}$;
33 | | The newly matched BS is removed from the corresponding $\mathcal{M}_{un,k}$;
34 | **end**
35 **end**
36 **return** the final matching Ψ^*.

stability cannot be applied in this case. We thus introduce the concept of a blocking pair, which is stricter than the traditional version.

DEFINITION 10.1 *Given a matching Ψ, where $\Psi(j_1) = (m_1,k_1)$ and $\Psi(m_2,k_2) = j_2$ (k_1 and k_2 are allowed to be equal), we denote the matching where user j_1 is matched with (m_2,k_2) while other pairs are unchanged as $\Psi' = \Psi \backslash \{(j_1,(m_1,k_1)),(j_2,(m_2,k_2))\} \cup \{(j_1,(m_2,k_2)),(j_2,(m_1,k_1))\}$. The pairs $(j_2,(m_1,k_1))$ and $(j_1,(m_2,k_2))$ are individual* rational blocking pairs *if (1)* $R^B_{m',j',k_i}(\Psi') > R^B_{m',j',k_i}(\Psi)$, $\forall j', (m',k_i) \in \Psi$, $i = 1,2$, *and (2)* $R^B_{m_1,j_2,k_1}(\Psi') > R^B_{m_1,j_1,k_1}(\Psi)$ *and* $R^B_{m_2,j_1,k_2}(\Psi') > R^B_{m_2,j_2,k_2}(\Psi)$. *In other words, the blocking pairs* $(j_1,(m_2,k_2))$ *and* $(j_2,(m_1,k_1))$ *can bring higher utility to all existing pairs with respect to subchannels* k_1 *and* k_2 *in* Ψ.

Given this definition, we discuss the stability and equilibrium based on different attitudes of the BS-subchannel units. The following statements stand:

- *When* $\rho_2 = 0$, *we have* group stability *in the sense that* j_1, j_2, (m_1,k_1), *and* (m_2,k_2) *are considered as a group. In other words, there does not exist individual-rational blocking pairs satisfying condition 2 in Definition 10.1.*
- *When* $\rho_2 \neq 0$, *instead of group stability, we have equilibrium. When there is no individual-rational blocking pair in a matching, any new matching pair* $(j, (m,k))$ *cannot improve the utility of the* (m,k) *unit without compromising the utility of other matched BS-subchannel units. The final matching then reaches an* equilibrium point.

Proof Suppose that we have $\Psi(j_1) = (m_1,k_1)$ and $\Psi(j_2) = (m_2,k_2)$ in the final matching Ψ^*. Consider the matching pairs $(j_1, (m_2,k_2))$ and $(j_2, (m_1,k_1))$.

For $\rho_2 = 0$ (case 1), there are three possible reasons why the prior two matching pairs do not show up in Ψ^*. The first reason is that in the initialization phase user j_1 is already matched with (m_1,k_1), and thus, j_1 has never been proposed in the matching process. If so, we can infer that either $|h^B_{j_1,m_1,k_1}| > |h^B_{j_2,m_1,k_1}|$ (i.e., user j_1 and BS m_1 are most preferred by subchannel k_1) or $|h^B_{j_2,m_2,k_2}| > |h^B_{j_1,m_2,k_2}|$. The second possible reason is that the pair $(j_1, (m_2,k_2))$ has never been proposed throughout the algorithm. This implies that user j_2 has already been matched before the pair $(j_1, (m_2,k_2))$ is considered by some (m,k) unit, and thus, we have $|h^B_{j_2,m_2,k_2}| > |h^B_{j_1,m_2,k_2}|$. The third reason is that the pair $(j_1, (m_2,k_2))$ was proposed once but rejected by user j_1. If so, we can infer that either $|h^B_{j_1,m_1,k_1}| > |h^B_{j_2,m_1,k_1}|$ (user j_2 is not matched yet) or $|h^B_{j_2,m_2,k_2}| > |h^B_{j_1,m_2,k_2}|$ (user j_2 has been matched earlier).

For the case where $|h^B_{j_1,m_1,k_1}| > |h^B_{j_2,m_1,k_1}|$, since the cochannel interference to BS m_1 keeps unchanged for $(j_1, (m_1,k_1))$ and $(j_2, (m_1,k_1))$ as shown in (10.2), we have $R^B_{m_1,j_1,k_1} > R^B_{m_1,j_2,k_1}$. Therefore, condition 2 in Definition 10.1 is not satisfied, implying that Ψ^* is group stable.

For $\rho_2 \neq 0$ (case 2), when $\rho_1 = 0$, the deduction is similar to case 1. We can always find an existing pair $(j',(m',k_2))$ over subchanel k_2 such that (m',k_2) has proposed to $(j_2,(m_2,k_2))$ earlier. This indicates that user j_1 will bring larger interference to BS m' than user j_2, and thus, $(j_1,(m_2,k_2))$ cannot bring higher utility to (m',k_2). We then say that Ψ^* is an equilibrium point.

When $\rho_1 \neq 0$, we take the pair $(j_1,(m_2,k_2))$ as an example. Similar to the three options in case 1, we can infer that there must exist an existing pair $(j',(m',k_2))$ in Ψ^* satisfying $|h^B_{j_2,m_2,k_2}|/|h^B_{j_2,m',k_2}| > |h^B_{j_1,m_2,k_2}|/|h^B_{j_2,m',k_2}|$. Therefore, we have either $|h^B_{j_2,m_2,k_2}| > |h^B_{j_1,m_2,k_2}|$ or $|h^B_{j_2,m',k_2}| < |h^B_{j_2,m',k_2}|$. In other words, either conditions 1 or 2 in Definition 10.1 is violated, implying that Ψ^* is an equilibrium point. □

In terms of computational complexity, the maximum number of iterations in the initialization phase is J, and the number of outer iterations in the matching phase is also proportional to J. For the worst case, in each outer iteration all matched pairs propose to the same user such that only one new matching pair is formed. The total number of outer iterations in this case is J. For the best case, in each iteration a new matching pair is accepted over a subchannel (i.e., K new pairs are recorded). Because there are J users in total, the number of required iterations is $\lceil J/K \rceil$. In practice, the iteration number varies between $\lceil J/K \rceil$ and J.

10.4 Algorithm Design for LEO-Based Backhaul Capacity Maximization

In this section, we convert the LBCO problem in (10.13) into a many-to-one matching problem with externalities. To depict the multiconnectivity and angle-sensitive nature of the UD-LEO networks, a novel swap-matching algorithm with continuous power control (SMPC) is proposed.

Matching Problem Formulation

Consider the set of TSTs, satellites, and available subchannels as $\mathcal{M}'$, $\mathcal{N}$, and $\mathcal{Q}$. We assume that each satellite n and subchannel q form a satellite-subchannel (SS) unit (n,q). A matching Φ is defined as a mapping between $\mathcal{M}'$ and $\mathcal{N} \times \mathcal{Q}$ such that for each TST m and each SS unit (n,q), we have (1) $\Phi(n,q) = m$ if and only if $(n,q) \in \Phi(m)$; (2) $\Phi(m) \subseteq (\mathcal{N} \times \mathcal{Q}) \cup \varnothing$ and $|\Phi(m)| \leq N_r$; and (3) $\Phi(n,q) \in \mathcal{M}'$ and $|\Phi(n,q)| \leq 1$. Specifically, each matching pair is denoted by $(m, (n,q))_{p^T_{m,n}}$, where $p^T_{m,n}$ is the transmit power of TST m over this link, satisfying constraint (10.9g). We aim to find a matching such that the weighted capacity $\sum_{m=M'+1}^{M} \lambda_m \tilde{C}_m$ can be maximized. The equivalent capacity can then be obtained by solving equations in (10.7).

Preference Relation

Following Section 10.3.3, the influence brought by a potential pair $(m',(n',q))$ with respect to the existing matched SS unit (n,q) is redefined as

$$w^{(n,q)}_{m',n',q} = v_{m',n'} \left[G^{m',n'}_{m',n'} \left| h^T_{m',n',q} \right|^2 \right]^{\rho_1} \Big/ \left[G^{m',n'}_{m',n} \left| h^T_{m',n,q} \right|^2 \right]^{\rho_2}. \tag{10.20}$$

Therefore, the preference relation $\succ_{(n,q)}$ can be given by

$$(m_1, (n_1,q)) \succ_{(n,q)} (m_2, (n_2,q)) \Leftrightarrow w^{(n,q)}_{m_1,n_1,q} > w^{(n,q)}_{m_2,n_2,q}, \quad \forall m_1, m_2, \forall n_1, n_2 \neq n. \tag{10.21}$$

Each subchannel's preference over different matchings is similar to that in Section 10.3.3. The utility of subchannel q is defined as $R^T_q = \sum_{m=M'+1}^{M} \lambda_m \sum_{n=1}^{N} R^T_{m,n,q}$ in this case.

Remark on the New Problem
Two new challenges have been posed, which are different from the TTO problem, rendering most existing matching algorithms [282–284] as well as Algorithm 10.1 no longer suitable, as illustrated next.

First, no mechanism to transmit power adjustments exists in each project-and-reject operation, as shown in Section 10.3.5. In addition, the complete preference list based on (10.20) over all possible power levels is difficult to construct.

Second, due to the angular sensitivity, the group stability in Algorithm 10.1 can no longer be guaranteed. Typically, in the matching Ψ obtained from Algorithm 10.1, the interference caused by user j to a given BS m is irrelevant of its matched BS. However, in the LBCO problem, as long as TST m switches its matched satellites from n_1 to n_2, the cochannel interference brought to any other satellite n' varies due to different off-axis antenna gains (i.e., $G_{m,n'}^{m,n_2} \neq G_{m,n'}^{m,n_1} \neq G_{m,n_2}^{m,n_2}$). By comparing (10.1) and (10.4), we can infer that the group stability does not hold in the LBCO problem even if $\rho_2 = 0$, implying that the propose-and-reject operation fails to depict such a dynamic matching structure change over each subchannel in the LBCO problem.

10.4.1 Algorithm Design

We next introduce the swap matching performed by each TST. Generally speaking, in a swap operation, a TST tends to swap its matches with another TST while keeping other TSTs' strategies unchanged.

DEFINITION 10.2 *Given a matching Φ with two existing matching pairs $(m_1,(n_1,q_1))_{p^T_{m_1,n_1,q_1}}$ and $(m_2,(n_2,q_2))_{p^T_{m_2,n_2,q_2}}$, a swap matching is defined as*

$$\Phi^{m_2,(n_2,q_2)}_{m_1,(n_1,q_1)} = \Phi \backslash \left\{(m_1,(n_1,q_1))_{p^T_{m_1,n_1,q_1}}, (m_2,(n_2,q_2))_{p^T_{m_2,n_2,q_2}}\right\} \cup \left\{(m_2,(n_1,q_1))_{p^T_{m_2,n_1,q_1}}, (m_1,(n_2,q_2))_{p^T_{m_1,n_2,q_2}}\right\}. \quad (10.22)$$

Both the TSTs and SS units are allowed to be virtual. There is no physical meaning for a virtual TST or SS unit, and the corresponding utility is zero.

Power Allocation in Swap Matchings
Based on this generalized definition, we discuss the power control strategy for five typical types of swap matchings. Other cases not mentioned here can be classified into one of five types.

In type 1 TST m_2 and SS unit (n_1,q_1) are virtual. The swap matching degrades to TST m_1 proposing to the unmatched SS unit (n_2,q_2) and $p^T_{m_2,n_2,q_2} = 0$. Note that by allocating power to the TST m_1 and satellite n_2 link, the SS units in $\Phi(m_1)$ may be influenced because the allocated power needs to be adjusted to satisfy the power constraint. Therefore, unlike traditional swap matching, more than two matching pairs may be involved during the swap matching in the LBCO problem.

To evaluate how the power control influences other SS units in $\Phi(m_1)$, we consider a pair $(m_1, (n,q))$ with $(n,q) \in \Phi(m_1)$. The utility of subchannel q is rewritten as a function of $p^T_{m_1,n,q}$:

$$R^T_q\left(p^T_{m_1,n,q}\right) = \lambda_{m_1}\log_2\left(1 + p^T_{m_1,n,q}\cdot\frac{G^{m_1,n}_{m_1 n}\left|h^T_{m_1,n,q}\right|^2}{I_{n,q}+\sigma^2}\right)$$
$$+\sum_{m'}\lambda_{m'}\sum_{n'\neq n_i}\log_2\left(1+\frac{S_{m',n',q}}{p^T_{m_1,n,q}G^{m_1,n}_{m_1,n'}\left|h^T_{m_1,n',q}\right|^2+I_{n',q}+\sigma^2}\right), \tag{10.23}$$

where $I_{n,q}$ (or $I_{n',q}$) is the interference received by satellite n_i (or n') over subchannel q_t and $S_{m',n',q}$ is the signal strength of the TST m' and satellite n' link. All three of these items are irrelevant to $p^T_{m_1,n,q}$.

We define the negative gradient of $R^T_q(p^T_{m_1,n,q})$ as $f(p^T_{m_1,n,q}) = -\nabla_{p^T_{m_1,n,q}}(R^T_q)$. It is observed that the larger $f(p^T_{m_1,n,q})$ is, the smaller influence $(m_1,(n_i,q_t))$ suffers from the reduction of its transmit power. Therefore, we select a pair $(m_1,(n^*,q^*))$ from $\Gamma(m_1)$ that is least affected by the swap matching:

$$\left(n^*,q^*\right) = \arg\max_{(n,q)\in\Gamma(m_1)} f\left(p^T_{m_1,n,q}\right). \tag{10.24}$$

To perform the swap matching networking, we consider maximizing the total utility of the involved subcahnnels via the power control of TST m_1. For convenience, we denote the available power budget of this swap matching as $P^{budget}_{m_1} = p_{m_1}(\Phi) + p^T_{m_1,n^*,q^*}(\Phi)$, where $p_{m_1}(\Phi)$ is the unallocated power of TST m_1. The power control (PC) problem for m_1 can be formulated as

$$\underline{(PC1:)} \quad \max_{\left\{p^T_{m_1,n_2,q_2},\,p^T_{m_1,n^*,q^*}\right\}} \quad R^T_{q_2}\left(p^T_{m_1,n_2,q_2}\right) + R^T_{q^*}\left(p^T_{m_1,n^*,q^*}\right)$$
$$s.t.\; p^T_{m_1,n_2,q_2} + p^T_{m_1,n^*,q^*} \leq P^{budget}_{m_1}. \tag{10.25}$$

Specifically, when $q^* = q_2$, the objective function $R^T_{q^*}\left(p^T_{m_1,n_2,q^*}, p^T_{m_1,n^*,q^*}\right)$ can be written as

$$R^T_{q^*}\left(p^T_{m_1,n_2,q^*}, p^T_{m_1,n^*,q^*}\right)$$
$$= \lambda_{m_1}\left[\log_2\left(1+\frac{p^T_{m_1,n^*,q^*}A}{I_{n^*,q^*}+p^T_{m_1,n_2,q^*}B+\sigma^2}\right)+\log_2\left(1+\frac{p^T_{m_1,n_2,q^*}C}{I_{n_2,q^*}+p^T_{m_1,n^*,q^*}D+\sigma^2}\right)\right]$$
$$+\sum_{m'\neq m_1}\lambda_{m'}\sum_{n'\neq n^*\neq n_2}\log_2\left(1+\frac{S_{m',n',q^*}}{I_{n',q^*}+p^T_{m_1,n^*,q^*}F_{m',n'}+p^T_{m_1,n_2,q^*}G_{m',n'}+\sigma^2}\right), \tag{10.26}$$

where $A \sim D$, $F_{m',n'}$, and $G_{m',n'}$ are all channel-relevant constants, and the detailed forms are omitted here. The PC problem of TST m_1 is then formulated as

$$\begin{aligned}(\underline{PC2}:) \quad & \max_{\left\{p^T_{m_1,n_2,q^*},\, p^T_{m_1,n^*,q^*}\right\}} \quad R^T_{q^*}\left(p^T_{m_1,n_2,q^*}, p^T_{m_1,n^*,q^*}\right) \\ & s.t.\ p^T_{m_1,n_2,q^*} + p^T_{m_1,n^*,q^*} \le P^{budget}_{m_1}. \end{aligned} \tag{10.27}$$

These two PC problems can be easily solved as follows:

For the solution to the first power control (PC1) problem, we first divide this problem into two PC3-type problems with the objective functions $R^T_{q_2}(p^T_{m_1,n_2,q_2})$ and $R^T_{q^*}(p^T_{m_1,n^*,q^*})$, respectively. The constraints over $p^T_{m_1,n_2}$ and $p^T_{m_1,n^*,q^*}$ are both set as $[0, P_T]$. Let $\nabla_{p^T_{m_1,n^*,q^*}} R^T_{q^*}(p^T_{m_1,n^*,q^*}) = \nabla_{p^T_{m_1,n_2,q_2}} R^T_{q_2}(p^T_{m_1,n_2,q_2}) = 0$ such that the extreme points can be found. We can then obtain the maximum values of the two PC3-type problems by traversing all extreme points and two boundary points (i.e., 0 and P_T). We denote the solutions as $p^{T*}_{m_1,n_2,q_2}$ and $p^{T*}_{m_1,n^*,q^*}$. If $p^{T*}_{m_1,n^*,q^*} + p^{T*}_{m_1,n_2,q_2} \le p^{budget}_{m_1}$, then we naturally obtain the optimal solution of PC1; otherwise, we set $p^T_{m_1,n^*,q^*} = p^{budget}_{m_1} - p^T_{m_1,n_2,q_2}$ and substitute it into the original objective function of (10.25). The problem then becomes a maximization problem with one continuous variable, which can be easily solved by searching the extreme points.

Next we have the solution to the PC2 problem. For brevity, we set $p_1 = p^T_{m_1,n^*,q^*}$ and $p_2 = p^T_{m_1,n_2,q^*}$ in (10.26). The objective function in (10.26) can be reformulated as

$$\begin{aligned} R^T_{q^*}(p_1, p_2) = & \Bigg[\lambda_{m_1} \log_2\left(A p_1 + B p_2 + I'_{n^*}\right) + \lambda_{m_1} \log_2\left(C p_2 + D p_1 + I'_{n_2}\right) \\ & + \sum_{m' \neq m_1} \lambda_{m'} \sum_{n'} \log_2\left(F_{m',n'} p_1 + G_{m',n'} p_2 + I'_{n'} + S_{m',n',q_2}\right) \Bigg] \\ & - \left[\lambda_{m_1} \log_2\left(B p_2 + I'_{n^*}\right) + \lambda_{m_1} \log_2\left(D p_1 + I'_{n_2}\right) \right. \\ & \left. + \sum\nolimits_{m' \neq m_1} \lambda_{m'} \sum\nolimits_{n'} \log_2\left(F_{m',n'} p_1 + G_{m',n'} p_2 + I'_{n'}\right)\right] \\ = & \, g_1(\boldsymbol{p}) - g_2(\boldsymbol{p}), \end{aligned} \tag{10.28}$$

where we set the vector $\boldsymbol{p} = (p_1, p_2)$ and all I' refers to the sum of interference and noise. Because each item in $g_1(\boldsymbol{p})$ and $g_2(\boldsymbol{p})$ is a concave function of $\boldsymbol{p}$, both $g_1(\boldsymbol{p})$ and $g_2(\boldsymbol{p})$ are concave. Therefore, $R^T_{q^*}(p_1, p_2)$ can be considered a difference of convex function. Becuase the constraint in PC2 represents a convex and close set, a classic difference of convex algorithm [285] can be utilized to solve the PC2 problem.

In type 2 swap matchings, $(m_2, (n_2, q_2))$ is a virtual pair. The swap matching degrades to the case where TST m_1 considers withdrawing some power resources allocated to the TST m_1 and satellite n_1 link over subchannel q_1. The power control problem for TST m_1 is formulated as

$$(\underline{PC3}:) \max_{p_{m_1,n_1,q_1}} R^T_{q_1}\left(p^T_{m_1,n_1}\right)$$
$$s.t.\ 0 \le p_{m_1,n_1,q_1} \le p_{m_1,n_1,q_1}(\Phi). \tag{10.29}$$

The solution can be found by traversing the extreme points and boundary points of the objective function.

In type 3 swap matchings, $m_1 = m_2$. In this case, TST m_1 adjusts its power allocated to two links. The power budget of m_1 is $P^{budget}_{m_1} = p^T_{m_1,n_1,q_1}(\Phi) + p^T_{m_1,n_2,q_2}(\Phi) + p_{m_1}(\Phi)$. When $q_1 \ne q_2$, the power control problem follows PC1, as shown in (10.25) with the objective function $R^T_{q_2}(p^T_{m_1,n_2,q_2}) + R^T_{q_1}(p^T_{m_1,n_1,q_1})$. In contrast, when $q_1 = q_2 = q$, this becomes a PC2-type problem as shown in (10.27) with the objective function $R^T_q(p^T_{m_1,n_2,q}, p^T_{m_1,n_1,q})$.

In type 4 swap matchings, no nodes are virtual or identical. TSTs m_1 and m_2 swap their matches over different subchannels with their power budgets unchanged. We evaluate the total utility of two involved subchannels such that each TST separately solves a PC3-type problem. The power budget for TST m_1 is $P^{budget}_{m_1} = p^T_{m_1,n_1,q_1}(\Phi) + p_{m_1}(\Phi)$, and the objective function is $R^T_{q_2}(p^T_{m_1,n_2,q_2})$ where the TST m_2 and satellite n_2 link has been removed. A symmetric problem can be formulated for TST m_2. After solving the prior two PC3 problems, the total utility is obtained by $R^T_{q_2}(p^{T,*}_{m_1,n_2,q_2}) + R^T_{q_1}(p^{T,*}_{m_2,n_1,q_1})$, where $p^{T,*}_{m_1,n_2,q_2}$ and $p^{T,*}_{m_2,n_1,q_1}$ are the solutions for two PC3 problems, respectively.

Lastly, in type 5 swap matchings, no nodes are virtual and only $q_1 = q_2 = q$. TSTs m_1 and m_2 swap their matches over subchannel q with their power budgets unchanged. This directly degrades to a PC2 problem, where the constraints are separated as those in the type 4 swap matching.

Feasibility and Validity of a Swap Matching

To maximize the total utility of the matching, each potential swap matching needs to be evaluated before being executed.

DEFINITION 10.3 *A swap matching* $\Phi^{m_2,(n_2,q_2)}_{m_1,(n_1,q_1)}$ *is* feasible *if and only if all constraints in* (10.13) *are not violated; that is, (1)* $|\theta_{m_1,n_2} - \theta_{m_1,n_i}| \ge \theta_{th}, \forall n_i \in \Phi(m_1), n_i \ne n_2$ *and* $|\theta_{m_2,n_1} - \theta_{m_2,n_{i'}}| \ge \theta_{th}, \forall n_{i'} \in \Phi(m_2), n_{i'} \ne n_1$*; (2)* $|\Phi(m_1)| \le N_r$ *and* $|\Phi(m_2)| \le N_r$*; and (3)* $|\Phi(n_1,q_1)| \le 1$ *and* $|\Phi(n_2,q_2)| \le 1$ *still stand after the swap matching. Based on the power strategy of two TSTs, a swap matching can only be* approved *if the total utility of the corresponding subchannels is improved:*

$$R^T_{q_1}\left(\Phi^{m_2,(n_2,q_2)}_{m_1,(n_1,q_1)}\right) + R^T_{q_2}\left(\Phi^{m_2,(n_2,q_2)}_{m_1,(n_1,q_1)}\right) > R^T_{q_1}(\Phi) + R^T_{q_2}(\Phi). \tag{10.30}$$

For case 1, $(m_2, (n_2, q_2))$ *is replaced by* $(m_1, (n^*, q^*))$.

Pruning Procedure for Swap-Matching Candidate Selection

In traditional swap-matching algorithms, every two matching pairs are searched for finding an approved swap matching as long as the matching structure changes. Each time it involves optimizing and comparing the utility before and after the candidate swap matching, thereby leading to a high complexity.

To address this issue, we add a gradient-based pruning procedure for preprocessing before the swap matching. Given a matching Φ, we construct a gradient matrix $\mathbf{W}_m$ of size $N \times Q$ for each TST m where the element $w_{n,q}^m$ is defined as

$$w_{n,q}^m = \begin{cases} -\nabla_{p_{m,n}^T} R_q^T \left(p_{m,n,q}^T\right), & \text{if } (m,(n,q)) \in \Phi, \\ -\nabla_{p_{m,n}^T} R_q^T \left(0^+\right), & \text{otherwise.} \end{cases} \tag{10.31}$$

The gradient matrix is updated after every approved swap matching $\Phi_{m_1,(n_1,q_1)}^{m_2,(n_2,q_2)}$. Instead of updating all elements, only those involving q_1 and q_2 need to be updated. The candidate swap matching can then be selected according the following pruning rules:

- Type 1 (or type 2) swap matching: TST m_1 should satisfy $w_{n_2,q_2}^{m_1} < 0$ (or $w_{n_1,q_1}^{m_1} > 0$).
- Type 3 swap matching: TST m_1 should satisfy $w_{n_2,q_2}^{m_1} \neq w_{n_1,q_1}^{m_1}$. Specifically, when (n_2,q_2) is an unmatched SS unit, it is required that $w_{n_2,q_2}^{m_1} < 0$ and $w_{n_2,q_2}^{m_1} < w_{n_1,q_1}^{m_1}$.
- Type 4 swap matching: It is required that $w_{n_2,q_2}^{m_1} > 0$ and $w_{n_1,q_1}^{m_2} > 0$ cannot hold at the same time ($w_{n_2,q_2}^{m_1}$ and $w_{n_1,q_1}^{m_2}$ need to be updated first based on $\Phi_{m_1,(n_1,q_1)}^{m_2,(n_2,q_2)}$).

By pruning those matching pairs violating these rules, we do not have to traverse all potential swap matchings. Because updating the gradient matrix is more convenient than solving the power control problems (PC1, PC2, or PC3), the computation task can be largely reduced.

Algorithm Description

Given the provided definitions, we propose SMPC algorithm to solve the LBCO problem, as shown in detail in Algorithm 10.2. For initialization, Algorithm 10.1 is adopted where each TST is only allowed to match one SS unit given fixed power P^T/N_r. The preference relation defined in Section 10.4 is utilized. The following swap-matching process contains multiple iterations in each of which an approved swap matching is found and executed. The iterations will not stop until there is no approved swap matching in the current matching.

10.4.2 Stability, Convergence, and Complexity

From Definition 10.3, each approved swap matching can bring higher utility to the whole system. Note that there exists an upper bound of the weighted capacity given λ due to the limited resources. Therefore, we can always find an approved matching after which the total utility does not increase any further. The convergence of the proposed SMPC algorithm can then be guaranteed.

We next present the definition of a *swap-stable* matching.

DEFINITION 10.4 *A matching Φ is swap-stable if there is no feasible and approved swap matching that can further improve the total utility.*

Algorithm 10.2 Swap-matching algorithm with power control (SMPC) for solving LBCO problem

Input: Sets of TSTs, satellites, and subchannels $\mathscr{M}$, $\mathscr{N}$, and $\mathscr{Q}$; constraint matrix $\boldsymbol{V}$.

Output: A swap stable matching Φ^*.

1 **Initialization**
2 Allocate a fixed transmit power P_T/N_r to each TST m;
3 Perform Algorithm 10.1 to obtain a one-to-one matching Φ;
4 Construct a gradient matrix $\boldsymbol{W}_m$ for each TST m;
5 Denote a virtual TST, satellite, and subchannel as m_0, n_0, and q_0, respectively;
6 Set $p_{m_1}^T(\Phi) = (1 - 1/N_r)P_T$ for each matched TST m_1;
7 Set $p_{m_2}^T(\Phi) = P_T$ for each unmatched TST m_2;
8 **Swap Matching Process**
9 **repeat**
10 Select a pair $(m, (n,q)) \notin \Phi$ satisfying pruning rules;
11 **if** $|\Phi(m)| < N_r$ *and* $\Phi_{m_0,(n,q)}^{m,(n_0,q_0)}$ *is feasible* **then**
12 Execute this swap matching if it is approved according to Definition 10.3;
13 Set $\Phi_{m_0,(n,q)}^{m,(n_0,q_0)}$ and update $p_m^T(\Phi)$;
14 Update $w_{n',q}^{m'}$, $\forall n' \in \mathscr{N}, m' \in \mathscr{M}'$;
15 **end**
16 Select a pair $(m, (n,q)) \in \Phi$ satisfying pruning rules;
17 **if** $\Phi_{m_0,(n_0,q_0)}^{m,(n,q)}$ *is feasible and can be approved* **then**
18 Execute this swap matching and set $\Phi = \Phi_{m_0,(n_0,q_0)}^{m,(n,q)}$;
19 Update $p_{m1}^T(\Phi)$ and $w_{n',q}^{m'}$, $\forall n' \in \mathscr{N}, m' \in \mathscr{M}$;
20 **end**
21 Select two pairs $(m_1, (n_1,q_1)), (m_2, (n_2,q_2)) \in \Phi$;
22 **if** *they fall into type 3 (or 4 or 5) swap matching satisfying the pruning rule* **then**
23 **if** $\Phi_{m_1,(n_1,q_1)}^{m_2,(n_2,q_2)}$ *is feasible and can be approved according to Definition 10.3* **then**
24 Execute this swap matching and set $\Phi = \Phi_{m_1,(n_1,q_1)}^{m_2,(n_2,q_2)}$;
25 Update $p_{m_1}^T(\Phi)$, $p_{m_2}^T(\Phi)$, $w_{n',q_1}^{m'}$, and $w_{n',q_2}^{m'}$, $\forall n' \in \mathscr{N}, m' \in \mathscr{M}$;
26 **end**
27 **end**
28 **until** *the total utility cannot be improved by any swap matching*;
29 **return** the final matching Φ^*.

According to Algorithm 10.2, the matching process will not stop until there is no approved swap matching in the current matching. Therefore, in the final matching, either there is no feasible swap matching, or the remained swap matching cannot improve the utility of the corresponding subchannels. This is naturally a swap-stable matching based on Definition 10.4. To further analyze the equilibrium of Algorithm 10.2, we present the following remark.

Remark When the TSTs are close enough to each other (i.e., they can be represented by the same geographic coordinates), the final matching is also an equilibrium when $\rho_2 = 0$.

Proof Given two matching pairs in Φ^*, say, $(m_1, (n_1,q))$ and $(m_2, (n_2,q))$ with equal transmit power, we check whether $(m_2, (n_1,q))$ and $(m_1, (n_2,q))$ can form blocking pairs. Because all TSTs share the same coordinates, we use the index m in the antenna coefficients G. Suppose that the prior two existing pairs are formed in phase 1. According to the preference relation in (10.20), we have either $\left|h_{m_1,n_1,q}\right|^2 > \left|h_{m_2,n_1,q}\right|^2$ or $\left|h_{m_2,n_2,q}\right|^2 > \left|h_{m_1,n_2,q}\right|^2$. We present the rates of satellite n_1 before and after blocking as before $\left(\text{i.e., } R_{n_1,q}^{before} = \log_2\left(1+\frac{G_{m,n_1}^{m,n_1}\left|h_{m_1,n_1,q}\right|^2}{I'_{m_1,q}+G_{m,n_1}^{m,n_1}\left|h_{m_2,n_1,q}\right|^2}\right)\right.$ and $\left.R_{n_1,q}^{after} = \log_2\left(1+\frac{G_{m,n_1}^{m,n_1}\left|h_{m_2,n_1,q}\right|^2}{I'_{m_1,q}+G_{m,n_1}^{m,n_1}\left|h_{m_1,n_1,q}\right|^2}\right)\right)$. Based on the channel conditions, we have either $R_{n_1,q}^{before} > R_{n_1,q}^{after}$ or $R_{n_2,q}^{before} > R_{n_2,q}^{after}$. Therefore, $(m_2, (n_1,q))$ and $(m_1, (n_2,q))$ are not blocking pairs because they cannot bring higher utility to all corresponding satellites and TSTs. Moreover, even if the prior two existing pairs are obtained from an approved swap matching, the earlier statement still holds because the swap matching further improves the utility of two matching pairs obtained from phase 1. □

Because the complexity of the initialization phase is similar to that in Section 10.3.7, we focus on analyzing the complexity of the swap-matching process. Because the complexity has a positive correlation with the number of potential swap matchings in a matching structure Φ, we focus on the latter to evaluate the complexity. Based on the size of three sets $\mathcal{M}'$, $\mathcal{N}$, and $\mathcal{Q}$, it is easy to obtain the number of type 1 (or 2) swap matchings as $NQ(M - M')$. For type 3 swap matchings, each TST with N_r links can construct $N_r(N_r - 1)/2$ swap matchings, and thus, the total number of swap matchings is $N_r(N_r - 1)(M - M')/2$. We now consider the type 4 and 5 swap matchings. For each TST m with N_r links, there exist at most $\max\{NQ - N_r, (M - M' - 1)N_r\}$ links not including m, each of which can form a swap matching with one of TST m's links. Therefore, the total number of swap matchings is

$$\begin{aligned} N_r \cdot & \sum_{i=1}^{M-M'-1} \max\left\{NQ - iN_r, \left(M - M' - i\right)N_r\right\} \\ &= \max\left\{2NQ - \left(M - M'\right)N_r, \left(M - M'\right)N_r\right\} \\ &\quad \cdot N_r\left(M - M' - 1\right)/2. \end{aligned} \tag{10.32}$$

In practice, the number of traversed swap matchings after pruning is much smaller than that of the earlier case, as shown in Fig. 10.3b.

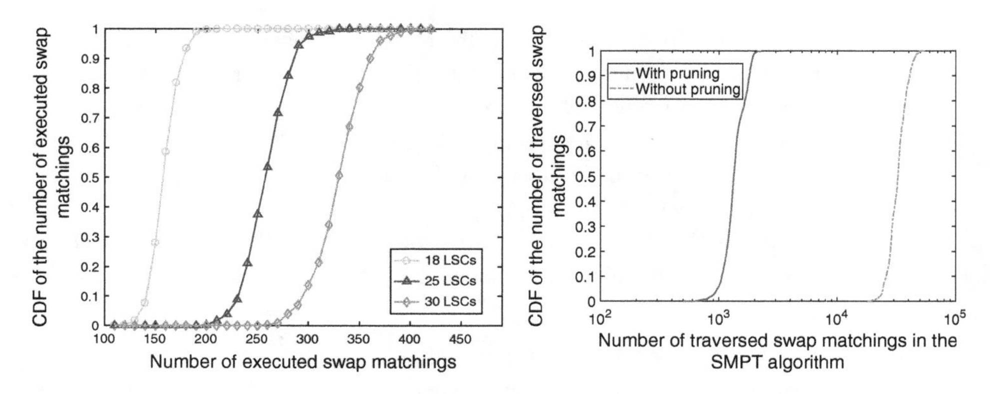

Figure 10.3 Cumulative distribution function (CDF) of (a) the total number of executed swap operations in the proposed SMPC algorithm and (b) the total number of traversed swap operations in the proposed SMPC algorithm with 18 TSTs

For reference, the complexity of the random matching algorithm is $O(M)$, and that of the greedy algorithm is $O(M^2)$.

10.5 Performance

In this section, we evaluate the performance of our proposed LEO-based integrated terrestrial-satellite (LITS) scheme. For comparison, the following schemes are also performed:

- **Ideal backhaul:** We set ideal backhaul for the macro cell and all small cells such that the sum rate obtained in this case is regarded as the upper bound.
- **Traditional terrestrial heterogeneous (TTH) networks:** All small cells are traditionally backhauled via wireless or wired links with limited capacity.
- **Nonintegrated terrestrial-satellite (NITS) networks:** This is the sum of the outcomes from two independent networks: (1) the traditional terrestrial network where J users access M' TSCs and (2) the LEO-based network where J users access $M - M'$ LSCs.

In our simulation, we set the radius of the macro cell as 1,000 m and that of each small cell as 200 m. The transmit power of each user is set as 23 dBm, the data generation rate of each user is 3,000 bytes/s, the noise density for terrestrial communications is -174 dBm/Hz, and the bandwidth for C-band communications is 20 MHz and that for Ka-band communications is 400 MHz. We set the numbers of subchannels to $K = 10 - 20$ and $Q = 10 - 15$ in the sense that the satellite networks adopt the broadband communications. The small-scale fading over the Ka-band is modeled as Rician fading with the K parameter set to seven, and that over the C-band is modeled as the Rayleigh fading. We adopt the Umi pathloss model in [286] for the terrestrial networks and the free-space pathloss model for the satellite networks. The backhaul capacities of the macro cell and each TSC are 150 Mbps and 20–30 Mbps, respectively. The numbers

Table 10.2 Simulation parameters

Parameter	Value
Maximum transmit power of each TST	2 W
Antenna type of the TST	60 cm equivalent aperture diameter
Noise figure over Ka-band	1.2 dB
Ka-band carrier frequency	30 GHz
Antenna gain	43.3 dBi
Lagrangian iteration parameter ϵ	10^{-7}
G/T parameter of each satellite	18.5 dB/K
Elevation angle [2]	35°

of TSCs and LSCs are both 25, and the number of LEO satellites is eight. We set the coordinates of satellites according to the satellite tool kit [287] such that the heights of satellites vary from 600 km to 1,200 km. The coordinates of the MBS are set to (116.38°E, 39.92°N). The number of equipped antennas [1] for each TST is $N_r = 2$. Other simulation parameters are set according to [253], as listed in Table 10.2.

We denote the random variables $\tilde{Y}_1$ and $\tilde{Y}_2$ as the total numbers of executed and attempted swap operations in the proposed SMPC algorithm, respectively. As mentioned in Section 10.4.2, $\tilde{Y}_2$ directly reflects the computational complexity, and $\tilde{Y}_1$ is proportional to the complexity. Figure 10.3a shows the cumulative distribution function (CDF) of $\tilde{Y}_1$, $\Pr(\tilde{Y}_1 \leq \tilde{y}_1)$, versus $\tilde{y}_1$ for different numbers of TSTs. Figure 10.3b shows the CDF of $\tilde{Y}_2$, $\Pr(\tilde{Y}_2 \leq \tilde{y}_2)$, versus $\tilde{y}_2$ for the cases where the pruning procedure is performed and not performed, respectively, given 18 TSTs. As the number of TSTs grows, the speed of convergence increases, especially when the pruning procedure is performed. This also reflects the low computational complexity of the proposed SMPC algorithm with the effective pruning procedure.

Figure 10.4a illustrates the sum rate of all users versus user density obtained by different schemes. In this chapter, the sum rate increases with the user density in the given area. It is observed that the performance of our proposed LITS scheme is close to the case with ideal backhaul and is much better than that of the TTH scheme and the NITS scheme. This implies that the LEO-based backhaul can significantly improve the system performance due to its high backhaul capacity. In addition, via the integration of the TSCs and the LSCs, the user scheduling and frequency resource allocation can be performed more efficiently compared to the nonintegrated scheme.

Figure 10.4b shows the total backhaul capacity versus the number of LEO satellites with different numbers of satellites that each TST can access, N_r. The figure shows the total backhaul capacity grows with the number of satellites, with diminishing returns, indicating the optimal density of satellites occurs at the inflection point. As N_r grows, the total backhaul capacity increases rapidly owning to the performance gain brought by the multiconnectivity of the TSTs.

In Fig. 10.4c, we also present how the projected area of a fixed number of LEO satellites influences the total backhaul capacity. As the projected area decreases, the distance between a TST and a satellite is shortened, bringing more high-quality back-

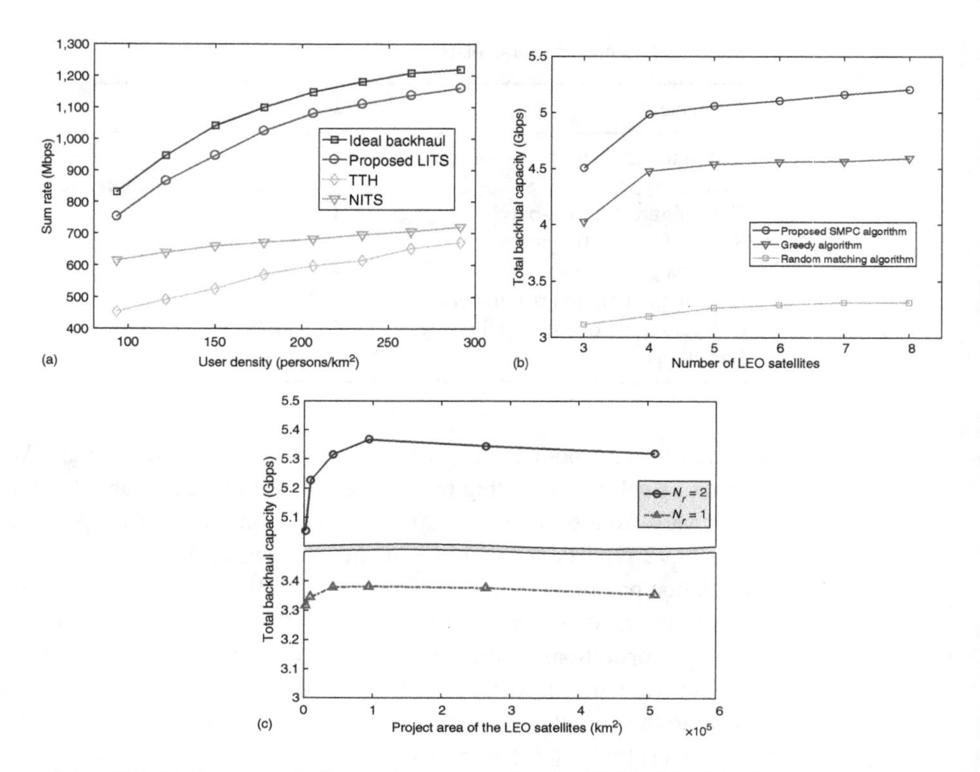

Figure 10.4 (a) Sum rate of all users versus user density. (b) Total backhaul capacity versus number of LEO satellites. (c) Total backhaul capacity versus projected area of the LEO satellites

haul links. When the projected area is small enough, the intersatellite interference dominates the backhaul capacity as the angular separation of any two satellites shrinks, leading to a decrease in backhaul capacity. This offers a reference for the LEO constellation deployment for backhaul capacity maximization.

Figure 10.5 presents how the traffic load influences the system performance in terms of the backhaul delay and the user scheduling strategies. In Fig. 10.5a, the backhaul delay is evaluated in both the LSCs and the TSCs with different traffic loads at each TST or SBS. Note that the backhaul capacity of the LSC is usually larger than that of the TSCs, leading to low transmission delay. However, the round-trip time of the LEO-based small cell is also much larger than that of the TSC due to the distance between the TST and the LEO satellites. Therefore, when the traffic load is small, the backhaul delay of the TSCs is lower than that of the LSCs. In contrast, when the traffic load is large enough, the advantage of the LEO-based backhaul is reflected because the round-trip time can be neglected compared to the transmission delay. This provides a criterion for the network to select the backhaul strategy based on the traffic load.

Figure 10.5b further evaluates how the traffic load influences the user scheduling strategy. The figure shows the proportion of users accessing the LEO-base small cells grows with the amount of data generated by each user. Here we ignore the terres-

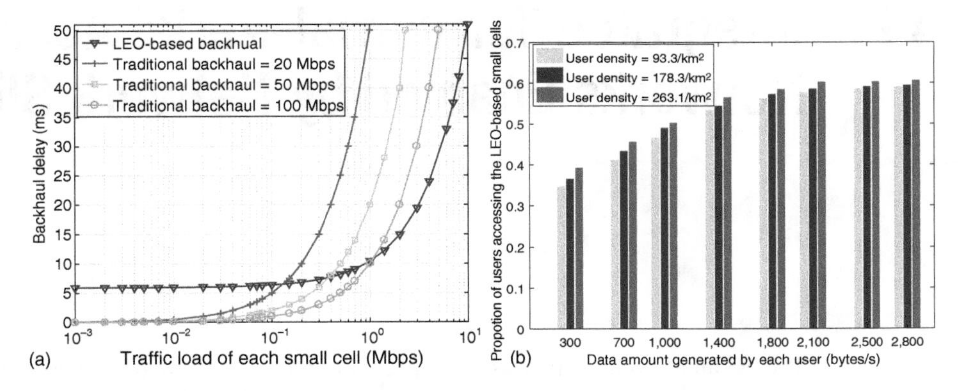

Figure 10.5 (a) Backhaul delay versus traffic load of each small cell. (b) Proportion of users accessing the LSCs versus amount of data generated by each user

trial transmission delay because it is too small. When the data generated by each user increases, the traffic load at each TST becomes larger, and thus, the equivalent backhaul capacity for each TST grows according to (10.6). Therefore, more users are scheduled to access the LSCs with a smaller delay. This is also consistent with Fig. 10.5a in the sense that the LEO-based backhaul has a large advantage over the traditional backhaul, especially with high traffic load.

10.6 Conclusion

In this chapter, we proposed the ultra-dense LEO-based integrated terrestrial-satellite network architecture where the LEO-based backhaul provides a new approach to extending traditional networks. We formulated an optimization problem to maximize the sum rate and the number of the accessed users subject to the simultaneously optimized backhaul capacity of each LSC. Such an intractable optimization problem was decomposed into two subproblems connected by the Lagrangian multiplexers. We then converted these two subproblems into matching problems and solved them separately by proposing two different low-complexity matching algorithms.

Two important conclusions can be obtained from the simulation results. First, it is advantageous to arrange the users to access the TSC for low propagation delay when the traffic load is small. As the traffic load grows, more users are inclined to access the LSC for lower transmission delay. Second, for a given number of LEO satellites in the visual range of TSTs, there exists an optimal satellite deployment for the total backhaul capacity maximization.

11 Integrated Terrestrial-Aerial Access Networks Enabled by Network Slicing

The 5G and beyond networks are expected to provide combined and advanced forms of massive machine-type communications (mMTC), enhanced mobile broadband (eMBB), and ultra-reliable low-latency communications (URLLC) services [288]. This necessitates innovative network solutions to meet the networks' stringent requirements in terms of high data rates, high reliability, low latency, and wide coverage for a massive number of devices. To address the insufficiency of current terrestrial networks, integrated terrestrial-aerial access networks have been proposed as a promising paradigm. Ground BSs, UAVs, and HAPs can coordinate with each other to satisfy heterogeneous service requirements owing to the wide coverage, large bandwidth, and flexible deployment of aerial vehicles [289].

To enable such an integration of aerial networks and terrestrial networks, network slicing (NS) has been proposed as an efficient method to deal with the heterogeneity of the system [290]. Using the network function virtualization (NFV) technique, the multidomain physical infrastructure resources of an integrated network maps them to dedicated logical networks in order to adapt to different use cases. Various virtual network functions (VNFs) are deployed at different physical nodes (such as the BS, UAV, and HAP) and tailored to support the heterogeneous service requirements of ground users. Each VNF is designed for one or several dedicated services. Once a VNF is deployed at an access point (AP), it provides corresponding service to all users connected to that AP. When more VNFs are deployed at these APs, a higher data rate can be achieved because more service types are supported by these APs, allowing for more flexible user scheduling. However, this also leads to extra compute resource consumption [291] at each AP. To depict the influence of such a VNF deployment, it is necessary to consider the compute resource utilization efficiency, which is the ratio of the sum rate to the compute resource consumption, measuring the trade-off between these two metrics.

The literature includes research on NS-enabled integrated terrestrial-aerial access networks. In [292], the authors designed a traffic offloading approach covering the requirements of two heterogeneous slices. The proposed offloading approach reduces the latency through resource allocation and UAV trajectory design. An integrated terrestrial-aerial vehicular network is considered in [293], and different machine learning methods are applied to the VNF deployment and resource scheduling. In [294], researchers propose an online control framework to dynamically slice the spectrum resource to maximize the system throughput. However, most existing works have not considered the influence of VNF deployment on the trade-off between the sum rate and resource consumption.

Against this background, in this chapter, we consider an NFV-enabled integrated terrestrial-aerial access network where the compute resource utilization efficiency is maximized by jointly optimizing the VNF deployment, spectrum, and power allocation of both aerial vehicles and terrestrial BSs. For the first time the mutual relation between user scheduling, VNF deployment, and compute resource consumption is constructed given such a heterogeneous terrestrial-aerial network structure. We then solve the problem by iteratively solving three subproblems based on the block coordinate descent (BCD) and successive convex approximation (SCA) algorithms. Via simulations, we reveal how the VNF deployment (including both the number of deployed VNFs and VNF-AP association) influences the system performance.

11.1 System Model

11.1.1 Scenario Description

As shown in Fig. 11.1, we consider an integrated terrestrial-aerial network consisting of N BSs, denoted by $\mathcal{N} = \{1,2,\ldots,N\}$, J UAVs denoted by $\mathcal{J} = \{1,2,\ldots,J\}$, M HAPs denoted by $\mathcal{M} = \{1,2,\ldots,M\}$, and I ground users distributed in an urban area. Base stations connect with ground users over the C-band frequency, while UAVs and HAPs connect with ground users over the Ka-band frequency. The network serves eMBB and URLLC users via downlink channels and mMTC users via uplink channels.

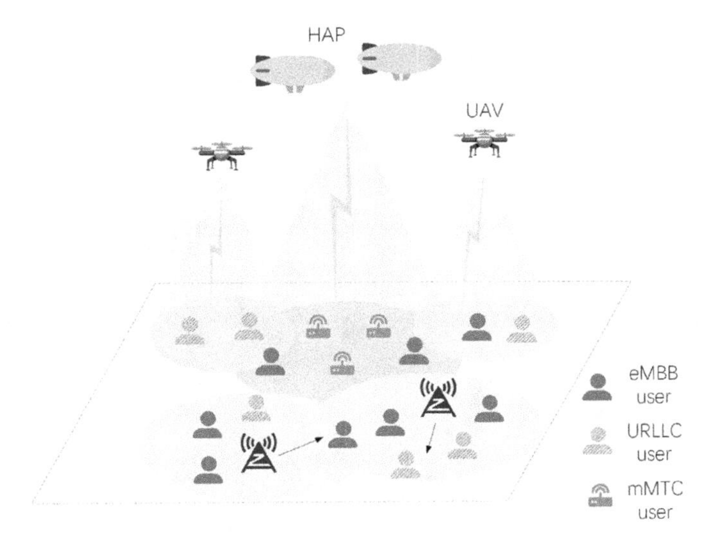

Figure 11.1 Integrated terrestrial-aerial access networks

11.1.2 Network-Slicing-Based System Architecture

To meet the different requirements of eMBB, URLLC, and mMTC services, we consider virtually isolating the network resources and functions using the concept of NS. The integrated network architecture, shown in Fig. 11.2, consists of three layers:

1. In the infrastructure layer, resources at APs such as compute, spectrum, and power resources compose a resource pool.
2. In the control layer, the slice controller is in charge of creating, maintaining, and releasing slices during their life cycles. The resource manager coordinates physical resources in the resource pool to deploy different VNFs. Logically, the network possesses three types of slices: eMBB, URLLC, and mMTC.
3. In the application layer, a variety of services and applications are performed based on the VNFs provided by the control layer.

11.2 NFV-Based Function Division and Transmission Scheme

As shown in Fig. 11.2, each physical node (i.e., BS, UAV, and HAP) is considered as a node of the NFV infrastructure, where different slices and VNFs can be deployed. The VNFs contained in each slice are customized to provide specific functions requested by users. In addition, a VNF utilizes the plug-and-play approach, meaning it can provide the predesigned function to all the users connected to the node once the VNF is installed on an AP.

The NFV-based transmission scheme has the following steps.

1. Each user sends its service request containing specific functions to the resource manager.

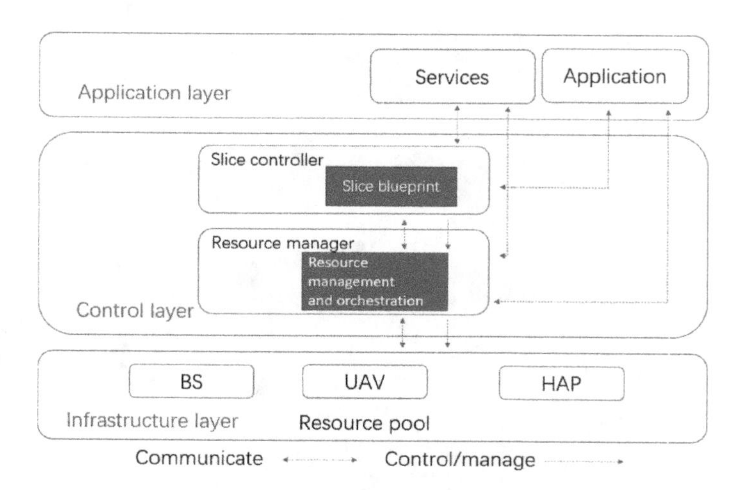

Figure 11.2 Network architecture of integrated terrestrial-aerial access networks

2. After collecting the requests from users, the resource manager sends a *slice request* to the slice controller.
3. Based on the slice request, the slice controller generates a network slice blueprint for creating, maintaining, and releasing the *network slice instance* during its life cycle.
4. The resource manager deploys VNFs and allocates physical resources to each slice instance at APs according to the blueprint.
5. Equipped with such customized slices, each AP then communicates with its associated users.

Note that any two eMBB (or URLLC/mMTC) slice instances may contain different VNFs required by users with various rate and latency demands. Containing different requests from users, each slice request of the slice controller corresponds to one type of service:

- An *eMBB slice request* is described as a tuple $\{I_s^e, R_s^e, F_s^e\}$ for a slice $s \in \mathcal{S}^e = \{1, \ldots, S^e\}$, where I_s^e is the number of eMBB users served by s, R_s^e denotes the minimum transmission rate of each eMBB user in s, and F_s^e is a tuple containing functions required by eMBB users, such as access control and mobility management.
- A *URLLC slice request* is denoted by a tuple $\{I_s^u, \tau_s^u, \varepsilon_s^u, F_s^u\}$ for a slice $s \in \mathcal{S}^u = \{1, \ldots, S^u\}$, where I_s^u is the number of URLLC users served by s, τ_s^u denotes the maximum transmission latency of an URLLC packet in s, ε_s^u is the decoding error probability of the data packet, and F_s^u is a tuple containing required functions, respectively.
- An *mMTC slice request* is given by $\{I_s^m, \gamma_s^m, F_s^m\}$ for a slice $s \in \mathcal{S}^m = \{1, \ldots, S^m\}$, where I_s^m is the number of mMTC users served by s, γ_s^m is the SNR threshold of an mMTC packet in s, and F_s^m contains required functions.

11.3 Joint Resource Allocation and VNF Deployment Optimization Problem Formulation

11.3.1 VNF Deployment and Resource Constraints

We first construct the relation between the VNF deployment and user scheduling. For function $f \in F_s^\kappa$, its consumed compute resource is denoted by C_f^κ, where $\kappa \in \{u, m, e\}$ corresponds to URLLC, mMTC, or eMBB slices. We use a binary variable $\beta_{i,f}^\kappa$ to indicate whether function f is requested by user i and $\alpha_{f,\chi}^\kappa$ to denote whether function f is installed at AP χ, where χ corresponds to BS n, UAV j, or HAP m. A binary variable $x_{i\chi,s}^\kappa$ is used to indicate whether user i is connected with the AP χ and served by the type-κ slice s.

The relation between the VNF deployment variable $\alpha_{f,\chi}^\kappa$ and the user-scheduling variable $x_{i\chi,s}^\kappa$ can be expressed by

$$\alpha_{f,\chi}^{\kappa} = Sgn\left(\sum_{i} x_{i\chi,s}^{\kappa} \beta_{i,f}^{\kappa}\right), \quad \forall f \in F_s^{\kappa}, \quad s \in S^{\kappa}, \tag{11.1}$$

where $Sgn(\cdot)$ is the sign function. The right side of (11.1) indicates whether there exists a user that requires function f at AP χ. If so, $\alpha_{f,\chi}^{\kappa} = 1$; otherwise, $\alpha_{f,\chi}^{\kappa} = 0$. This constraint guarantees that the required functions of each user are deployed at at least one AP.

We next show the physical resource constraints. For user fairness, we assume that a user can be served by just one AP:

$$\sum_{\chi} x_{i\chi,s}^{\kappa} = 1, \quad \forall i \in I_s^{\kappa}, s \in S^{\kappa}. \tag{11.2}$$

For each AP χ, the maximum transmit power constraint is given by

$$\sum_{i,s} x_{i\chi,s}^{e} p_{i,\chi}^{e} + x_{i\chi,s}^{u} p_{i,\chi}^{u} + x_{i\chi,s}^{m} p_{i,\chi}^{m} \leq p_{\chi}, \quad \forall \chi, \tag{11.3}$$

where p_{χ} denotes the total transmit power of AP χ. In addition, the spectrum resource constraint is given by

$$\sum_{i,s} x_{i\chi,s}^{e} W_{i,\chi}^{e} + x_{i\chi,s}^{u} W_{i,\chi}^{u} + x_{i\chi,s}^{m} W_{i,\chi}^{m} \leq W_{\chi}, \quad \forall \chi, \tag{11.4}$$

where W_{χ} is the bandwidth allocated to the AP χ.

11.3.2 UAV Trajectory Constraints

For convenience, we denote the locations of user i and AP χ as $\mathbf{z}_i$ and $\mathbf{z}_{\chi}$, respectively. The channel between user i and AP χ is given by

$$h_{i,\chi}^{\kappa} \triangleq \frac{g_{i,\chi}^{\kappa}}{\| \mathbf{z}_i - \mathbf{z}_{\chi} \|^{2\upsilon}}, \tag{11.5}$$

where $g_{i,\chi}^{\kappa}$ depicts the small-scale fading, and υ is the pathloss parameter.

Specifically, due to the mobility of UAVs, any two consecutive points of the UAV trajectory are subject to its speed. Thus, we have

$$\| \mathbf{z}_j(t) - \mathbf{z}_j(t-1) \|^2 \leq e_{\max}^2, \tag{11.6}$$

where $\mathbf{z}_j(t)$ is the position of UAV j at slot t, and $e_{\max}$ is UAV's maximum flight distance during a time slot. In addition, for collision avoidance, the distance between any two UAVs at each slot should not be less than a safe distance:

$$\| \mathbf{z}_j(t) - \mathbf{z}_k(t) \|^2 \geq d_{\min}^2, \quad \forall t, \tag{11.7}$$

where $d_{\min}$ is the minimum safe distance.

11.3.3 Data-Rate Constraints

For eMBB user i, the achievable data rate $R^e_{i,s}$ is given by

$$R^e_{i,s} = \sum_{\chi} x^e_{i\chi,s} W^e_{i,\chi} \log_2\left(1 + \frac{p^e_{i,\chi}|h^e_{i,\chi}|^2}{N_0 W^e_{i,\chi}}\right), \quad \forall i \in I^e_s, \quad s \in S^e, \tag{11.8}$$

where $W^e_{i,\chi}$ denotes the bandwidth allocated to user i, $p^e_{i,\chi}$ denotes the transmit power of the AP χ and user i link, and N_0 denotes the noise power. Each user i's minimum data-rate requirement should be satisfied:

$$R^e_{i,s} \geq R^e_s, \quad \forall i \in I^e_s, \quad s \in S^e. \tag{11.9}$$

For URLLC service, the following delay constraint should be satisfied:

$$R^u_{i,s} \geq \frac{b^u_i}{\tau^u_s}, \quad \forall i \in I^u_s, \quad s \in S^u, \tag{11.10}$$

where $R^u_{i,s}$ is the achievable rate of user i served by slice s. For a quasistatic Rayleigh fading channel, we use the rate expression in finite blocklength regime to approximate $R^u_{i,s}$ as

$$R^u_{i,s} = \sum_{\chi} x^u_{i\chi,s} W^u_{i,\chi}\left(\log_2\left(1 + \frac{p^u_{i,\chi}|h^u_{i,\chi}|^2}{N_0 W^u_{i,\chi}}\right) - \sqrt{\frac{C_{i,\chi}}{n_i}} Q^{-1}(\varepsilon^u_s)\log_2 e\right),$$
$$\forall i \in I^u_s, \quad s \in S^u, \tag{11.11}$$

where $W^u_{i,\chi}$ denotes the bandwidth allocated to user i, $p^u_{i,\chi}$ denotes the transmit power of the user i and AP χ link, $C_{i,\chi} = 1 - \frac{1}{\left(1+\frac{p^u_{i,\chi}|h^u_{i,\chi}|^2}{N_0 W^u_{i,\chi}}\right)^2}$ is the *channel dispersion*, n_i is the length of codeword block, and $Q^{-1}(\cdot)$ is the inverse of the Gaussian Q-function.

For mMTC user i, the received SNR at AP χ is

$$\gamma^m_{i,s} = \sum\nolimits_{\chi} x^m_{i\chi,s} \frac{p^m_{i,\chi}|h^m_{i,\chi}|^2}{N_0 W^m_{i,\chi}}, \quad \forall i \in I^m_s, \quad s \in S^m, \tag{11.12}$$

where $W^m_{i,\chi}$ denotes the bandwidth allocated to user i, and $p^m_{i,\chi}$ denotes the transmit power of the user i and AP χ link.

By setting the received SNR threshold as γ^m_s, we have

$$\gamma^m_{i,s} \geq \gamma^m_s, \quad \forall i \in I^m_s, \quad s \in S^m. \tag{11.13}$$

The data rate of mMTC user i with respect to slice s is

$$R^m_{i,s} = \sum_{\chi} x^m_{i\chi,s} W^m_{i,\chi} \log_2\left(1 + \frac{p^m_{i,\chi}|h^m_{i,\chi}|^2}{N_0 W^m_{i,\chi}}\right). \tag{11.14}$$

Note that the metric of mMTC is the number of accessed users rather than high data rates. We thus set an upper bound on the data rate to avoid resource waste:

$$R^m_{i,s} \leq R^m_s, \quad \forall i \in I^m_s, \quad s \in S^m. \tag{11.15}$$

11.4 Joint Resource Allocation and VNF Deployment Optimization Problem Formulation

Our goal is to maximize the compute resource utilization efficiency subject to the heterogeneous requirements of users.

For eMBB, URLLC, and mMTC services, the compute resource utilization efficiency $E^{(\kappa)}$ can be expressed as

$$E^{(\kappa)} = \frac{\sum_{\chi,i} x^{\kappa}_{i\chi,s} R^{\kappa}_{i,s}}{\sum_{\chi,f} C^{\kappa}_{f} \alpha^{\kappa}_{f,\chi}}, \quad \kappa \in \{e,m,u\}. \tag{11.16}$$

For each time slot, we can formulate the problem as follows (the subscript t is omitted):

$$\begin{aligned} \max_{\boldsymbol{\alpha},\mathbf{W},\mathbf{P},\mathbf{z}} \quad & \sum_{\kappa\in\{e,m,u\}} E^{(\kappa)}, \\ \text{subject to} \quad & (11.1)-(11.4), (11.6), (11.7), (11.9), (11.10), (11.13), (11.15) \\ & \alpha^{\kappa}_{f,\chi} \in \{0,1\}, \end{aligned} \tag{11.17}$$

where $\boldsymbol{\alpha}$ denotes a tuple of VNF deployment variable $\alpha^{\kappa}_{f,\chi}$, $\mathbf{W}$ denotes a tuple of the spectrum $W^{\kappa}_{i,\chi}$, $\mathbf{P}$ denotes a tuple of transmit power $p^{\kappa}_{i,\chi}$, and $\mathbf{z}$ denotes a tuple of the UAVs' positions.

11.5 Iterative VNF Deployment and Resource Allocation Algorithm

In this section, to solve (11.17) efficiently, we divide it into three subproblems: VNF deployment, spectrum allocation, and power allocation. At each time slot, we solve them iteratively to obtain the suboptimal solution, and then we update the UAVs' positions at the next time slot for better system performance.

11.5.1 BCD-Based VNF Deployment Algorithm

Given the spectrum and power resource allocation, we propose a BCD-based method to solve the VNF deployment subproblem. Instead of directly optimizing the VNF deployment, we solve the subproblem with respect to user scheduling and then obtain VNF deployment according to (11.1). As shown in Algorithm 11.1, we first initialize $x^{\kappa}_{i\chi,s}$ randomly. In each iteration, variable $x^{\kappa}_{i\chi,s}$ will be optimized while keeping other $x^{\kappa}_{l\chi,s} (l \in \{1,2,\ldots,I^{\kappa}_{s}\}\backslash i)$ unchanged. We set λ as the solution to (11.17):

$$\lambda = \max \sum_{\kappa\in\{e,m,u\}} E^{(\kappa)}. \tag{11.18}$$

We then update λ, and the optimal VNF deployment is obtained as well. The results will reach convergence after a sufficient number of iterations.

Algorithm 11.1 BCD-based VNF deployment

Data: Current optimal user scheduling $\mathbf{x}$, current user scheduling $\mathbf{x}'$, current optimal function value λ, current function value λ', current optimal VNF deployment $\boldsymbol{\alpha}$.

1 **Initialization**: Each user i obtains current optimal solution $\mathbf{x}$ and the optimal function value λ;
2 **while** *stopping criterion is not met yet* **do**
3 　**for** *each user i* **do**
4 　　**for** *each AP* χ **do**
5 　　　Set $x_{i\chi,s} = 1$;
6 　　　Set other $x_{ia,s} = 0$, where a denotes APs except χ;
7 　　　**if** *current function value* $\lambda' \geq \lambda$ **then**
8 　　　　Set $\lambda = \lambda'$;
9 　　　　Let current solution $\mathbf{x}'$ be the optimal solution $\mathbf{x}$ and $\boldsymbol{\alpha}$ is obtained by (11.1);
10 　　　**else**
11 　　　　Recover $\mathbf{x}'$, set $\mathbf{x}' = \mathbf{x}$;
12 　　　**end**
13 　　**end**
14 　**end**
15 **end**

11.5.2 SCA-Based Spectrum Allocation Algorithm

In (11.17), we fix the VNF deployment and the allocation of power resource. Note that the objective function is still nonconvex, making it nontrivial to be solved directly. We apply the SCA method to approximate both the objective function and the nonconvex constraint (11.11) such that the spectrum allocation subproblem can be transformed into a tractable convex optimization one. Through a series of iterative steps as shown in Algorithm 11.2, we can obtain the solution to this subproblem.

Note that the nonconvexity of the objective function and constraints is mainly caused by the the data-rate expression of URLLC users. Thus, we just need to perform approximation of (11.11), which is given by

$$\tilde{R}_{i,s}^{u,(t)} = R_{i,s}^{u,(t)} + \left(\nabla R_{i,s}^{u,(t)}\right)^H (\mathbf{W} - \mathbf{W}^{(t)}) + \tau \|\mathbf{W} - \mathbf{W}^{(t)}\|^2, \tag{11.19}$$

where τ is a constant.

After approximating the nonconvex objective function and constraints via the Taylor expansion, the original problem can be transformed into a convex optimization one:

$$\begin{aligned} \max_{\mathbf{W}} \quad & E^{(e)} + \tilde{E}^{(u)} + E^{(m)}, \\ \text{subject to} \quad & (11.4), (11.9), (11.13), (11.15) \\ & \tilde{R}_{i,s}^{u,(t)} \geq \frac{b_i^u}{\tau_s^u}, \quad \forall i \in I_s^u, \quad s \in S^u, \end{aligned} \tag{11.20}$$

Algorithm 11.2 SCA-based spectrum allocation

Data: Stepsize γ, approximation constant τ, iteration index t.
1 **Initialization**: Each user i obtains last optimal $W^{e,(0)}_{i\chi,s}$, $W^{u,(0)}_{i\chi,s}$, and $W^{m,(0)}_{i\chi,s}$;
2 $\gamma = 0.1$, $\tau = 0.1$;
3 Set $t = 0$;
4 **while** *stopping criterion is not met yet* **do**
5 Calculate $R^{u,(t)}_{i,s}$ with $\mathbf{W}^{(t)}$;
6 Obtain the approximation $\tilde{R}^{u,(t)}_{i,s}$ with (11.19);
7 Set $\tilde{\mathbf{W}}^{(t)}$ to be the solution of the convex optimization problem (11.20);
8 Set $\mathbf{W}^{(t+1)} = \gamma\tilde{\mathbf{W}}^{(t)} + (1-\gamma)\mathbf{W}^{(t)}$;
9 Set $t = t+1$;
10 **end**

Algorithm 11.3 SCA-based power allocation

Data: Stepsize γ, approximation constant τ, iteration index t.
1 **Initialization**: Each user i obtains last optimal $P^{e,(0)}_{i\chi,s}$, $P^{u,(0)}_{i\chi,s}$, and $P^{m,(0)}_{i\chi,s}$;
2 $\gamma = 0.1$, $\tau = 0.1$.;
3 Set $t = 0$;
4 **while** *stopping criterion is not met yet* **do**
5 Calculate $R^{u,(t)}_{i,s}$ with $\mathbf{P}^{(t)}$;
6 Obtain the approximation $\tilde{R}^{u,(t)}_{i,s}$ with (11.21);
7 Set $\tilde{\mathbf{P}}^{(t)}$ to be the solution of the convex optimization problem (11.22);
8 Set $\mathbf{P}^{(t+1)} = \gamma\tilde{\mathbf{P}}^{(t)} + (1-\gamma)\mathbf{P}^{(t)}$;
9 Set $t = t+1$;
10 **end**

where $\tilde{E}^{(u)} = \frac{\sum_{\chi,i} x^u_{i\chi,s}\tilde{R}^u_{i,s}}{\sum_{\chi,f} C^u_f \alpha^u_{f,\chi}}$. Therefore, we can find the optimal spectrum resource allocation.

11.5.3 SCA-Based Algorithm to Obtain Optimal Power Allocation

Given the VNF deployment and the spectrum resource allocation, we now focus on the power-allocation subproblem, which is also nonconvex due to the existence of (11.11). Similarly, we use the SCA method to solve it.

The approximation of the data rate of URLLC services shown in (11.11) can be given by

$$\tilde{R}^{u,(t)}_{i,s} = R^{u,(t)}_{i,s} + \left(\nabla R^{u,(t)}_{i,s}\right)^H (\mathbf{P} - \mathbf{P}^{(t)}) + \tau\|\mathbf{P} - \mathbf{P}^{(t)}\|^2, \tag{11.21}$$

where τ is a constant.

The power-allocation subproblem can then be transformed into a convex one:

$$\begin{aligned} \max_{P} \quad & E^{(e)} + \tilde{E}^{(u)} + E^{(m)}, \\ \text{subject to} \quad & (11.3), (11.9), (11.13), (11.15), \\ & \tilde{R}_{i,s}^{u,(t)} \geq \frac{b_i^u}{\tau_s^u}, \quad \forall i \in I_s^u, \quad s \in S^u, \end{aligned} \tag{11.22}$$

where $\tilde{E}^{(u)} = \frac{\sum_{\chi,i} x_{i\chi,s}^u \tilde{R}_{i,s}^u}{\sum_{\chi,f} C_f^u \alpha_{f,\chi}^u}$. We can then find the optimal solution for the power-allocation subproblem.

11.5.4 Overall Algorithm Description

In this section, we describe the overall algorithm to solve (11.17), where the three subproblems are solved iteratively at each time slot. In the ith iteration at time slot t, we first obtain $\lambda(\boldsymbol{\alpha}^{(i)}, \mathbf{W}^{(i-1)}, \mathbf{P}^{(i-1)})$ by performing Algorithm 11.1 and then $\lambda(\boldsymbol{\alpha}^{(i)}, \mathbf{W}^{(i)}, \mathbf{P}^{(i-1)})$ and $\lambda(\boldsymbol{\alpha}^{(i)}, \mathbf{W}^{(i)}, \mathbf{P}^{(i)})$ via Algorithms 11.2 and 11.3 in sequence. When the gap between two adjacent iterations, $\lambda(\boldsymbol{\alpha}^{(i)}, \mathbf{W}^{(i)}, \mathbf{P}^{(i)})$ and $\lambda(\boldsymbol{\alpha}^{(i-1)}, \mathbf{W}^{(i-1)}, \mathbf{P}^{(i-1)})$, is smaller than ϵ, the whole algorithm converges to a suboptimal solution. Afterward, we calculate the optimal position of the UAVs for the next time slot $t+1$ by solving (11.17) given the VNF deployment and resource-allocation scheme. The whole process continues within the considered time period.

11.5.5 Convergence Analysis

PROPOSITION 11.1 *The proposed overall algorithm converges after a limited number of iterations.*

Proof For the VNF-deployment subproblem, because we iteratively maximize the objective function (11.17), the objective value does not decrease after each iteration. In other words, in the ith iteration, the efficiency $\lambda(\boldsymbol{\alpha}^{(i)}, \mathbf{W}^{(i-1)}, \mathbf{P}^{(i-1)})$ is no less than $\lambda(\boldsymbol{\alpha}^{(i-1)}, \mathbf{W}^{(i-1)}, \mathbf{P}^{(i-1)})$. Note that the convergence of the SCA algorithm can be guaranteed according to [295]. Thus, $\lambda(\boldsymbol{\alpha}^{(i)}, \mathbf{W}^{(i)}, \mathbf{P}^{(i-1)})$ is no less than $\lambda(\boldsymbol{\alpha}^{(i)}, \mathbf{W}^{(i-1)}, \mathbf{P}^{(i-1)})$. Similarly, $\lambda(\boldsymbol{\alpha}^{(i)}, \mathbf{W}^{(i)}, \mathbf{P}^{(i)})$ is no less than $\lambda(\boldsymbol{\alpha}^{(i)}, \mathbf{W}^{(i)}, \mathbf{P}^{(i-1)})$. Because the resources are limited, the compute resource utilization efficiency has an upper bound. It will stop increasing after a limited number of iterations, and then the algorithm converges. □

11.6 Performance Evaluation

In this section, we provide numerical results to evaluate our proposed scheme for the integrated terrestrial-aerial access network in terms of the compute resource utilization efficiency and sum rate.

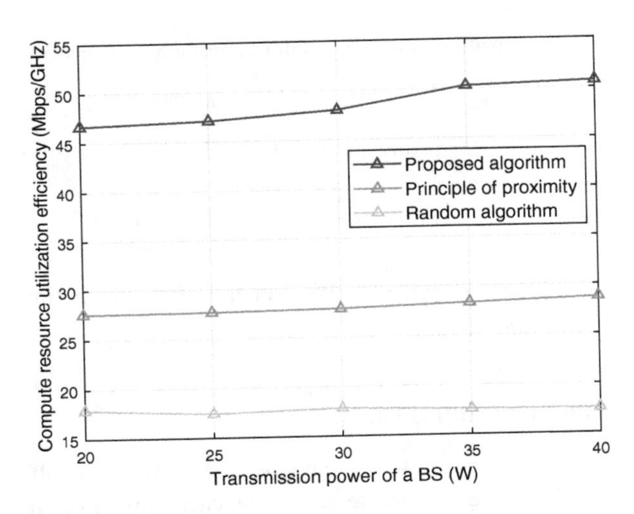

Figure 11.3 Compute resource utilization efficiency versus BS transmission power

For the simulations, all terrestrial nodes and users are uniformly distributed in a rectangular area of 2 km × 2 km. We follow the typical parameter settings in [234], and the heights of UAVs and HAPs are set as 50 m and 10 km, respectively. The power spectral density of AWGN N_0 is -173 dBm/Hz, and the maximum transmission power of each BS is 46 dBm. For eMBB users, the required minimum transmission rate is 10 Mbps. For URLLC users, we set the length of a codeword block as $n_i = 168$, the codeword decoding error probability as $\varepsilon = 10^{-7}$, the transmission latency requirement $\tau = 5$ ms, and the packet length $b = 5{,}000$. For mMTC users, the maximum transmission rate is 1 Mbps, and the minimum SNR is -2 dB. The bandwidths of the terrestrial and aerial networks are set as 45 MHz and 100 MHz, respectively.

To evaluate the algorithm proposed in Section 11.5, we compare it with two benchmarks: (1) the principle of proximity algorithm where each user associates with the AP that provides the best channel quality and (2) a random algorithm where each user randomly associates with one AP and the node deploys the VNFs accordingly. Power and spectrum allocation are performed via the SCA-based methods in the two benchmarks. The results are presented in Fig. 11.3, where the compute resource utilization efficiency increases with the transmit power of the terrestrial BS based on the proposed algorithm, which performs significantly better than the benchmarks This also reflects the superiority of our VNF deployment scheme in maximizing the compute efficiency.

Figure 11.4 shows the user-scheduling results in the integrated terrestrial-aerial access networks, which also reflect the corresponding VNF deployment. We observe that mMTC users prefer connecting with the HAPs rather than the BSs and UAVs. In this way, the advantage of the HAP's vast coverage can be fully exploited to serve a massive number of mMTC users. For the URLLC users, the percentage of user BS associations is the highest compared to the UAV and HAP counterparts. This is because when terrestrial BSs are accessible, the short propagation distance between

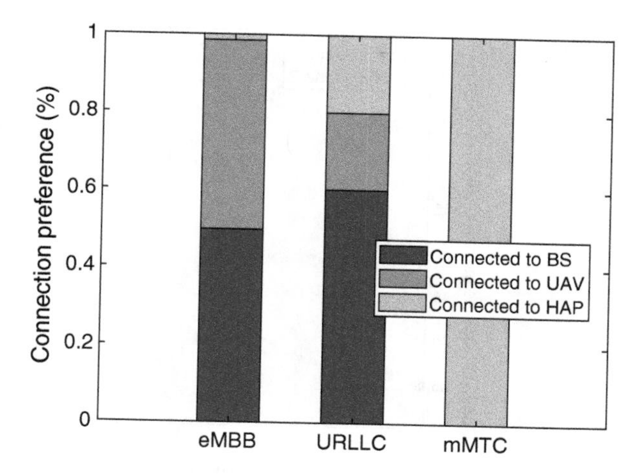

Figure 11.4 User scheduling results of different types of users

users and the BS can better guarantee the low latency requirements. Note that most eMBB users receive data from BSs and UAVs, which achieve a larger data rate than the HAP. In this case, it is recommended that the BSs and UAVs should be equipped with eMBB and URLLC related VNFs, whereas the mMTC related VNFs should be provided by the HAP.

Note that each VNF contains different functional modules. By selecting different modules to construct the VNFs, we have different VNF deployment schemes. For a finer-grained VNF construction where each VNF only contains a small number of modules, the total number of constructed VNFs to be deployed is larger than the coarse VNF construction where each VNF contains many modules. To evaluate how such VNF construction influences the system performance, we consider two extreme cases. In case 1, all modules of each VNF are requested by users (i.e., these are general or basic function modules). In case 2, only one module of each VNF is requested by users (i.e., it is an exclusive or personalized module).

For these cases, Fig. 11.5 shows the compute resource utilization efficiency vsersus the number of constructed VNFs. We observe that for the basic modules (case 1) a smaller number of constructed VNFs improves the efficiency, implying a coarse VNF construction where basic modules are collected into several VNFs is preferable. Because VNF configuration requires extra compute resources, collecting these basic modules into fewer VNFs saves those additional compute resources for configuring VNFs, thereby improving compute resource utilization efficiency. In case 2, the efficiency metric grows with the number of constructed VNFs, showing that a fine-grained VNF construction is desired when we construct VNFs based on exclusive modules. It is not wise to package them into one VNF because these modules are dedicated and users may not need all of them, which leads to a waste of compute resources. These two remarks provide useful guidelines for VNF construction, depending on whether the modules are generalized or personalized.

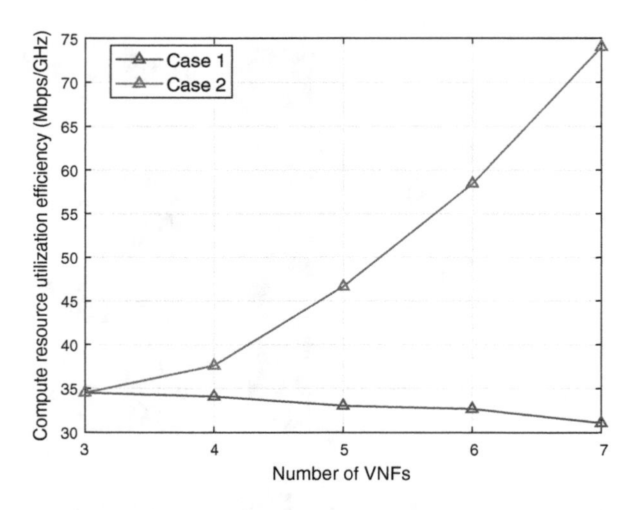

Figure 11.5 Compute resource utilization efficiency versus the number of constructed VNFs

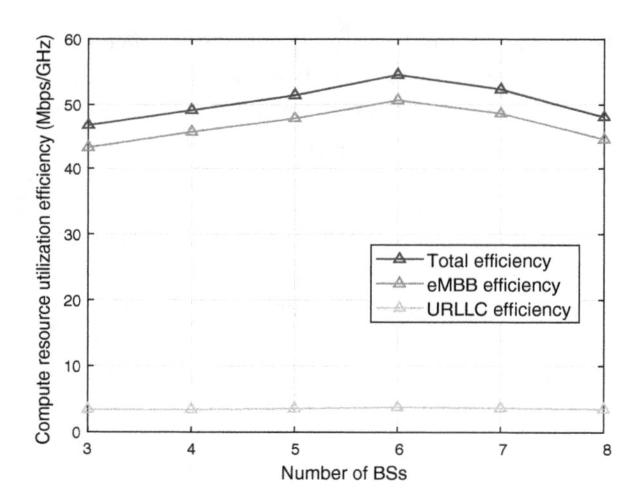

Figure 11.6 Compute resource utilization efficiency versus the number of BSs

Figure 11.6 shows the influence of the number of BSs on compute resource utilization efficiency. The efficiency first increases with the number of BSs because the transmission power grows, resulting in a higher sum rate. It then decreases with the number of BSs because extra BSs require higher compute resource consumption for VNF deployment, which dominates the efficiency metric. This indicates that there exists an optimal number of BSs for the considered area of interest to maximize the compute resource utilization efficiency. Figure 11.6 also shows that the performance of eMBB users is most sensitive to the number of BSs because eMBB users have stringent requirements on data rates.

11.7 Conclusion

In this chapter, we considered an integrated terrestrial-aerial access network enabled by NS. To achieve the trade-off between the sum rate and VNF compute resource consumption, we formulated a compute resource utilization efficiency maximization problem where VNF deployment, spectrum, and power allocation are jointly optimized. We constructed the mapping between VNF deployment and user scheduling and then decoupled the problem into three subproblems, which are solved iteratively using BCD and SCA methods. The effectiveness of the proposed algorithm is verified by simulation results. It was also proved that there exists an optimal number of BSs to maximize the compute resource utilization efficiency.

Notes

Chapter 3. UAVs Serving as Flying Infrastructures

1 The length of each time slot can be a few seconds because our algorithm has a high efficiency.

2 The power consumption of wireless transmission can be ignored compared to that of the engines of the UAV [59].

3 Small-scale fading is ignored because we only use average SNR to determine which UAV or MBS the user should connect to.

4 We assume that each potential user has an independent probability of becoming an active user at each moment. Therefore, the number of active users in a certain area should be a binomial distributed random variable. When the number of users in each area is large enough (>100), a Poisson distribution is a good approximation of binomial distribution without a loss of accuracy.

5 In addition, by deleting unnecessary values in Table 3.2, we can resolve a whole contract with $M = 300$ in only one second.

6 Based on Fig. 3.3, although *Social Welfare* = *Revenue of the MBS Manager* + *Total Profits of all the UAVs*, we can also express it as *Social Welfare* = *Total Utilities of all the UAVs* − *Cost of the MBS Manager*, as given in (3.32).

7 The same method can also be applied to the decode-and-forward (DF) relay. For more details, please refer to [67].

8 With total power constraint, the optimal power solution that minimizes the outage probability can be obtained with various MD-UAV and UAV-BS distances ratios. It also guarantees that the maximum power efficiency can be reached with different transmission distance scenarios [68].

Chapter 4. UAVs Serving as Flying Users

1 We assume that the collected sensory data of each UAV in a cycle can be converted into a single sensory data frame with the same length.

2 For example, the UAV's transmission will interfere with its sensing if the UAV tries to sense electromagnetic signals in the frequency bands that are adjacent to its transmission frequency.

3 This mode corresponds to the direct mode in the centralized sense-and-send protocol in Section 4.1.2.

4 This mode corresponds to the relay mode in the centralized sense-and-send protocol in Section 4.1.2.

5 We consider the UAVs rotary-wing UAVs, which can hover in the air for some time slots. The rotary-wing UAVs can move with the speed of $[0, v_{\max}]$ in any time slot.

6 The value of γ_{th} is set according to the QoS in the specific network.

7 The uplink sum rate is the sum of the U2N and CU transmission rate, and U2U transmission is not included in the uplink sum rate. Therefore, the objective function does not contain the U2U

rate. Instead, we set a minimum threshold for each U2U link to guarantee the success of the U2U transmissions.

8 The dataset can be found at *https://github.com/YyzHarry/AQI_Dataset*.

Chapter 8. Ultra-dense LEO Satellite Networks

1 In general, LEO satellites are deployed in multiple circular orbital planes (at altitudes ranging from 200 km to 2,000 km). The number of orbital planes and the number of satellites on each orbital plane are designed specifically to meet the communication coverage requirements (such as global coverage).

2 Without loss of generality, we set $\varphi_0 = 0$.

3 Note that the distance between the satellite and the terrestrial terminal station increases with the visual range and the satellite formation altitude. Considering the maximum communication distance of each satellite and terrestrial terminal station, for simplicity, we assume that τ first remains constant and then decreases linearly with respect to the visual range and the satellite formation altitude.

4 In the special single-antenna satellite case, $MK_t > NLK_r$ only holds when M is larger than causing a 800, which is difficult to implement in practice.

5 This is different from reconfigurable intelligent surfaces (RISs) widely used as passive relays owing to their reflection characteristic. Furthermore, the feeds of an RHS are embedded in the metasurface, whereas those of an RIS are on the outside of the metasurface [211].

6 The efficiency of an RHS element η is determined by its size and physical structure. For a RHS element with a given size and physical structure, its efficiency η is a constant [211, 214].

7 When the number of RHS elements does not satisfy $\eta \cdot N_x N_y \leq \frac{8}{3}$, all the power of the reference wave will be radiated into free space when the reference wave has not propagated to the edge of the RHS, such that the elements at the edge of the RHS are redundant and ineffective. In this book, we only focus on the effective RHS elements satisfying $\eta \cdot N_x N_y \leq \frac{8}{3}$.

8 In general, the material for fabricating RHSs experiences little loss. Thus, the power loss during the wave propagation can be neglected, and the value of α is usually small [210].

9 The initial position of each LEO satellite (i.e., each satellite's altitude H, zenith angle $\{\theta_q^{(1)}, \theta_q^{(2)}\}$, and azimuth angle $\{\varphi_q^{(1)}, \varphi_q^{(2)}\}$) can be obtained by conventional received-signal-strength-based techniques.

10 Because the satellite moves in a uniform circular motion, given the position of the satellite in two adjacent time slots, the angular velocity of the satellite can be obtained, based on which the position of the satellite in the next time slot can be derived; that is, the satellite's position at time slot t can be derived from its positions at time slots $(t-1)$ and $(t-2)$.

11 Without loss of generality, we assume that each satellite receives one data stream from the UT.

12 The number of RF chains in the hybrid beamforming architecture must be no smaller than the number of data streams[223]. In other words, the number of the feeds should also be no smaller than the number of active data streams (i.e., $L \geq Q$).

13 It is common to only consider large-scale fading in LEO satellite communication networks [189, 224]. The main reason is that the distance between the UT and the LEO satellite is large, and the large-scale fading has a major impact on the change in signal strength, whereas the influence of the small-scale fading can be neglected.

14 Because the distance between the satellite and the UT is much larger than the physical dimensions of both the RHS and the receive antennas, $|\mathbf{d}_{n_x,n_y}^{q,k}|$ can be approximated by the distance between the satellite and the origin, which is denoted as $|\mathbf{d}_q|$.

15 Because the holographic beamforming scheme is developed on the premise that the satellites' positions are known (i.e., the directions of transmission are fixed), the effective aperture of the element A_q is a fixed constant.

16 For convenience, we consider the uniform linear phased array at each satellite where the element spacing is d_s. We denote the angle of arrival of the signals with respect to the antenna array at satellite q as ϕ_q, and thus, $\Psi_{q,k} = (k-1)d_s \sin\phi_q$.

Chapter 9. Ultra-dense LEO Satellite Constellation Design

1 The backhaul requirement means that the backhaul capacity of each UT should be larger than a predefined threshold (such as 100 Mbps).

2 The coverage ratio requirement means that the satellite constellation's coverage ratio on the Earth averaged over time should be larger than a predefined threshold (such as 90%).

3 Because the value of Δ_T is artificially set, we can choose a proper value for Δ_T to guarantee that T is an integer.

4 It is common to consider large-scale fading in LEO satellite networks [251, 252]. On the one hand, the distance between the UT and LEO satellite is far enough that the influence of the motion of LEO satellites on the distance can be neglected for short periods. On the other hand, because the channel state information outdates frequently due to the long propagation delay of the UT-satellite link, the large-scale fading has a major impact on the change in signal strength, whereas the influence of the small-scale fading can be neglected. In addition, we consider a LEO satellite constellation design that satisfies the backhaul requirement of the UT and provided seamless global coverage over a long period, which has low real-time requirements. The influence of small-scale fading is averaged out, and thus, the small-scale fading is not taken into account.

5 The polar-orbit constellation is one of the most representative satellite constellations. Iridium [258] and OneWeb [1] both adopt a polar-orbit constellation for LEO satellite deployment to provide global coverage. The advantage of the polar-orbit constellation is that it can strengthen the effective coverage of polar regions greatly. It can also can save more satellites and outperform the non-polar-orbit constellation in terms of global coverage.

6 The proposed 3D constellation optimization algorithm can be extended to a multilayer satellite network because the proposed three criteria for LEO satellite constellation design have no limits on the flying altitudes of the satellites.

7 In Fig. 9.8, we set $\theta_{\min} = 16^\circ$.

8 Note that according to (9.28), each combination of h and $\theta_{\min}$ corresponds to multiple combinations of N_0 and M. As shown in Fig. 9.8, the number of deployed LEO satellites in the optimized constellation has a close relation to the initial LEO satellite constellation. For the sake of fairness, we choose the initial LEO satellite constellation with the minimum number of satellites for each combination of h and $\theta_{\min}$.

9 In general, the satellites' maximum radial offset is about 5 km [262, 263].

10 The Telesat constellation consists of no fewer than 117 LEO satellites distributed on two sets of orbital planes. The first set consists of 6 polar orbits with at least 12 satellites on each orbital plane. The orbital inclination is 99.5° and the altitude is 1,000 km. The second set consists of no less than 5 inclined orbits with a minimum of 10 satellites on each orbital plane. The orbital inclination is 37.4° and the altitude is 1,200 km [264], which can better cover the low-latitude area on the Earth. To increase the fairness of the comparison, we consider a modified Telesat constellation where the inclination of the inclined orbits is 45° in order to achieve more even coverage of the Earth.

11 Assume that the UTs are equipped with the GNSS receivers and are able to estimate the position of the satellite, and thus, the Doppler shift of LEO satellites can be significantly compensated [265, 266].

12 The network topology is viewed as static in each time slot with a 1 second duration.

Chapter 10. Integrating Terrestrial-Satellite Networks into 5G and Beyond

1 The proposed architecture is independent of the satellite altitude and constellation and thus can be directly extended to the MEO and GEO satellite systems. The data-routing and signaling-control problems need to be considered for the multilayer system, which is not the focus of this chapter. Readers may find an initial work focusing on contact-graph-based routing in [272].

2 To save signaling costs, we adopt the semipersistent scheduling where the LEO backhaul links are updated every few time slots.

3 The channel state information (CSI) here can be obtained by adopting the widely utilized training data based CSI estimation techniques [274, 276]. For reference, the probability density function (PDF) of the channel coefficient is given as $f_{\left|h^T_{m,n,q}\right|^2}(x) = \alpha e^{-\beta x} {}_1F_1\left(m'; 1; \delta x\right)$, where the parameters α, β, m', and δ can be found in [277].

Bibliography

[1] "OneWeb non-geostationary satellite system (Attachment A)," Tech. Rep., Federal Communications Commissions, 2016.

[2] "Spacex non-geostationary satellite system," Tech. Rep., Federal Communications Commissions, 2016.

[3] O. Kodheli, E. Lagunas, N. Maturo, S. K. Sharma, B. Shankar, J. F. M. Montoya, J. C. M. Duncan, D. Spano, S. Chatzinotas, S. Kisseleff, et al., "Satellite communications in the new space era: A survey and future challenges," *IEEE Communications Surveys & Tutorials*, vol. 23, no. 1, pp. 70–109, 2021.

[4] M. Centenaro, C. E. Costa, F. Granelli, C. Sacchi, and L. Vangelista, "A survey on technologies, standards and open challenges in satellite IoT," *IEEE Communications Surveys & Tutorials*, vol. 23, no. 3, pp. 1693–1720, 2021.

[5] G. K. Kurt, M. G. Khoshkholgh, S. Alfattani, A. Ibrahim, T. S. Darwish, M. S. Alam, H. Yanikomeroglu, and A. Yongacoglu, "A vision and framework for the high altitude platform station (HAPs) networks of the future," *IEEE Communications Surveys & Tutorials*, vol. 23, no. 2, pp. 729–779, 2021.

[6] X. Cao, P. Yang, M. Alzenad, X. Xi, D. Wu, and H. Yanikomeroglu, "Airborne communication networks: A survey," *IEEE Journal on Selected Areas in Communications*, vol. 36, no. 9, pp. 1907–1926, 2018.

[7] A. Fotouhi, H. Qiang, M. Ding, M. Hassan, L. G. Giordano, A. Garcia-Rodriguez, and J. Yuan, "Survey on UAV cellular communications: Practical aspects, standardization advancements, regulation, and security challenges," *IEEE Communications Surveys & Tutorials*, vol. 21, no. 4, pp. 3417–3442, 2019.

[8] X. Zhu and C. Jiang, "Integrated satellite-terrestrial networks toward 6G: Architectures, applications, and challenges," *IEEE Internet of Things Journal*, vol. 9, no. 1, pp. 437–461, 2022.

[9] T. Wei, W. Feng, Y. Chen, C.-X. Wang, N. Ge, and J. Lu, "Hybrid satellite-terrestrial communication networks for the maritime internet of things: Key technologies, opportunities, and challenges," *IEEE Internet of Things Journal*, vol. 8, no. 11, pp. 8910–8934, 2021.

[10] J. Liu, Y. Shi, Z. M. Fadlullah, and N. Kato, "Space-air-ground integrated network: A survey," *IEEE Communications Surveys & Tutorials*, vol. 20, no. 4, pp. 2714–2741, 2018.

[11] H. Cui, J. Zhang, Y. Geng, Z. Xiao, T. Sun, N. Zhang, J. Liu, Q. Wu, and X. Cao, "Space-air-ground integrated network (SAGIN) for 6G: Requirements, architecture and challenges," *China Communications*, vol. 19, no. 2, pp. 90–108, 2022.

[12] "More than 50 billion connected devices," white paper, Ericsson, Feb. 2011.

[13] R. Deng, B. Di, S. Chen, S. Sun, and L. Song, "Ultra-dense LEO satellite offloading for terrestrial networks: How much to pay the satellite operator?" *IEEE Transactions on Wireless Communications*, vol. 19, no. 10, pp. 6240–6254, 2020.

[14] Q. Di, Y. Wang, A. Zanobetti, Y. Wang, P. Koutrakis, C. Choirat, F. Dominici, and J. D. Schwartz, "Air pollution and mortality in the Medicare population," *New England Journal of Medicine*, vol. 376, no. 26, pp. 2513–2522, 2017.

[15] M. Casoni, C. A. Grazia, M. Klapez, N. Patriciello, A. Amditis, and E. Sdongos, "Integration of satellite and LTE for disaster recovery," *IEEE Communications Magazine*, vol. 53, no. 3, pp. 47–53, 2015.

[16] S. Chandrasekharan, K. Gomez, A. Al-Hourani, S. Kandeepan, T. Rasheed, L. Goratti, L. Reynaud, D. Grace, I. Bucaille, T. Wirth, et al., "Designing and implementing future aerial communication networks," *IEEE Communications Magazine*, vol. 54, no. 5, pp. 26–34, 2016.

[17] C. A. Wargo, G. C. Church, J. Glaneueski, and M. Strout, "Unmanned aircraft systems (UAS) research and future analysis," in *Proc. IEEE Aerospace Conference*, 2014, pp. 1–16.

[18] L. Gupta, R. Jain, and G. Vaszkun, "Survey of important issues in UAV communication networks," *IEEE Communications Surveys & Tutorials*, vol. 18, no. 2, pp. 1123–1152, 2016.

[19] J. Wang, C. Jiang, Z. Han, Y. Ren, R. G. Maunder, and L. Hanzo, "Taking drones to the next level: Cooperative distributed unmanned-aerial-vehicular networks for small and mini drones," *IEEE Vehicular Technology MagazIne*, vol. 12, no. 3, pp. 73–82, 2017.

[20] N. H. Motlagh, T. Taleb, and O. Arouk, "Low-altitude unmanned aerial vehicles-based internet of things services: Comprehensive survey and future perspectives," *IEEE Internet of Things Journal*, vol. 3, no. 6, pp. 899–922, 2016.

[21] Y. Yang, Z. Zheng, K. Bian, L. Song, and Z. Han, "Real-time profiling of fine-grained air quality index distribution using UAV sensing," *IEEE Internet of Things Journal*, vol. 5, no. 1, pp. 186–198, 2018.

[22] T. Kersnovski, F. Gonzalez, and K. Morton, "A UAV system for autonomous target detection and gas sensing," in *2017 IEEE Aerospace Conference*, 2017, pp. 1–12.

[23] B. H. Y. Alsalam, K. Morton, D. Campbell, and F. Gonzalez, "Autonomous UAV with vision based on-board decision making for remote sensing and precision agriculture," in *2017 Aerospace Conference*, 2017, pp. 1–12.

[24] T. Zhao, D. Doll, D. Wang, and Y. Chen, "A new framework for UAV-based remote sensing data processing and its application in almond water stress quantification," in *2017 International Conference on Unmanned Aircraft Systems (ICUAS)*, 2017, pp. 1794–1799.

[25] S. Zhang, H. Zhang, Q. He, K. Bian, and L. Song, "Joint trajectory and power optimization for UAV relay networks," *IEEE Communications Letters*, vol. 22, no. 1, pp. 161–164, 2018.

[26] H. Zhang, L. Song, Z. Han, and H. V. Poor, "Cooperation techniques for a cellular internet of unmanned aerial vehicles," *IEEE Wireless Communications*, vol. 26, no. 5, pp. 167–173, 2019.

[27] O. B. Sezer, E. Dogdu, and A. M. Ozbayoglu, "Context-aware computing, learning, and big data in internet of things: A survey," *IEEE Internet of Things Journal*, vol. 5, no. 1, pp. 1–27, 2017.

[28] S. Zhang, H. Zhang, B. Di, and L. Song, "Cellular UAV-to-X communications: Design and optimization for multi-UAV networks," *IEEE Transactions on Wireless Communications*, vol. 18, no. 2, pp. 1346–1359, 2019.

[29] T. Tozer and D. Grace, "High-altitude platforms for wireless communications," *Electronics & Communication Engineering Journal*, vol. 13, no. 3, pp. 127–137, 2001.

[30] D. Grace and M. Mohorcic, *Broadband communications via high altitude platforms*, John Wiley & Sons, 2011.

[31] A. Mohammed, A. Mehmood, F.-N. Pavlidou, and M. Mohorcic, "The role of high-altitude platforms (HAPs) in the global wireless connectivity," *Proc. of the IEEE*, vol. 99, no. 11, pp. 1939–1953, 2011.

[32] "Recommendation ITU-RF.1500, preferred characteristics of systems in the FS using high altitude platforms operating in the bands 47.2–47.5 GHz and 47.9–48.2 GHz," International Telecommunications Union, 2000.

[33] G. Avdikos, G. Papadakis, and N. Dimitriou, "Overview of the application of high altitude platform (HAP) systems in future telecommunication networks," in *2008 10th International Workshop on Signal Processing for Space Communications*, 2008, pp. 1–6.

[34] Google, "Google loon project," [Online], https://x.company/loon/.

[35] Capanina, "Capanina project, 2004–2007," [Online], www.capanina.org.

[36] M. Alzenad, M. Z. Shakir, H. Yanikomeroglu, and M.-S. Alouini, "FSO-based vertical backhaul/fronthaul framework for 5G+ wireless networks," *IEEE Communications Magazine*, vol. 56, no. 1, pp. 218–224, 2018.

[37] "SpaceX non-geostationary satellite system–Attachment A," SpaceX, 2017.

[38] R. Barnett, "OneWeb non-geostationary satellite system: Attachment ałtechnical information to supplement schedule S," Tech. Rep. SAT-MOD-20180319-00022, Federal Communications Commission, 2018.

[39] Y. Henri, "The OneWeb satellite system," *Handbook of small satellites: Technology, design, manufacture, applications, economics and regulation*, Springer, pp. 1–10, 2020.

[40] M. Á. Vázquez, L. Blanco, X. Artiga, and A. Pérez-Neira, "Hybrid analog-digital transmit beamforming for spectrum sharing satellite-terrestrial systems," in *2016 IEEE 17th International Workshop on Signal Processing Advances in Wireless Communications (SPAWC)*, 2016, pp. 1–5.

[41] X. Artiga, J. Nunez-Martinez, A. Perez-Neira, G. J. L. Vela, J. M. F. Garcia, and G. Ziaragkas, "Terrestrial-satellite integration in dynamic 5G backhaul networks," in *2016 8th Advanced Satellite Multimedia Systems Conference and 14th Signal Processing for Space Communications Workshop (ASMS/SPSC)*, 2016, pp. 1–6.

[42] E. Lagunas, S. Chatzinotas, and B. Ottersten, "Fair carrier allocation for 5G integrated satellite-terrestrial backhaul networks," in *2018 25th International Conference on Telecommunications (ICT)*, 2018, pp. 617–622.

[43] Y. Wang, J. Zhang, X. Zhang, P. Wang, and L. Liu, "A computation offloading strategy in satellite terrestrial networks with double edge computing," in *2018 IEEE International Conference on Communication Systems (ICCS)*, 2018, pp. 450–455.

[44] B. Di, L. Song, Y. Li, and H. V. Poor, "Ultra-dense LEO: Integration of satellite access networks into 5G and beyond," *IEEE Wireless Communications*, vol. 26, no. 2, pp. 62–69, 2019.

[45] X. Wang, M. Chen, T. Taleb, A. Ksentini, and V. C. Leung, "Cache in the air: Exploiting content caching and delivery techniques for 5G systems," *IEEE Communications Magazine*, vol. 52, no. 2, pp. 131–139, 2014.

[46] R. Amorim, H. Nguyen, P. Mogensen, I. Z. Kovács, J. Wigard, and T. B. Sørensen, "Radio channel modeling for UAV communication over cellular networks," *IEEE Wireless Communications Letters*, vol. 6, no. 4, pp. 514–517, 2017.

[47] D. W. Matolak and R. Sun, "Unmanned aircraft systems: Air-ground channel characterization for future applications," *IEEE Vehicular Technology Magazine*, vol. 10, no. 2, pp. 79–85, 2015.

[48] Y. Zeng, R. Zhang, and T. J. Lim, "Wireless communications with unmanned aerial vehicles: Opportunities and challenges," *IEEE Communications Magazine*, vol. 54, no. 5, pp. 36–42, 2016.

[49] A. Al-Hourani, S. Kandeepan, and S. Lardner, "Optimal LAP altitude for maximum coverage," *IEEE Wireless Communications Letters*, vol. 3, no. 6, pp. 569–572, 2014.

[50] M. Mozaffari, W. Saad, M. Bennis, and M. Debbah, "Drone small cells in the clouds: Design, deployment and performance analysis," in *Proc. IEEE Global Communications Conference*, 2015, pp. 1–6.

[51] S. Rohde and C. Wietfeld, "Interference aware positioning of aerial relays for cell overload and outage compensation," in *Proc. IEEE Vehicular Technology Conference*, 2012, pp. 1–5.

[52] A. Merwaday and I. Guvenc, "UAV assisted heterogeneous networks for public safety communications," in *Proc. IEEE Wireless Communications and Networking Conference Workshops*, 2015, pp. 329–334.

[53] V. Sharma, M. Bennis, and R. Kumar, "UAV-assisted heterogeneous networks for capacity enhancement," *IEEE Communications Letters*, vol. 20, no. 6, pp. 1207–1210, 2016.

[54] M. Mozaffari, W. Saad, M. Bennis, and M. Debbah, "Optimal transport theory for cell association in UAV-enabled cellular networks," *IEEE Communications Letters*, vol. 21, no. 9, pp. 2053–2056, 2017.

[55] S. Jangsher and V. O. Li, "Resource allocation in moving small cell network," *IEEE Transactions on Wireless Communications*, vol. 15, no. 7, pp. 4559–4570, 2016.

[56] A. Valcarce, T. Rasheed, K. Gomez, S. Kandeepan, L. Reynaud, R. Hermenier, A. Munari, M. Mohorcic, M. Smolnikar, and I. Bucaille, "Airborne base stations for emergency and temporary events," *Personal Satellite Services*, Springer, pp. 49–58, 2013.

[57] P. Bolton and M. Dewatripont, *Contract theory*, MIT Press, 2005.

[58] Z. Hu, Z. Zheng, L. Song, T. Wang, and X. Li, "UAV offloading: Spectrum trading contract design for UAV-assisted cellular networks," *IEEE Transactions on Wireless Communications*, vol. 17, no. 9, pp. 6093–6107, 2018.

[59] E. W. Frew and T. X. Brown, "Airborne communication networks for small unmanned aircraft systems," *Proc. of the IEEE*, vol. 96, no. 12, pp. 2008–2027, 2008.

[60] L. Gao, X. Wang, Y. Xu, and Q. Zhang, "Spectrum trading in cognitive radio networks: A contract-theoretic modeling approach," *IEEE Journal on Selected Areas in Communications*, vol. 29, no. 4, pp. 843–855, 2011.

[61] P. Toth and S. Martello, *Knapsack problems: Algorithms and computer implementations*, Wiley, 1990.

[62] M. Alzenad, A. El-Keyi, F. Lagum, and H. Yanikomeroglu, "3-D placement of an unmanned aerial vehicle base station (UAV-BS) for energy-efficient maximal coverage," *IEEE Wireless Communications Letters*, vol. 6, no. 4, pp. 434–437, 2017.

[63] J. Lyu, Y. Zeng, R. Zhang, and T. J. Lim, "Placement optimization of UAV-mounted mobile base stations," *IEEE Communications Letters*, vol. 21, no. 3, pp. 604–607, 2016.

[64] M. Erdelj, E. Natalizio, K. R. Chowdhury, and I. F. Akyildiz, "Help from the sky: Leveraging UAVs for disaster management," *IEEE Pervasive Computing*, vol. 16, no. 1, pp. 24–32, 2017.

[65] D. H. Choi, S. H. Kim, and D. K. Sung, "Energy-efficient maneuvering and communication of a single UAV-based relay," *IEEE Transactions on Aerospace and Electronic Systems*, vol. 50, no. 3, pp. 2320–2327, 2014.

[66] Y. Zeng, R. Zhang, and T. J. Lim, "Throughput maximization for UAV-enabled mobile relaying systems," *IEEE Transactions on Communications*, vol. 64, no. 12, pp. 4983–4996, 2016.

[67] S. Zeng, H. Zhang, K. Bian, and L. Song, "UAV relaying: Power allocation and trajectory optimization using decode-and-forward protocol," in *Proc. IEEE International Conference on Communications Workshops*, 2018, pp. 1–6.

[68] S. Salari, M. Z. Amirani, I.-M. Kim, D. I. Kim, and J. Yang, "Distributed beamforming in two-way relay networks with interference and imperfect CSI," *IEEE Transactions on Wireless Communications*, vol. 15, no. 6, pp. 4455–4469, 2016.

[69] S. Zhang, B. Di, L. Song, and Y. Li, "Sub-channel and power allocation for non-orthogonal multiple access relay networks with amplify-and-forward protocol," *IEEE Transactions on Wireless Communications*, vol. 16, no. 4, pp. 2249–2261, 2017.

[70] I. S. Gradshteyn and I. M. Ryzhik, *Table of integrals, series, and products*, Academic Press, 2014.

[71] H. Zhang, Y. Liao, and L. Song, "D2D-U: Device-to-device communications in unlicensed bands for 5G system," *IEEE Transactions on Wireless Communications*, vol. 16, no. 6, pp. 3507–3519, 2017.

[72] S. Hayat, E. Yanmaz, and R. Muzaffar, "Survey on unmanned aerial vehicle networks for civil applications: A communications viewpoint," *IEEE Communications Surveys & Tutorials*, vol. 18, no. 4, pp. 2624–2661, 2016.

[73] Z. Zheng, A. K. Sangaiah, and T. Wang, "Adaptive communication protocols in flying ad hoc network," *IEEE Communications Magazine*, vol. 56, no. 1, pp. 136–142, 2018.

[74] "Enhanced LTE support for aerial vehicles (release 15)," Tech. Rep. 36.777 V0.4.0, 3GPP, 2017.

[75] V. V. Shakhov and I. Koo, "Experiment design for parameter estimation in probabilistic sensing models," *IEEE Sensors Journal*, vol. 17, no. 24, pp. 8431–8437, 2017.

[76] G. Demange, D. Gale, and M. Sotomayor, "Multi-item auctions," *Journal of Political Economy*, vol. 94, no. 4, pp. 863–872, 1986.

[77] J. Hu, H. Zhang, and L. Song, "Reinforcement learning for decentralized trajectory design in cellular UAV networks with sense-and-send protocol," *IEEE Internet of Things Journal*, vol. 6, no. 4, pp. 6177–6189, 2019.

[78] "Study on channel model for frequencies from 0.5 to 100 GHz," Tech. Rep. 38.901, version 14.0.0, 3GPP, 2017.

[79] S. O. Rice, "Mathematical analysis of random noise," *Bell System Technical Journal*, vol. 23, no. 3, pp. 282–332, 1944.

[80] J. Marcum, "Table of q functions," Tech. Report RM-339-PR, Rand Corporation, 1950.

[81] R. M. Corless, G. H. Gonnet, D. E. Hare, D. J. Jeffrey, and D. E. Knuth, "On the lambert *w* function," *Advances in Computational Mathematics*, vol. 5, no. 1, pp. 329–359, 1996.

[82] R. S. Sutton and A. G. Barto, *Reinforcement learning: An introduction*, MIT Press, 2018.

[83] D. Gu and E. Yang, "Multiagent reinforcement learning for multi-robot systems: A survey," Tech. Rep., Department of Computer Science, University of Essex, 2004.

[84] C. J. C. H. Watkins, "Learning from delayed rewards," doctoral thesis, University of Cambridge, 1989.

[85] M. Bowling, "Multiagent learning in the presence of agents with limitations," Tech. Rep., School of Computer Science, Carnegie Mellon University, 2003.

[86] W. Uther and M. Veloso, "Adversarial reinforcement learning," Tech. Rep., Carnegie Mellon University, 1997.

[87] C. Claus and C. Boutilier, "The dynamics of reinforcement learning in cooperative multiagent systems," in *Proc. Conference on Artificial intelligence/Innovative Applications of Artificial Intelligence*, 1998, pp. 746–752.

[88] T. Jaakkola, M. I. Jordan, and S. P. Singh, "Convergence of stochastic iterative dynamic programming algorithms," *Neural Computation*, vol. 6, no. 6, pp. 703–710, 1994.

[89] S. P. Singh, M. J. Kearns, and Y. Mansour, "Nash convergence of gradient dynamics in general-sum games," in *Proc. Conference on Uncertainty in Artificial Intelligence*, 2000, pp. 541–548.

[90] S. Zhang, H. Zhang, and L. Song, "Beyond D2D: Full dimension UAV-to-everything communications in 6G," *IEEE Transactions on Vehicular Technology*, to be published.

[91] F. Tang, Z. M. Fadlullah, N. Kato, F. Ono, and R. Miura, "Ac-poca: Anticoordination game based partially overlapping channels assignment in combined UAV and D2D-based networks," *IEEE Transactions on Vehicular Technology*, vol. 67, no. 2, pp. 1672–1683, 2018.

[92] G. Gui, M. Liu, F. Tang, N. Kato, and F. Adachi, "6G: Opening new horizons for integration of comfort, security and intelligence," *IEEE Wireless Communications*, to be published.

[93] F. Wu, H. Zhang, J. Wu, and L. Song, "Cellular UAV-to-device communications: Trajectory design and mode selection by multi-agent deep reinforcement learning," *IEEE Transactions on Communications*, to be published.

[94] "Spatial channel model for multiple input multiple output (MIMO) simulations (release 6)," Tech Rep., TS 25. 996, 3GPP, 2018.

[95] M. J. Garey and D. S. Johnson, *Computers and intractability: A guide to the theory of NP-completeness*, W.H. Freeman & Company, 1979.

[96] M. Sipser, *Introduction to the theory of computation*, Cengage Learning, 2012.

[97] M. Grant, S. Boyd, and Y. Ye, "Cvx: MATLAB software for disciplined convex programming," 2009.

[98] D. A. Plaisted, "Some polynomial and integer divisibility problems are NP-hard," in *Proc. Annual Symposium on Foundations of Computer Science*, 1976, pp. 246–267.

[99] D. Li and X. Sun, *Nonlinear integer programming*, Springer Science & Business Media, 2006.

[100] World Health Organization, "7 million premature deaths annually linked to air pollution," *Air Quality & Climate Change*, vol. 22, no. 1, pp. 53–59, 2014.

[101] Y. Li, Y. Zhu, W. Yin, Y. Liu, G. Shi, and Z. Han, "Prediction of high resolution spatial-temporal air pollutant map from big data sources," in *Proc. International Conference on Big Data Computing and Communications*, 2015.

[102] B. Zou, J. G. Wilson, F. B. Zhan, and Y. Zeng, "Air pollution exposure assessment methods utilized in epidemiological studies," *Journal of Environmental Monitoring*, vol. 11, no. 3, pp. 475–490, 2009.

[103] "Beijing municipal environmental monitoring center," [Online], www.bjmemc.com.cn/, March 2017.

[104] Y. Cheng, X. Li, Z. Li, S. Jiang, Y. Li, J. Jia, and X. Jiang, "Aircloud: A cloud-based air-quality monitoring system for everyone," in *Proc. ACM Conference on Embedded Network Sensor Systems*, 2014, pp. 251–265.

[105] D. Hasenfratz, O. Saukh, C. Walser, C. Hueglin, M. Fierz, T. Arn, J. Beutel, and L. Thiele, "Deriving high-resolution urban air pollution maps using mobile sensor nodes," *Pervasive and Mobile Computing*, vol. 16, no. 2, pp. 268–285, 2015.

[106] N. Nikzad, N. Verma, C. Ziftci, E. Bales, N. Quick, P. Zappi, K. Patrick, S. Dasgupta, I. Krueger, T. Š. Rosing, et al., "Citisense: Improving geospatial environmental assessment of air quality using a wireless personal exposure monitoring system," in *Proc. ACM Wireless Health*, 2012, pp. 1–8.

[107] Y. Gao, W. Dong, K. Guo, X. Liu, Y. Chen, X. Liu, J. Bu, and C. Chen, "Mosaic: A low-cost mobile sensing system for urban air quality monitoring," in *Proc. IEEE International Conference on Computer Communications*, 2016, pp. 1–9.

[108] D. Bisht, S. Tiwari, U. Dumka, A. Srivastava, P. Safai, S. Ghude, D. Chate, P. Rao, K. Ali, T. Prabhakaran, et al., "Tethered balloon-born and ground-based measurements of black carbon and particulate profiles within the lower troposphere during the foggy period in Delhi, India," *Science of the Total Environment*, vol. 573, pp. 894–905, 2016.

[109] Y. Hu, G. Dai, J. Fan, Y. Wu, and H. Zhang, "Blueaer: A fine-grained urban pm2. 5 3D monitoring system using mobile sensing," in *Proc. IEEE International Conference on Computer Communications*, 2016, pp. 1–9.

[110] T. N. Quang, C. He, L. Morawska, L. D. Knibbs, and M. Falk, "Vertical particle concentration profiles around urban office buildings," *Atmospheric Chemistry and Physics*, vol. 12, no. 11, pp. 5017–5030, 2012.

[111] F. M. Rubino, L. Floridia, M. Tavazzani, S. Fustinoni, R. Giampiccolo, and A. Colombi, "Height profile of some air quality markers in the urban atmosphere surrounding a 100 m tower building," *Atmospheric Environment*, vol. 32, no. 20, pp. 3569–3580, 1998.

[112] C. Borrego, H. Martins, O. Tchepel, L. Salmim, A. Monteiro, and A. I. Miranda, "How urban structure can affect city sustainability from an air quality perspective," *Environmental Modelling & Software*, vol. 21, no. 4, pp. 461–467, 2006.

[113] Y. Zheng, F. Liu, and H.-P. Hsieh, "U-air: When urban air quality inference meets big data," in *Proc. ACM Conference on Knowledge Discovery and Data Mining*, 2013, pp. 1436–1444.

[114] H.-X. Xu, G. Li, S.-L. Yang, and X. Xu, "Modeling and simulation of haze process based on Gaussian model," in *International Computer Conference on Wavelet Actiev Media Technology and Information Processing*, 2014, pp. 68–74.

[115] M. Cameletti, R. Ignaccolo, and S. Bande, "Comparing spatio-temporal models for particulate matter in Piemonte," *Environmetrics*, vol. 22, no. 8, pp. 985–996, 2011.

[116] C. Zhao, M. van Heeswijk, and J. Karhunen, "Air quality forecasting using neural networks," in *Proc. IEEE Symposium Series on Computational Intelligence (SSCI)*, 2016, pp. 1–7.

[117] M. Cai, Y. Yin, and M. Xie, "Prediction of hourly air pollutant concentrations near urban arterials using artificial neural network approach," *Transportation Research Part D: Transport and Environment*, vol. 14, no. 1, pp. 32–41, 2009.

[118] M. Gardner and S. Dorling, "Neural network modelling and prediction of hourly NO_x and NO_2 concentrations in urban air in London," *Atmospheric Environment*, vol. 33, no. 5, pp. 709–719, 1999.

[119] M. M. Dedovic, S. Avdakovic, I. Turkovic, N. Dautbasic, and T. Konjic, "Forecasting pm10 concentrations using neural networks and system for improving air quality," in *Proc. XI International Symposium on Telecommunications*, 2016, pp. 1–6.

[120] "Beijing municipal environmental protection bureau," [Online], www.bjepb.gov.cn/, March 2017.

[121] "Technology laser pm2.5 sensor, air quality sensor," [Online], www.plantower.com/en/.

[122] "Da-jiang Innovations Science and Technology Co., Ltd. (DJI), Phantom 3 professional," [Online], www.dji.com/cn/phantom-3-pro.

[123] Y. Yang, Z. Zheng, K. Bian, Y. Jiang, L. Song, and Z. Han, "Arms: A fine-grained 3D AQI realtime monitoring system by UAV," in *Proc. IEEE Global Communications Conference*, 2017, pp. 1–6.

[124] R. V. Hogg, J. McKean, and A. T. Craig, *Introduction to mathematical statistics*, Pearson Education, 2005.

[125] M. Cameletti, F. Lindgren, D. Simpson, and H. Rue, "Spatio-temporal modeling of particulate matter concentration through the SPDE approach," *AStA Advances in Statistical Analysis*, vol. 97, no. 2, pp. 109–131, 2013.

[126] D. Middleton, "Modelling air pollution transport and deposition," in *IEE Colloquium on Pollution of Land, Sea and Air: An Overview for Engineers*, 1995, pp. 1–11.

[127] J. M. Stockie, "The mathematics of atmospheric dispersion modeling," *Siam Review*, vol. 53, no. 2, pp. 349–372, 2011.

[128] S. Brusca, F. Famoso, R. Lanzafame, S. Mauro, A. M. C. Garrano, and P. Monforte, "Theoretical and experimental study of Gaussian plume model in small scale system," *Energy Procedia*, vol. 101, pp. 58–65, 2016.

[129] F. Tang, B. Mao, Z. M. Fadlullah, N. Kato, O. Akashi, T. Inoue, and K. Mizutani, "On removing routing protocol from future wireless networks: A real-time deep learning approach for intelligent traffic control," *IEEE Wireless Communications*, vol. 25, no. 1, pp. 154–160, 2017.

[130] A. Al-Molegi, M. Jabreel, and B. Ghaleb, "STF-RNN: Space time features-based recurrent neural network for predicting people next location," in *Proc. IEEE Symposium Series on Computational Intelligence*, 2016, pp. 1–7.

[131] S. M. Carroll and B. W. Dickinson, "Construction of neural nets using the radon transform," in *Proc. IEEE International Joint Conference on Neural Networks*, 1989, pp. 607–611.

[132] G. Cybenko, "Approximation by superpositions of a sigmoidal function," *Mathematics of Control, Signals and Systems*, vol. 2, no. 4, pp. 303–314, 1989.

[133] K.-I. Funahashi, "On the approximate realization of continuous mappings by neural networks," *Neural Networks*, vol. 2, no. 3, pp. 183–192, 1989.

[134] D. P. Bertsekas, *Nonlinear programming*, Athena Scientific, 1999.

[135] R. Penrose, "A generalized inverse for matrices," *Mathematical Proc. Cambridge Philosophical Society*, vol. 51, no. 3, pp. 406–413, 1955.

[136] R. MacAusland, "The Moore-Penrose inverse and least squares," *Math 420: Advanced Topics in Linear Algebra*, pp. 1–10, 2014.

[137] D. E. Goldberg and R. Lingle, "Alleles, loci, and the traveling salesman problem," in *Proc. International Conference on Genetic Algorithms and Their Applications*, 1985, pp. 154–159.

[138] S. Karapantazis and F. Pavlidou, "Broadband communications via high-altitude platforms: A survey," *IEEE Communications Surveys and Tutorials*, vol. 7, no. 1, pp. 2–31, 2005.

[139] C. Zhu, Y. Li, M. Zhang, Q. Wang, and W. Zhou, "Optimal HAP deployment and power control for space-air-ground IoRT networks," in *2021 IEEE Wireless Communications and Networking Conference (WCNC)*, 2021, pp. 1–7.

[140] D. Xu, X. Yi, Z. Chen, C. Li, C. Zhang, and B. Xia, "Coverage ratio optimization for HAP communications," in *2017 IEEE 28th Annual International Symposium on Personal, Indoor, and Mobile Radio Communications (PIMRC)*, 2017, pp. 1–5.

[141] S. Tang, D. Yan, P. You, S. Yong, and S. Xu, "Multiobjective optimization deployment of HAP broadband communication networks," in *2017 IEEE 9th International Conference on Communication Software and Networks (ICCSN)*, 2017, pp. 436–442.

[142] P. He, N. Cheng, and J. Cui, "Handover performance analysis of cellular communication system from high altitude platform in the swing state," in *2016 IEEE International Conference on Signal and Image Processing (ICSIP)*, 2016, pp. 407–411.

[143] X. Wang, L. Li, and W. Zhou, "The effect of HAPs unstable movement on handover performance," in *2019 28th Wireless and Optical Communications Conference (WOCC)*, 2019, pp. 1–5.

[144] I.-M. Kim, Z. Yi, D. Kim, and W. Chung, "Improved opportunistic beamforming in ricean channels," *IEEE Transactions on Communications*, vol. 54, no. 12, pp. 2199–2211, 2006.

[145] D. Xu, Y. Sun, D. W. K. Ng, and R. Schober, "Robust resource allocation for UAV systems with UAV jittering and user location uncertainty," in *2018 IEEE Globecom Workshops (GC Wkshps)*, 2018, pp. 1–6.

[146] W. Yuan, C. Liu, F. Liu, S. Li, and D. W. K. Ng, "Learning-based predictive beamforming for UAV communications with jittering," *IEEE Wireless Communications Letters*, vol. 9, no. 11, pp. 1970–1974, 2020.

[147] Z. Ling, F. Hu, H. Zhang, Z. Han, and H. V. Poor, "Distributionally robust optimization for peak age of information minimization in e-health IoT," in *Proc. IEEE International Conference on Communications (ICC)*, 2021, pp. 1–6.

[148] Z. Ling, F. Hu, H. Zhang, and Z. Han, "Age of information minimization in healthcare IoT using distributionally robust optimization," *IEEE Internet of Things Journal*, vol. 9, no. 17, pp. 16154–16167, 2022.

[149] Z. Lian, L. Jiang, C. He, and D. He, "User grouping and beamforming for HAP massive MIMO systems based on statistical-eigenmode," *IEEE Wireless Communications Letters*, vol. 8, no. 3, pp. 961–964, 2019.

[150] Q. Zhang, Q. Xi, C. He, and L. Jiang, "User clustered opportunistic beamforming for stratospheric communications," *IEEE Communications Letters*, vol. 20, no. 9, pp. 1832–1835, 2016.

[151] W. Saad, M. Bennis, and M. Chen, "A vision of 6G wireless systems: Applications, trends, technologies, and open research problems," *IEEE Network*, vol. 34, no. 3, pp. 134–142, 2020.

[152] B. Li, Z. Fei, C. Zhou, and Y. Zhang, "Physical-layer security in space information networks: A survey," *IEEE Internet of Things Journal*, vol. 7, no. 1, pp. 33–52, 2020.

[153] Z. Jia, M. Sheng, J. Li, D. Niyato, and Z. Han, "LEO-satellite-assisted UAV: Joint trajectory and data collection for Internet of remote things in 6G aerial access networks," *IEEE Internet of Things Journal*, vol. 8, no. 12, pp. 9814–9826, 2021.

[154] Y. Zhao, W. Zhai, J. Zhao, T. Zhang, S. Sun, D. Niyato, and K.-Y. Lam, "A comprehensive survey of 6G wireless communications," https://arxiv.org/abs/2101.03889, Feb. 2021.

[155] Z. Zhang, Y. Xiao, Z. Ma, M. Xiao, Z. Ding, X. Lei, G. K. Karagiannidis, and P. Fan, "6G wireless networks: Vision, requirements, architecture, and key technologies," *IEEE Vehicular Technology Magazine*, vol. 14, no. 3, pp. 28–41, 2019.

[156] D. Zhou, M. Sheng, J. Luo, R. Liu, J. Li, and Z. Han, "Collaborative data scheduling with joint forward and backward induction in small satellite networks," *IEEE Transactions on Communications*, vol. 67, no. 5, pp. 3443–3456, 2019.

[157] Y. Wang, M. Sheng, W. Zhuang, S. Zhang, N. Zhang, R. Liu, and J. Li, "Multi-resource coordinate scheduling for Earth observation in space information networks," *IEEE Journal on Selected Areas in Communications*, vol. 36, no. 2, pp. 268–279, 2018.

[158] A. Golkar and I. L. i Cruz, "The federated satellite systems paradigm: Concept and business case evaluation," *Acta Astronautica*, vol. 111, pp. 230–248, 2015.

[159] D. Zhou, M. Sheng, B. Li, J. Li, and Z. Han, "Distributionally robust planning for data delivery in distributed satellite cluster network," *IEEE Transactions on Wireless Communications*, vol. 18, no. 7, pp. 3642–3657, 2019.

[160] Y. Pochet and L. A. Wolsey, *Production planning by mixed integer programming*, Springer, 2006.

[161] A. Lodi, *Mixed integer programming computation*, Springer, pp. 619–645, 2010.

[162] G. Nemhauser and L. Wolsey, *Computational complexity*, John Wiley & Sons, pp. 114–145, 2014.

[163] R. Rahmaniani, T. G. Crainic, M. Gendreau, and W. Rei, "The Benders decomposition algorithm: A literature review," *European Journal of Operational Research*, vol. 259, no. 3, pp. 801–817, 2017.

[164] L. Fan, J. Wang, R. Jiang, and Y. Guan, "Min-max regret bidding strategy for thermal generator considering price uncertainty," *IEEE Transactions on Power Systems*, vol. 29, no. 5, pp. 2169–2179, 2014.

[165] A. Ibrahim, O. A. Dobre, T. M. N. Ngatched, and A. G. Armada, "Bender's decomposition for optimization design problems in communication networks," *IEEE Network*, vol. 34, no. 3, pp. 232–239, 2020.

[166] Y. Yu, X. Bu, K. Yang, H. K. Nguyen, and Z. Han, "Network function virtualization resource allocation based on joint benders decomposition and ADMM," *IEEE Transactions on Vehicular Technology*, vol. 69, no. 2, pp. 1706–1718, 2020.

[167] H. Wu and H. Lu, "Delay and power tradeoff with consideration of caching capabilities in dense wireless networks," *IEEE Transactions on Wireless Communications*, vol. 18, no. 10, pp. 5011–5025, 2019.

[168] M. Sheng, Y. Wang, J. Li, R. Liu, D. Zhou, and L. He, "Toward a flexible and reconfigurable broadband satellite network: Resource management architecture and strategies," *IEEE Wireless Communications*, vol. 24, no. 4, pp. 127–133, 2017.

[169] C. Luo, S. Guo, S. Guo, L. T. Yang, G. Min, and X. Xie, "Green communication in energy renewable wireless mesh networks: Routing, rate control, and power allocation,"

IEEE Transactions on Parallel and Distributed Systems, vol. 25, no. 12, pp. 3211–3220, 2014.

[170] Z. Jia, M. Sheng, J. Li, D. Zhou, and Z. Han, "Joint HAP access and LEO satellite backhaul in 6G: Matching game-based approaches," *IEEE Journal on Selected Areas in Communications*, vol. 39, no. 4, pp. 1147–1159, 2021.

[171] F. Dong, H. Li, X. Gong, Q. Liu, and J. Wang, "Energy-efficient transmissions for remote wireless sensor networks: An integrated HAP/satellite architecture for emergency scenarios," *Sensors*, vol. 15, no. 9, pp. 22266–22290, 2015.

[172] A. I. Aravanis, B. Shankar M. R., P. Arapoglou, G. Danoy, P. G. Cottis, and B. Ottersten, "Power allocation in multibeam satellite systems: A two-stage multi-objective optimization," *IEEE Transactions on Wireless Communications*, vol. 14, no. 6, pp. 3171–3182, 2015.

[173] D. Zhou, M. Sheng, J. Li, C. Xu, R. Liu, and Y. Wang, "Toward high throughput contact plan design in resource-limited small satellite networks," in *2016 IEEE 27th Annual International Symposium on Personal, Indoor, and Mobile Radio Communications (PIMRC)*, 2016, pp. 1–6.

[174] F. Granelli, C. Costa, J. Zhang, R. Bassoli, and F. H. P. Fitzek, "Design of an on-demand agile 5G multi-access edge computing platform using aerial vehicles," *IEEE Communications Standards Magazine*, vol. 4, no. 4, pp. 34–41, 2020.

[175] N. Kato, Z. M. Fadlullah, F. Tang, B. Mao, S. Tani, A. Okamura, and J. Liu, "Optimizing space-air-ground integrated networks by artificial intelligence," *IEEE Wireless Communications*, vol. 26, no. 4, pp. 140–147, 2019.

[176] X. Zhu, C. Jiang, L. Kuang, N. Ge, and J. Lu, "Non-orthogonal multiple access based integrated terrestrial-satellite networks," *IEEE Journal on Selected Areas in Communications*, vol. 35, no. 10, pp. 2253–2267, 2017.

[177] Z. Zhang, W. Zhang, and F.-H. Tseng, "Satellite mobile edge computing: Improving QoS of high-speed satellite-terrestrial networks using edge computing techniques," *IEEE Network*, vol. 33, no. 1, pp. 70–76, 2019.

[178] X. Cao, B. Yang, C. Huang, C. Yuen, M. Di Renzo, D. Niyato, and Z. Han, "Reconfigurable intelligent surface-assisted aerial-terrestrial communications via multi-task learning," *IEEE Journal on Selected Areas in Communications*, vol. 39, no. 10, pp. 3035–3050, 2021.

[179] C. Zhang, C. Jiang, L. Kuang, J. Jin, Y. He, and Z. Han, "Spatial spectrum sharing for satellite and terrestrial communication networks," *IEEE Transactions on Aerospace and Electronic Systems*, vol. 55, no. 3, pp. 1075–1089, 2019.

[180] G. Bianchi, "Performance analysis of the IEEE 802.11 distributed coordination function," *IEEE Journal on Selected Areas in Communications*, vol. 18, no. 3, pp. 535–547, 2000.

[181] B. Yang, B. Li, Z. Yan, D. J. Deng, and M. Yang, "Performance analysis of multi-channel MAC with single transceiver for the next generation WLAN," *Journal of Network and Computer Applications*, vol. 146, p. 102408, 2019.

[182] "Study on using satellite access in 5G (stage 1)," Tech. Rep., TR 22.822 Release 16, 3GPP, 2018.

[183] R. Radhakrishnan, W. W. Edmonson, F. Afghah, R. M. Rodriguez-Osorio, F. Pinto, and S. C. Burleigh, "Survey of inter-satellite communication for small satellite systems: Physical layer to network layer view," *IEEE Communications Surveys & Tutorials*, vol. 18, no. 4, pp. 2442–2473, 2016.

[184] L. Zhang, L. Zhu, and C. Ju, "Generalized mimo channel model and its capacity analysis in formation flying satellite communication systems," in *2011 6th International ICST Conference on Communications and Networking in China (CHINACOM)*, 2011, pp. 1079–1082.

[185] R. T. Schwarz, A. Knopp, D. Ogermann, C. A. Hofmann, and B. Lankl, "Optimum-capacity MIMO satellite link for fixed and mobile services," in *2008 International ITG Workshop on Smart Antennas*, 2008, pp. 209–216.

[186] X. Jia, T. Lv, F. He, and H. Huang, "Collaborative data downloading by using inter-satellite links in LEO satellite networks," *IEEE Transactions on Wireless Communications*, vol. 16, no. 3, pp. 1523–1532, 2017.

[187] B. Devillers, A. Pérez-Neira, and C. Mosquera, "Joint linear precoding and beamforming for the forward link of multi-beam broadband satellite systems," in *2011 IEEE Global Telecommunications Conference (GLOBECOM 2011)*, 2011, pp. 1–6.

[188] L. Cao, Y. Chen, Z. Zhang, H. Li, and A. K. Misra, "Predictive smooth variable structure filter for attitude synchronization estimation during satellite formation flying," *IEEE Transactions on Aerospace and Electronic Systems*, vol. 53, no. 3, pp. 1375–1383, 2017.

[189] R. Schwarz, A. Knopp, B. Lankl, D. Ogermann, and C. Hofmann, "Optimum-capacity MIMO satellite broadcast system: Conceptual design for LoS channels," in *2008 4th Advanced Satellite Mobile Systems*, 2008, pp. 66–71.

[190] G. B. Shaw, D. Miller, and D. Hastings, "Generalized characteristics of communication, sensing, and navigation satellite Systems," *Journal of Spacecraft and Rockets*, vol. 37, no. 6, pp. 801–811, 2000.

[191] G. Maral, M. Bousquet, and Z. Sun, *Satellite communications systems: Systems, techniques and technology*, John Wiley & Sons, 2020.

[192] J. Lee and N. Jindal, "High SNR analysis for MIMO broadcast channels: Dirty paper coding versus linear precoding," *IEEE Transactions on Information Theory*, vol. 53, no. 12, pp. 4787–4792, 2007.

[193] A. Goldsmith, S. A. Jafar, N. Jindal, and S. Vishwanath, "Capacity limits of MIMO channels," *IEEE Journal on selected areas in Communications*, vol. 21, no. 5, pp. 684–702, 2003.

[194] J. N. Franklin, *Matrix theory*, Courier Corporation, 2012.

[195] M. Arti, "Channel estimation and detection in satellite communication systems," *IEEE Transactions on Vehicular Technology*, vol. 65, no. 12, pp. 10173–10179, 2016.

[196] M. R. Bhatnagar and M. Arti, "On the closed-form performance analysis of maximal ratio combining in shadowed-rician fading LMS channels," *IEEE Communications Letters*, vol. 18, no. 1, pp. 54–57, 2013.

[197] S. Renukadevi, S. Rajarajan, R. Vedanayagi, and J. Raja, "Antenna designs for amateur band low earth orbit (LEO) satellites–A review," *International Journal for Research in Engineering Application and Management*, vol. 4, no. 5, pp. 254–259, 2018.

[198] J. W. D. Ben Noble, *Applied linear algebra*, Prentice-Hall, 1988.

[199] L. N. Trefethen and D. Bau III, *Numerical linear algebra*, vol. 50, Siam, 1997.

[200] I. N. Bronshtein and K. A. Semendyayev, *Handbook of mathematics*, Springer Science & Business Media, 2013.

[201] J. W. Tukey, "On the distribution of the fractional part of a statistical variable," *Mathematical Collection*, vol. 4, no. 3, pp. 561–562, 1938.

[202] G. Casella and R. L. Berger, *Statistical inference*, Cengage Learning, 2021.

[203] B. Di, H. Zhang, L. Song, Y. Li, and G. Y. Li, "Ultra-dense LEO: Integrating terrestrial-satellite networks into 5G and beyond for data offloading," *IEEE Transactions on Wireless Communications*, vol. 18, no. 1, pp. 47–62, 2018.

[204] "Study on new radio (NR) to support non-terrestrial networks," Tech. Rep., TR 38.811, 3GPP, 2017.

[205] J. G. Andrews, S. Buzzi, W. Choi, S. V. Hanly, A. Lozano, A. C. Soong, and J. C. Zhang, "What will 5G be?" *IEEE Journal on Selected Areas in Communications*, vol. 32, no. 6, pp. 1065–1082, 2014.

[206] R. Deng, B. Di, H. Zhang, L. Kuang, and L. Song, "Ultra-dense LEO satellite constellations: How many LEO satellites do we need?" *IEEE Transactions on Wireless Communications*, vol. 20, no. 8, pp. 4843–4857, 2021.

[207] D. Wang, Y. Zhang, H. Wei, X. You, X. Gao, and J. Wang, "An overview of transmission theory and techniques of large-scale antenna systems for 5G wireless communications," *Science China Information Sciences*, vol. 59, no. 8, pp. 1–18, 2016.

[208] M. E. Badawe, T. S. Almoneef, and O. M. Ramahi, "A true metasurface antenna," *Scientific Reports*, vol. 6, no. 1, pp. 1–8, 2016.

[209] R.-B. R. Hwang, "Binary meta-hologram for a reconfigurable holographic metamaterial antenna," *Scientific Reports*, vol. 10, no. 1, pp. 1–10, 2020.

[210] D. R. Smith, O. Yurduseven, L. P. Mancera, P. Bowen, and N. B. Kundtz, "Analysis of a waveguide-fed metasurface antenna," *Physical Review Applied*, vol. 8, no. 5, p. 054048, 2017.

[211] B.-J. Che, F.-Y. Meng, Y.-L. Lyu, Y.-Q. Zhao, and Q. Wu, "Reconfigurable holographic antenna with low sidelobe level based on liquid crystals," *Journal of Physics D: Applied Physics*, vol. 53, no. 31, p. 315302, 2020.

[212] R. Deng, B. Di, H. Zhang, Y. Tan, and L. Song, "Reconfigurable holographic surface: Holographic beamforming for metasurface-aided wireless communications," *IEEE Transactions on Vehicular Technology*, vol. 70, no. 6, pp. 6255–6259, 2021.

[213] "Holographic beamforming and phased arrays," Tech. Rep., Pivotal Commware, 2019.

[214] T. Sleasman, M. F. Imani, W. Xu, J. Hunt, T. Driscoll, M. S. Reynolds, and D. R. Smith, "Waveguide-fed tunable metamaterial element for dynamic apertures," *IEEE Antennas and Wireless Propagation Letters*, vol. 15, pp. 606–609, 2015.

[215] M. C. Johnson, S. L. Brunton, N. B. Kundtz, and N. J. Kutz, "Extremum-seeking control of the beam pattern of a reconfigurable holographic metamaterial antenna," *Journal of the Optical Society of America A*, vol. 33, no. 1, pp. 59–68, 2016.

[216] Y. B. Li, L. L. Li, B. G. Cai, Q. Cheng, and T. J. Cui, "Holographic leaky-wave metasurfaces for dual-sensor imaging," *Scientific Reports*, vol. 5, no. 1, pp. 1–7, 2015.

[217] M. Boyarsky, T. Sleasman, L. Pulido-Mancera, T. Fromenteze, A. Pedross-Engel, C. M. Watts, M. F. Imani, M. S. Reynolds, and D. R. Smith, "Synthetic aperture radar with dynamic metasurface antennas: A conceptual development," *Journal of the Optical Society of America A*, vol. 34, no. 5, pp. A22–A36, 2017.

[218] R. Deng, B. Di, H. Zhang, D. Niyato, Z. Han, H. V. Poor, and L. Song, "Reconfigurable holographic surfaces for future wireless communications," *IEEE Wireless Communications*, vol. 28, no. 6, pp. 126–131, 2021.

[219] J. L. Gómez-Tornero, F. D. Quesada-Pereira, and A. Álvarez-Melcón, "Analysis and design of periodic leaky-wave antennas for the millimeter waveband in hybrid waveguide-planar technology," *IEEE Transactions on Antennas and Propagation*, vol. 53, no. 9, pp. 2834–2842, 2005.

[220] K. Sidibeh and T. Vladimirova, "Wireless communication in LEO satellite formations," in *2008 NASA/ESA Conference on Adaptive Hardware and Systems*, 2008, pp. 255–262.

[221] J. Sedin, L. Feltrin, and X. Lin, "Throughput and capacity evaluation of 5G new radio non-terrestrial networks with LEO satellites," in *GLOBECOM 2020-2020 IEEE Global Communications Conference*, 2020, pp. 1–6.

[222] X. Gao, L. Dai, Y. Zhang, T. Xie, X. Dai, and Z. Wang, "Fast channel tracking for terahertz beamspace massive mimo systems," *IEEE Transactions on Vehicular Technology*, vol. 66, no. 7, pp. 5689–5696, 2016.

[223] F. Sohrabi and W. Yu, "Hybrid digital and analog beamforming design for large-scale antenna arrays," *IEEE Journal of Selected Topics in Signal Processing*, vol. 10, no. 3, pp. 501–513, 2016.

[224] R. T. Schwarz and A. Knopp, "MIMO capacity of co-located satellites in longitude separation," in *IEEE International Conference on Communications (ICC)*, 2019, pp. 1–7.

[225] F. Rusek, D. Persson, B. K. Lau, E. G. Larsson, T. L. Marzetta, O. Edfors, and F. Tufvesson, "Scaling up MIMO: Opportunities and challenges with very large arrays," *IEEE Signal Processing Magazine*, vol. 30, no. 1, pp. 40–60, 2012.

[226] L. Liang, W. Xu, and X. Dong, "Low-complexity hybrid precoding in massive multiuser MIMO systems," *IEEE Wireless Communications Letters*, vol. 3, no. 6, pp. 653–656, 2014.

[227] U. Erez, S. Shamai, and R. Zamir, "Capacity and lattice strategies for canceling known interference," *IEEE Transactions on Information Theory*, vol. 51, no. 11, pp. 3820–3833, 2005.

[228] L.-N. Tran, M. Juntti, M. Bengtsson, and B. Ottersten, "Weighted sum rate maximization for MIMO broadcast channels using dirty paper coding and zero-forcing methods," *IEEE Transactions on Communications*, vol. 61, no. 6, pp. 2362–2373, 2013.

[229] X. Feng and Z. Zhang, "The rank of a random matrix," *Applied Mathematics and Computation*, vol. 185, no. 1, pp. 689–694, 2007.

[230] W. Mei and R. Zhang, "Aerial-ground interference mitigation for cellular-connected UAV," *IEEE Wireless Communications*, vol. 28, no. 1, pp. 167–173, 2021.

[231] W. Mei and R. Zhang, "Cooperative downlink interference transmission and cancellation for cellular-connected UAV: A divide-and-conquer approach," *IEEE Transactions on Communications*, vol. 68, no. 2, pp. 1297–1311, 2019.

[232] L. You, K.-X. Li, J. Wang, X. Gao, X.-G. Xia, and B. Ottersten, "Massive MIMO transmission for LEO satellite communications," *IEEE Journal on Selected Areas in Communications*, vol. 38, no. 8, pp. 1851–1865, 2020.

[233] F. Xu, K. Wu, and X. Zhang, "Periodic leaky-wave antenna for millimeter wave applications based on substrate integrated waveguide," *IEEE Transactions on Antenna and Propagation*, vol. 58, no. 2, pp. 340–347, 2009.

[234] "Study on new radio access technology: Radio frequency (RF) and coexistence aspect release 14," Tech. Rep., TR 38.803, 3GPP, 2017.

[235] M. Werner, A. Jahn, E. Lutz, and A. Bottcher, "Analysis of system parameters fo LEO/ICO-satellite communication networks," *IEEE Journal on Selected areas in Communications*, vol. 13, no. 2, pp. 371–381, 1995.

[236] M. Richharia, *Mobile satellite communications: principles and trends*, John Wiley & Sons, 2014.

[237] D. C. Beste, "Design of satellite constellations for optimal continuous coverage," *IEEE Transactions on Aerospace and Electronic Systems*, vol. AES-14, no. 3, pp. 466–473, 1978.

[238] C.-J. Wang, "Structural properties of a low earth orbit satellite constellation: The Walker Delta Network," in *Proc. MILCOM'93-IEEE Military Communications Conference*, 1993, vol. 3, pp. 968–972.

[239] D. Mortari and M. P. Wilkins, "Flower constellation set theory. Part I: Compatibility and phasing," *IEEE Transactions on Aerospace and Electronic Systems*, vol. 44, no. 3, pp. 953–962, 2008.

[240] M. P. Wilkins and D. Mortari, "Flower constellation set theory part II: Secondary paths and equivalency," *IEEE Transactions on Aerospace and Electronic Systems*, vol. 44, no. 3, pp. 964–976, 2008.

[241] D. Mortari, M. De Sanctis, and M. Lucente, "Design of flower constellations for telecommunication services," *Proc. IEEE*, vol. 99, no. 11, pp. 2008–2019, 2011.

[242] J. Jiang, S. Yan, and M. Peng, "Regional LEO satellite constellation design based on user requirements," in *2018 IEEE/CIC International Conference on Communications in China (ICCC)*, 2018, pp. 855–860.

[243] I. Meziane-Tani, G. Métris, G. Lion, A. Deschamps, F. T. Bendimerad, and M. Bekhti, "Optimization of small satellite constellation design for continuous mutual regional coverage with multi-objective genetic algorithm," *International Journal of Computational Intelligence Systems*, vol. 9, no. 4, pp. 627–637, 2016.

[244] C. Dai, G. Zheng, and Q. Chen, "Satellite constellation design with multi-objective genetic algorithm for regional terrestrial satellite network," *China Communications*, vol. 15, no. 8, pp. 1–10, 2018.

[245] T. Savitri, Y. Kim, S. Jo, and H. Bang, "Satellite constellation orbit design optimization with combined genetic algorithm and semianalytical approach," *International Journal of Aerospace Engineering*, vol. 2017, 2017.

[246] Z. Liu, W. Guo, W. Hu, and M. Xia, "Delay minimization for progressive construction of satellite constellation network," *IEEE Communications Letters*, vol. 19, no. 10, pp. 1718–1721, 2015.

[247] Z. Qu, G. Zhang, H. Cao, and J. Xie, "LEO satellite constellation for internet of things," *IEEE Access*, vol. 5, pp. 18391–18401, 2017.

[248] X. Zhu and Y. Gao, "Comparison of intelligent algorithms to design satellite constellations for enhanced coverage capability," in *2017 10th International Symposium on Computational Intelligence and Design (ISCID)*, 2017, vol. 2, pp. 223–226.

[249] B. Di, L. Song, Y. Li, and H. V. Poor, "Ultra-dense LEO: Integration of satellite access networks into 5G and beyond," *IEEE Wireless Communications*, vol. 26, no. 2, pp. 62–69, 2019.

[250] O. Montenbruck, E. Gill, and F. Lutze, "Satellite orbits: Models, methods, and applications," *Applied Mechanics Reviews*, vol. 55, no. 2, pp. B27–B28, 2002.

[251] J. López-Fernández, J. F. Paris, and E. Martos-Naya, "Bivariate Rician shadowed fading model," *IEEE Transactions on Vehicular Technology*, vol. 67, no. 1, pp. 378–384, 2017.

[252] R. Deng, B. Di, S. Chen, S. Sun, and L. Song, "Ultra-dense LEO satellite offloading for terrestrial networks: How much to pay the satellite operator?" *IEEE Transactions on Wireless Communications*, vol. 19, no. 10, pp. 6240–6254, 2020.

[253] "Study on new radio (NR) to support non terrestrial networks (Release 15)," Tech Rep. TR 38.811 (V0.3.0), 3GPP, Dec. 2017.

[254] M. Haenggi, *Stochastic geometry for wireless networks*, Cambridge University Press, 2012.

[255] G. Grimmett and D. Stirzaker, *Probability and random processes*, Oxford University Press, 2020.

[256] W. W. Bell, *Special functions for scientists and engineers*, Courier Corporation, 2004.

[257] A. Jeffrey and H. H. Dai, *Handbook of mathematical formulas and integrals*, Elsevier, 2008.

[258] Y. Jia and Z. Peng, "The analysis and simulation of communication network in iridium system based on opnet," in *2010 2nd IEEE International Conference on Information Management and Engineering*, 2010, pp. 68–72.

[259] X. Wang, G. Xing, Y. Zhang, C. Lu, R. Pless, and C. Gill, "Integrated coverage and connectivity configuration in wireless sensor networks," in *Proc. 1st International Conference on Embedded Networked Sensor Systems*, 2003, pp. 28–39.

[260] D. G. Luenberger and Y. Ye, *Linear and nonlinear programming*, vol. 2, Springer, 1984.

[261] P. E. Zadunaisky, "Small perturbations on artificial satellites as an inverse problem," *IEEE Transactions on Aerospace and Electronic Systems*, vol. 39, no. 4, pp. 1270–1276, 2003.

[262] P. Ligong, F. Zhuren, L. Ganhua, and H. Minzhang, "Relative motion model of satellites formation flying base on the influence of the j 2 perturbation," in *2009 IEEE International Conference on Robotics and Biomimetics (ROBIO)*, 2009, pp. 308–313.

[263] X. Cao, P. Zheng, and S. Zhang, "Atmospheric drag perturbation effect on the deployment of tether-assisted deorbit system," in *2009 International Conference on Mechatronics and Automation*, 2009, pp. 4316–4321.

[264] I. Del Portillo, B. G. Cameron, and E. F. Crawley, "A technical comparison of three low earth orbit satellite constellation systems to provide global broadband," *Acta Astronautica*, vol. 159, pp. 123–135, 2019.

[265] O. Kodheli, A. Guidotti, and A. Vanelli-Coralli, "Integration of satellites in 5G through LEO constellations," in *2017 IEEE Global Communications Conference (GLOBECOM 2017)*, 2017, pp. 1–6.

[266] I. Ali, N. Al-Dhahir, and J. E. Hershey, "Doppler characterization for LEO satellites," *IEEE Transactions on Communications*, vol. 46, no. 3, pp. 309–313, 1998.

[267] S. Singh, M. N. Kulkarni, A. Ghosh, and J. G. Andrews, "Tractable model for rate in self-backhauled millimeter wave cellular networks," *IEEE Journal on Selected Areas in Communications*, vol. 33, no. 10, pp. 2196–2211, 2015.

[268] "Propagation data and prediction methods required for the design of earth-space telecommunication systems," P Series, Recommendation ITU-R, P.618–13, 2017.

[269] D. C. Beste, "Design of satellite constellations for optimal continuous coverage," *IEEE Transactions on Aerospace and Electronic Systems*, no. 3, pp. 466–473, 1978.

[270] "SANSA project," [Online], 2016, http://sansa-h2020.eu/.

[271] E. Papapetrou, S. Karapantazis, G. Dimitriadis, and F.-N. Pavlidou, "Satellite handover techniques for LEO networks," *International Journal on Satellite Communications and Networking*, vol. 22, no. 2, pp. 231–245, 2004.

[272] W. Shi, D. Gao, H. Zhou, Q. Xu, and C. H. Foh, "Traffic aware inter-layer contact selection for multi-layer satellite terrestrial network," in *2017 IEEE Global Communications Conference (GLOBECOM 2017)*, 2017, pp. 1–7.

[273] A. Abdi, W. C. Lau, M.-S. Alouini, and M. Kaveh, "A new simple model for land mobile satellite channels: First-and second-order statistics," *IEEE Transactions on Wireless Communications*, vol. 2, no. 3, pp. 519–528, 2003.

[274] H. Chaouech and R. Bouallegue, "Channel estimation and detection for multibeam satellite communications," in *IEEE Asia Pacific Conference on Circuits and Systems (APCCAS)*, 2010, pp. 366–369.

[275] M. Arti, "Imperfect CSI based AF relaying in hybrid satellite-terrestrial cooperative communication systems," in *IEEE International Conference on Communication Workshop (ICCW)*, 2015, pp. 1681–1686.

[276] M. Arti, "Channel estimation and detection in hybrid satellite-terrestrial communication systems," *IEEE Transactions on Vehicular Technology*, vol. 65, no. 7, pp. 5764–5771, 2016.

[277] M. R. Bhatnagar and M. Arti, "Performance analysis of AF based hybrid satellite-terrestrial cooperative network over generalized fading channels," *IEEE Communications Letters*, vol. 17, no. 10, pp. 1912–1915, 2013.

[278] "Radio Regulations (Edition of 2012, part 2)," International Telecommunication Union, 2012.

[279] S. Boyd and L. Vandenberghe, *Convex optimization*, Cambridge University Press, 2004.

[280] M. Pycia and M. Yenmez, "Matching with externalities," [Online], 2015, https://papers.ssrn.com/sol3/papers.cfm?abstract_id=2475468.

[281] A. Mumcu and I. Saglam, "Stable one-to-one matchings with externalities," *Mathematical Social Sciences*, vol. 60, no. 2, pp. 154–159, 2010.

[282] D. F. Manlove, *Algorithmics of matching under preferences*, vol. 2, World Scientific, 2013.

[283] B. Di, L. Song, and Y. Li, "Sub-channel assignment, power allocation, and user scheduling for non-orthogonal multiple access networks," *IEEE Transactions on Wireless Communications*, vol. 15, no. 11, pp. 7686–7698, 2016.

[284] B. Di, L. Song, Y. Li, and G. Y. Li, "Non-orthogonal multiple access for high-reliable and low-latency V2X communications in 5G systems," *IEEE Journal on Selected Areas in Communications*, vol. 35, no. 10, pp. 2383–2397, 2017.

[285] P. Parida and S. S. Das, "Power allocation in OFDM based NOMA systems: A DC programming approach," in *Globecom Workshops (GC Wkshps)*, 2014, pp. 1026–1031.

[286] "Guidelines for evaluation of radio interface technologies for IMT-Advanced," Tech. Rep. ITU-R M.2135, 2009.

[287] "Satellite tool kit," [Online], https://softadvice.informer.com/Satellite_Tool_Kit_9.html.

[288] P. Popovski, K. F. Trillingsgaard, O. Simeone, and G. Durisi, "5G wireless network slicing for eMBB, URLLC, and MMTC: A communication-theoretic view," *IEEE Access*, vol. 6, pp. 55765–55779, 2018.

[289] S. Zhang, W. Quan, J. Li, W. Shi, P. Yang, and X. Shen, "Air-ground integrated vehicular network slicing with content pushing and caching," *IEEE Journal on Selected Areas in Communications*, vol. 36, no. 9, pp. 2114–2127, 2018.

[290] X. Foukas, G. Patounas, A. Elmokashfi, and M. K. Marina, "Network slicing in 5G: Survey and challenges," *IEEE Communications Magazine*, vol. 55, no. 5, pp. 94–100, 2017.

[291] L. Popokh, J. Su, S. Nair, and E. Olinick, "Illumicore: Optimization modeling and implementation for efficient VNF placement," in *2021 International Conference on Software, Telecommunications and Computer Networks (SoftCOM)*, 2021, pp. 1–7.

[292] L. Zhang, W. Abderrahim, and B. Shihada, "Heterogeneous traffic offloading in space-air-ground integrated networks," *IEEE Access*, vol. 9, pp. 165462–165475, 2021.

[293] H. Wu, J. Chen, C. Zhou, W. Shi, N. Cheng, W. Xu, W. Zhuang, and X. S. Shen, "Resource management in space-air-ground integrated vehicular networks: SDN control and AI algorithm design," *IEEE Wireless Communications*, vol. 27, no. 6, pp. 52–60, 2020.

[294] F. Lyu, P. Yang, H. Wu, C. Zhou, J. Ren, Y. Zhang, and X. Shen, "Service-oriented dynamic resource slicing and optimization for space-air-ground integrated vehicular networks," *IEEE Transactions on Intelligent Transportation Systems*, vol. 23, no. 7, pp. 7469–7483, 2021.

[295] M. Razaviyayn, "Successive Convex Approximation: Analysis and Applications," doctoral dissertation, University of Minnesota, 2014.

Index

Printed in the USA
CPSIA information can be obtained
at www.ICGtesting.com
LVHW081536111123
763669LV00008B/167

9 781108 837934